PHOTOCATALYSIS

Principles and Applications

PHOTOCATALYSIS
Principles and Applications

Rakshit Ameta
Suresh C. Ameta

CRC Press
Taylor & Francis Group
Boca Raton London New York

CRC Press is an imprint of the
Taylor & Francis Group, an **informa** business

CRC Press
Taylor & Francis Group
6000 Broken Sound Parkway NW, Suite 300
Boca Raton, FL 33487-2742

First issued in paperback 2019

© 2017 by Taylor & Francis Group, LLC
CRC Press is an imprint of Taylor & Francis Group, an Informa business

No claim to original U.S. Government works

ISBN-13: 978-0-4822-5493-8 (hbk)
ISBN-13: 978-0-367-87063-8 (pbk)

Visit the Taylor & Francis Web site at
http://www.taylorandfrancis.com

and the CRC Press Web site at
http://www.crcpress.com

Contents

Preface

Photocatalysis, reactions carried out in the presence of a semiconductor and light, is rapidly becoming one of the most active areas of chemical research, with applications in areas such as electrochemistry, medicine, and environmental chemistry. This book, *Photocatalysis: Principles and Applications*, stresses the development of various types of photocatalytic semiconductors, including binary, ternary, quaternary, and composite, and their modifications by metallization, sensitization, and doping to enhance their photocatalytic activities. In addition to describing the principles and mechanisms of photocatalysis, it also discusses other possible applications of photocatalysis such as use as antifouling agents, controlling air pollution by degrading contaminants present in the environment, self-cleaning of glasses and tiles in the presence of light/artificial light, green composites, wastewater treatment, hydrogen generation, and inactivation of microorganisms. This book also describes medical applications, photosplitting of water, photoreduction of carbon dioxide and mimicking photosynthesis.

- Introduces the basic principle of photocatalysis
- Provides an overview of the types of semiconductors, their immobilization, and modifications to make them more active
- Gives possible applications of photocatalysis in wastewater treatment and strategies to combat against different kinds of pollution such as water, air, and soil
- Discusses inactivation of different kinds of microorganisms
- Covers medical applications

Authors

Rakshit Ameta, PhD, is currently an Associate Professor and Head, Department of Chemistry, PAHER University, Udaipur, India. He has about ten years of experience in teaching and research. He has published more than 70 research papers. Seven students have been awarded PhDs under his supervision and presently six more are working. He served as a member of the executive council of the Indian Chemical Society, Kolkata and Indian Council of Chemists, Agra.

Suresh C. Ameta, PhD, is a Professor of Chemistry and Dean, Faculty of Science, PAHER University, Udaipur, India. He has about 43 years' experience in teaching and research. He has about 400 research papers to his credit, and he has guided successfully 79 students to their PhD theses. He has received many awards, including Lifetime Achievement Awards from the Indian Chemical Society, Kolkata (2011) and the Indian Council of Chemists, Agra (2015).

1 Introduction

1.1 HISTORY

The Greek word photocatalysis is a combination of two words: photo ("phos" means light) and catalysis ("katalyo" means to break apart, decompose). Generally, the term photocatalysis is used to describe a process where light is used to excite a photocatalyst and the rate of chemical reaction is accelerated without involving the photocatalyst. Although the term photocatalysis is quite confusing and there has been a long debate on the definition of this term, the International Union of Pure and Applied Chemistry (IUPAC) has finally decided that the term photocatalysis is reserved for the reactions carried out in the presence of a semiconductor and light. Plotnikow (1936) mentioned photocatalysis in his book entitled *Allgemeine Photochemie*. Almost four decades later, some researchers actively started conducting surface studies on photocatalysts such as TiO_2 and ZnO. In the meantime, some workers also thought on the possibility of using sunlight as the energy source. Fujishima and Honda (1972) conducted the photolysis of water using a semiconductor electrode (TiO_2) in a photoelectrochemical cell, which gave momentum to the field of photocatalysis. This was considered as the real beginning of this field. In the 1980s and 1990s, many efforts were made to understand the fundamental process and improve the photocatalytic efficiency of titania. In the last few years, semiconductor materials have been used as a photocatalyst for air and water remediation, mineralizing hazardous organic pollutants, and industrial and health applications (Rengel et al. 2012).

1.2 PHOTOCATALYSIS

A photoinduced reaction, which is accelerated in the presence of a catalyst (semiconductor), is termed photocatalysis. There are two types of photocatalysis: homogeneous and heterogeneous.

- **Homogeneous photocatalysis**
  When the reactant and photocatalyst exist in the same phase, the reaction is called homogeneous photocatalysis. Coordination compounds, dyes, natural pigments, and so on are the most common examples of homogenous photocatalysts.
- **Heterogeneous photocatalysis**
  In this type, the reactant and photocatalyst exist in different phases. Transition metal chalcogenides are the most common examples of heterogeneous photocatalysts, which have some unique characteristics.

Photocatalytic reactions are initiated by the absorption of a photon with appropriate energy that is equal to or higher than the band gap energy of the photocatalyst. The absorbed photon creates a charge separation as the electron is elevated from the valence band (VB) of a semiconductor to the conduction band (CB), creating a hole (h^+) in the VB. This excited electron can reduce any substrate or react with electron acceptors such as O_2 present on the semiconductor surface or dissolved in water, reducing it to superoxide radical anion $O_2^{-•}$. On the other hand, the hole can oxidize the organic molecule to form R^+, or react with ^-OH or H_2O, oxidizing them to $^•OH$ radicals.

Other highly oxidant species such as peroxide radicals are also responsible for the heterogeneous photodecomposition of organic substrates. ^-OH is quite a strong oxidizing agent which can oxidize most of the azo dyes and other pollutants to the minerals as end products (Konstantinou and Albanis 2014). The field of photocatalysis has been reviewed by different researchers from time to time (Fox and Dulay 1993; Fujishima et al. 2000; Ameta et al. 2003; Reloez et al. 2012).

1.3 SEMICONDUCTING MATERIALS

Currently, semiconductor photocatalysis is becoming one of the most active areas of research and has been studied in different streams such as catalysis; photochemistry; electrochemistry; inorganic and organic chemistries; physical, and polymer, environmental chemistry; and so on. Mainly binary semiconductors such as TiO_2, ZnO, Fe_2O_3, CdS, and ZnS have been used as photocatalysts because of a favorable combination of their electronic structure, light absorption properties, charge transport characteristics, and excited-state lifetime.

Apart from the binary chalcogenides, some ternary chalcogenides have also been a subject of investigation; these include $SrZrO_3$, $PbCrO_4$, $CuInS_2$, Cu_2SnS_3, and so on (Tell et al. 1971; Guo et al. 2014; Chen et al. 2015; Miseki et al. 2015).

Very little work has been carried out on the use of quaternary oxides and sulfides in comparison to binary and ternary chalcogenides. Some such photocatalysts are Bi_2AlVO_7, Cu_2ZnSnS_4, $FeZn_2Cu_3O_{6.5}$, and so on (Luan et al. 2009; Reshak et al. 2014; Kumawat et al. 2015).

1.4 MODIFICATIONS

Usually semiconductors providing promising solutions for environmental pollution problems and solar energy crisis are selected as photocatalysts. Considering the benefits and limitations of these photocatalytic materials, some researchers have attempted to enhance photocatalytic activity of these materials using various techniques. Different strategies have been used from time to time, such as surface and interface modification by controlling morphology and particle size, composite or coupling materials, transition metal doping, nonmetal doping, codoping (metal–metal, metal–nonmetal, nonmetal–nonmetal), noble metal deposition, and surface sensitization by organic dye and metal complexes, to enhance the photocatalytic properties.

1.4.1 DOPING

The addition of impurities to a very pure substance is known as doping, which is divided into the following two categories: (1) cationic doping and (2) anionic doping. In cationic doping, the semiconductor is doped with cations, for example Al, Cu, V, Cr, Fe, Ni, Co, Mn, and so on, while in anionic doping, anions are used, for example N, S, F, C, and so on. Each type of dopant has its own unique impact on crystal lattice of the photocatalyst. Metal and nonmetal ion doping on the surface of a photocatalyst increases its photoresponsiveness to the visible region by creating new energy levels (or impurity state) between the VB and CB to reduce its band gap. The electrons excited by light are shifted from the impurity state to the CB.

The photocatalytic activity of different nanoparticles such as Fe-doped TiO_2, WO_3/ZnO, and Fe-doped CeO_2 was examined by Siriwong et al. (2012). Metal-doped $SrTiO_3$ photocatalyst was prepared by Chen et al. (2012) for water splitting while Zhang et al. (2013) explained the effect of nonmetal dopants such as B, C, N, F, P, and S as anions on electronic structures of $SrTiO_3$. Anandan et al. (2012) studied the photocatalytic activity of TiO_2 and ZnO after doping by rare-earth metal La. Maeda and Yamada (2007) also studied doping of Cu, Al, and Fe with TiO_2 semiconductor.

Codoping of metal and nonmetal also increases the photocatalytic activity of the photocatalysts. Codoping of Cr + N in ZnO, Cu + Al codoped ZnO, Ga + N codoped TiO_2, and W + C codoped TiO_2 nanowires have been reported by various researchers (Wu et al. 2011; Li et al. 2012; Cho et al. 2013; Nibret et al. 2015).

1.4.2 COMPOSITES/COUPLING

Coupling of semiconductors or composite is another method to make photocatalysts effective in the visible light for different applications. Here, a large band gap semiconductor is coupled with a small band gap semiconductor having a more negative CB level. As a result, the electrons of CB can be injected from the small band gap semiconductor to the large band gap semiconductor. The dye sensitization process is similar to this method, but the only difference is that the electrons move from one semiconductor to another. Hydrogen production by coupled SnO_2, CdS, CdS/Pt–TiO_2, and NiS/Zn_xCd_{1-x}S/reduced graphene oxide has been studied (Gurunathan et al. 1997; Park et al. 2011; Zhang et al. 2014).

1.4.3 METALLIZATION

Noble metals such as Ag, Au, Pt, Ni, Cu, Rh, Pd, and so on, have been used to improve the photocatalytic activity of a semiconductor. This process decreases the possibility of electron–hole recombination, and causes efficient charge separation and higher rates of photocatalytic reaction. Noble metals having these properties can assist in electron transfer, leading to higher photocatalytic activity.

1.4.4 Dye Sensitization

Dye sensitization is a promising method for surface modification of photocatalysts to utilize the visible light for energy conversion. Dyes have redox properties and visible light sensitivity, which can be used in solar cells as well as in photocatalytic systems. When dyes are exposed to the visible light, they can inject electrons to the CB of semiconductors to start a catalytic reaction.

To convert absorbed light directly into electrical energy with higher efficiency in solar cells or through generation of hydrogen, a fast electron injection and slow backward reaction are the major requirements.

1.5 APPLICATIONS

Different semiconductors are used in the water and air purification, self-cleaning, self-sterilization, antifogging, antimicrobial activity, and so on. In these areas, the TiO_2 photocatalyst has attracted much attention because of its high catalytic efficiency, chemical stability, economy, low toxicity, and good compatibility with traditional construction materials. It is also useful for damaging microorganisms, for example bacteria and viruses, and even in inactivating some cancer cells, as well as for the photosplitting of water to produce hydrogen gas, the fuel of the future.

1.5.1 Water Treatment

Many binary and ternary semiconductors have been used as photocatalysts in wastewater treatment. TiO_2 and ZnO photocatalysts have been quite commonly used in wastewater purification. ZnO is an excellent photocatalytic oxidation material widely used in wastewater treatment in industries such as pharmacy, printing and dyeing, paper and pulp, and so on. TiO_2 nanotubes (TNTs) are the most promising photocatalysts for photocatalytic decontamination of water. Benjwal et al. (2015) reported that the graphene oxide–TiO_2/Fe_3O_4 based ternary nanocomposites have potential applications in wastewater treatment.

1.5.2 Removing Trace Metals

Some trace elements such as Hg, Cr, Pb, and other metals are extremely hazardous to human health. These toxic metals can be successfully removed, even at lower level concentrations such as parts per million, by heterogeneous photocatalysis to maintain water quality and human health.

1.5.3 Water Splitting

A number of oxides, sulfides, and selenides have been prepared as photocatalysts for water-splitting reactions. Nanosized TiO_2, many coupled semiconductor $CaFe_2O_4$/TiO_2, heterojunction WO_3/$BiVO_4$, core/shell nanofibers CdS/ZnO, and so on, offer a promising way to produce hydrogen from water (Su et al. 2011; Yang et al. 2013; Reddy et al. 2014).

1.5.4 SELF-CLEANING FUNCTIONS

The TiO$_2$ photocatalyst has attracted much attention as a photofunctional material, because cleaning glass and tile surfaces involves high-energy depletion, chemical detergents, and high costs. The organic and inorganic molecules remain adsorbed and easily degraded on the TiO$_2$-based self-cleaning surface. Then, it can be washed with water due to the high hydrophilicity of titania film. This function of titania is effective only when the number of incident solar photons per unit time is greater than the rate of adsorption of the organic pollutants on the surface. The best use of TiO$_2$ self-cleaning is in construction and as coating materials for walls in buildings, as these materials are exposed to sunlight and natural rainfall.

1.5.5 ANTIFOGGING

A superhydrophilic technology has an extremely wide range of applications, and antifogging is one of them. Generally, the fogging on the surfaces of mirrors and glasses occurs when humid air comes in contact with these surfaces to form many small droplets of water, and, as a result, the light is scattered. No water drops are formed on a highly hydrophilic surface, but a uniform thin film of water is formed, which prevents fogging. Once the surface turns into a highly hydrophilic state, it remains unchanged for several days or at least one week. Various types of glass products, mirrors, and eyeglasses have been prepared using this new technology with simple processing at a low cost. Many Japanese-made cars are being furnished with antifogging superhydrophilic side view mirrors (Spasiano et al. 2015).

1.5.6 ANTIBACTERIAL AND CANCER TREATMENT

The antibacterial effects of TiO$_2$ photocatalysis were observed by Matsunaga et al. (1985). They reported this novel concept of photochemical sterilization by using TiO$_2$ semiconductor under metal halide lamp irradiation. Different microbial toxins (such as lipopolysaccharide endotoxin, brevetoxins, microcystins, etc.) are inactivated by TiO$_2$ and it also killed a wide range of organisms, including bacteria, fungi, algae, viruses, and even cancer cells. Titania and doped TiO$_2$ coatings are being used to inhibit not only the reproduction of bacteria, but also to simultaneously decompose the bacterial cells under mild conditions. Silver is a good antibacterial material and therefore Yu et al. (2011) prepared neat TiO$_2$ and Ag–TiO$_2$ composite nanofilms for antimicrobial applications under ultraviolet (UV) illumination. One of the most common applications of TiO$_2$ photocatalytic material using its antimicrobial property is in interior paints. The mixture of TiO$_2$ and nano-ZnO shows the best photocatalytic antimicrobial effect. A photocatalyst fixation technology was used by the Japanese Arc-Flash Company, which sprayed photocatalyst directly on the surface. In these coatings, titania nanoparticles were used as the main component; these coatings can effectively sterilize and sanitize environments such as hospitals, residential kitchens, schools, floors, and so on, and killed bacteria with 98% efficiency, improving hygiene standards (Spasiano et al. 2015).

1.5.7 SELF-STERILIZATION

TiO_2 photocatalyst is chemically activated by light energy, and decomposes dirt and different types of typical pathogenic organisms such as bacteria, viruses, fungi, protozoa, and so on, from water, air, surfaces, and biological hosts under UV irradiation. Because this process is self-sterilizing TiO_2, being a physically and chemically stable and safe material, has been used as an additive for food and cosmetics. Sekiguchi et al. (2007) prepared TiO_2-coated silicon catheters. They are easily sterilized under light sources and used safely in cultured cell experiments, animal experiments, and clinical uses. Nakamura et al. (2007) also developed a self-sterilizing lancet coated with a TiO_2 photocatalytic monolayer, which showed a potential application for monitoring blood glucose in diabetes.

As nonrenewable energy sources are depleting at a great pace, there is an urgent need for such an energy source, which could prove to be inexhaustible. Photocatalysis is an emerging field of research that employs the energy of the Sun (the ultimate energy source) not only in splitting water into hydrogen and oxygen, but for different water purification and treatment processes. Photocatalysis finds applications in varied fields such as decontamination of wastewaters, solar energy conversion, photogeneration of hydrogen by splitting water, self-cleaning glasses and surfaces, antimicrobial activity, photoreduction of carbon dioxide, artificial photosynthesis, and so on. Photocatalysis also has the potential to be used in processes like self-sterilization of tiles, tents, buildings, and so on, and also in the treatment of cancer and different viral infections.

However, currently this process has certain limitations such as limited use of the solar spectrum, high cost in some cases, solubility, suspension, and so on. Efforts are being made to overcome these demerits by different modifications of the photocatalysts. Recently, use of nanoparticles of photocatalytic materials has added some newer dimensions, and the time is not far off when photocatalysis will prove to be an emerging green technology.

REFERENCES

Ameta, S. C., R. Chaudhary, R. Ameta, and J. Vardia. 2003. Photocatalysis: A promising technology for waste water treatment. *J. Indian Chem. Soc.* 80: 257–265.

Anandan, S., Y. Ikuma, and V. Murugesan. 2012. Highly active rare-earth metal La-doped photocatalysis: Fabrication, characterization, and their photocatalytic activity. *Int. J. Photoenergy*, 2012. doi:org/10.1155/2012/921412.

Benjwal, P., M. Kumar, P. Chamoli, and K. K. Kar. 2015. Enhanced photocatalytic degradation of methylene blue and adsorption of arsenic (III) by reduced graphene oxide (rGO)-metal oxide (TiO_2/Fe_3O_4) based nanocomposites. *RSC Adv.* 5: 73249–73260.

Chen, H.-C., C.-W. Huang, J. C. S. Wu, and S.-T. Lin. 2012. Theoretical investigation of the metal-doped $SrTiO_3$ photocatalysts for water splitting. *J. Phys. Chem. C.* 116(14): 7897–7903.

Chen, O. J., J. Chen, T. Wang, Z. Li, and X. Dou. 2015. The photovoltaic properties of novel narrow band gaps Cu_2SnS_3 films prepared by spray pyrolysis method. *RSC Adv.* 5(37): 28885–28891.

Cho, I. S., C. H. Lee, Y. Feng, M. Logar, P. M. Rao, L. Cai et al. 2013. Co-doping titanium dioxide nanowires with tungsten and carbon for enhanced photoelectrochemical performance. *Nat. Commun.* 4: 1723. doi:10.1038/ncomms2729.

Fox, M. A. and M. T. Dulay. 1993. Heterogeneous photocatalysis. *Chem. Rev.* 93(1): 341–357.

Fujishima, A. and K. Honda. 1972. Electrochemical photolysis of water at a semiconductor electrode. *Nature.* 238(5358): 37–38.

Fujishima, A., T. N. Rao, and D. A. Tryk. 2000. Titanium dioxide photocatalysis. *J. Photochem. Photobiol. C: Photochem. Rev.* 1: 1–21.

Guo, Z., B. Sa, B. Pathak, J. Zhou, R. Ahuja, and Z. Sun. 2014. Band gap engineering in huge-gap semiconductor $SrZrO_3$ for visible light photocatalysis. *Int. J. Hydrogen Energy.* 39(5): 2042–2048.

Gurunathan, K., P. Maruthamuthu, and V. C. Sastri. 1997. Photocatalytic hydrogen production by dye-sensitized Pt/SnO_2 and $Pt/SnO_2/RuO_2$ in aqueous methyl viologen solution. *Int. J. Hydrogen energy.* 22(1): 57–62.

Konstantinou, I. K. and T. A. Albanis. 2014. TiO_2-assisted photocatalytic degradation of azo dyes in aqueous solution: Kinetic and mechanistic investigations: A review. *Appl. Catal. B Environ.* 49: 1–14.

Kumawat, P., M. Joshi, R. Ameta, and S. C. Ameta. 2015. Photocatalytic degradation of basic fuchsim over quaternary oxide iron zinc cuprate ($FeZn_2Cu_3O_{6.5}$). *Adv. Appl. Sci. Res.* 6(7): 209–215.

Li, X. B., Q. Liu, X. Y. Jiang, and J. Huang. 2012. Enhanced photocatalytic activity of Ga-N co-doped anatase TiO_2 for water decomposition to hydrogen. *Int. J. Electrochem. Sci.* 7: 11519–11527.

Luan, J., W. Zhao, J. Feng, H. Cai, Z. Zheng, B. Pan, X. Wu, Z. Zou, and Y. Li. 2009. Structural, photophysical and photocatalytic properties of novel Bi_2AlVO_7. *J. Hazard. Mater.* 164(2–3): 781–789.

Maeda, M. and T. Yamada. 2007. Photocatalytic activity of metal-doped titanium oxide films prepared by sol-gel process. *J. Phys. Conf. Ser.* 61: 755–759.

Matsunaga, T., R. Tomoda, T. Nakajima, and H. Wake. 1985. Photoelectrochemical sterilization of microbial cells by semiconductor powders. *FEMS Microbiol. Lett.* 29(1–2): 211–214.

Miseki, Y., O. Kitao, and K. Sayama. 2015. Photocatalytic water oxidation over $PbCrO_4$ with 2.3 eV band gap in IO_3^-/I^- redox mediator under visible light. *RSC Adv.* 5(2): 1452–1455.

Nakamura, H., M. Tanaka, S. Shinohara, M. Gotoh, and I. Karube. 2007. Development of a self-sterilizing lancet coated with a titanium dioxide photocatalytic nano-layer for self-monitoring of blood glucose. *Biosens. Bioelectron.* 22(9–10): 1920–1925.

Nibret, A., O. P. Yadav, I. Diaz, and A. M. Taddesse. 2015. Cr-N co-doped ZnO nanoparticles: Synthesis, characterization and photocatalytic activity for degradation of thymol blue. *Bull. Chem. Soc. Ethiop.* 29(2): 247–258.

Park, H., Y. K. Kim, and W. Choi. 2011. Reversing CdS preparation order and its effects on photocatalytic hydrogen production of $CdS/Pt-TiO_2$ hybrids under visible light. *J. Phys. Chem. C.* 115(13): 6141–6148.

Plotnikow, J. 1936. *Allgemeine-photochemie*. Berlin and Leipzig: Walter De Gruyter & Co.

Reddy, P. A. K., B. Srinivas, V. D. Kumari, M. V. Shankar, M. Subrahmanyam, and J. S. Lee. 2014. $CaFe_2O_4$ sensitized hierarchical TiO_2 photo composite for hydrogen production under solar light irradiation. *Chem. Eng. J.* 247: 152–160.

Reloez, M., N. T. Nolan, S. C. Pillai, M. K. Seery, P. Falaras, A. G. Kontos et al. 2012. A review on the visible light active titanium dioxide photocatalysts for environmental applications. *Appl. Catal. B: Environ.* 125: 331–349.

Rengel, V. C., S. V. Khedkar, and N. K. J. Thanvi. 2012. Photocatalytic oxidation and reactors– A review. *Int. J. Adv. Eng. Technol.* 3(4): 31–35.

Reshak, A. H., K. Nouneh, I. V. Kityk, J. Bila, S. Auluck, H. Kamarudin, and Z. Sekkat. 2014. Structural, electronic and optical properties in earth-abundant photovoltaic absorber of Cu_2ZnSnS_4 and $Cu_2ZnSnSe_4$ from DFT calculations. *Int. J. Electrochem. Sci.* 9: 955–974.

Sekiguchi, Y., Y. Yao, Y. Ohko, K. Tanaka, T. Ishido, A. Fujishima, and Y. Kubota. 2007. Self-sterilizing catheters with titanium dioxide photocatalyst thin films for clean intermittent catheterization: Basics and study of clinical use. *Int. J. Urol.* 14(5): 426–430.

Siriwong, C., N. Wetchakun, B. Inceesungvorn, D. Channei, T. Samerjai, and S. Phanichphant. 2012. Doped-metal oxide nanoparticles for use as photocatalysts. *Prog. Cryst. Growth Charact. Mater.* 58: 145–163.

Spasiano, D., R. Marotta, S. Malato, P. Fernandez-Ibanez, and I. D. Somma. 2015. Solar photocatalysis: Materials, reactors, some commercial and pre-industrialized applications. A comprehensive approach. *Appl. Catal. B Environ.* 170: 90–123.

Su, J., L. Guo, N. Bao, and C. A. Grimes. 2011. Nanostructured $WO_3/BiVO_4$ heterojunction films for efficient photoelectrochemical water splitting. *Nano Lett.* 11(5): 1928–1933.

Tell, B., J. L. Shay, and H. M. Kasper. 1971. Electrical properties, optical properties and band structures of $CuGaS_2$ and $CuInS_2$. *Phys. Rev. B.* 4(8): 2463–2471.

Wu, S. Z., H. L. Yang, X. G. Xu, J. Miao, and Y. Jiang. 2011. Ferromagnetism studies of Cu-doped and (Cu, Al) co-doped ZnO thin films. *J. Phys. Conf. Ser.* 263: 012022.

Yang, G., W. Yan, Q. Zhang, S. Shen, and S. Ding. 2013. One dimensional CdS/ZnO core/shell nanofibers via single-spinneret electrospinning: Tunable morphology and efficient photocatalytic hydrogen production. *Nanoscale.* 5(24): 12432–12439.

Yu, B., K. M. Leung, Q. Guo, W. M. Lau, and J. Yang. 2011. Synthesis of $Ag\text{-}TiO_2$ composite nano thin film for antimicrobial application, *Nanotechnology.* 22(11): 115603.

Zhang, C., Y. Jia, Y. Jing, Y. Yao, J. Ma, and J. Sun. 2013. Effect of non-metal elements (B, C, N, F, P, S) mono-doping as anions on electronic structure of $SrTiO_3$. *Comput. Mater. Sci.* 79: 69–74.

Zhang, J., L. Qi, J. Ran, J. Yu, and S. Z. Qiao. 2014. Ternary $NiS/Zn_xCd_{1-x}S$/reduced graphene oxide nanocomposites for enhanced solar photocatalytic H_2-production activity. *Adv. Energy Mater.* 4(10). doi:10.1002/aenm.201301925.

2 Photocatalysis

2.1 INTRODUCTION

Chemical-based industries have played a significant role in human civilization, but extensive anthropogenic and industrial activities have introduced large quantities of chemicals in the environment, causing potential harm to ecosystems. These pollutants are becoming a major source of environmental contamination. As the international environmental standards are becoming more and more stringent, many research studies have been focused on the treatment of wastewater containing different contaminants. However, because of the complexity and variety of these pollutants, it has become rather difficult to find a unique treatment procedure that entirely covers the effective elimination of all types of compounds. Physical methods such as flocculation, reverse osmosis, and adsorption on activated charcoal are nondestructive and merely transfer the pollutant to other media, thus giving rise to secondary pollution.

Heterogeneous semiconductor photocatalysis has been extensively studied. It is a promising approach for degradation of a large number of organic pollutants and it is found to be cost effective as well. Among many semiconductors, TiO_2 has been one of the most commonly used and investigated photocatalysts ever since Fujishima and Honda (1972) used TiO_2 photoanode in combination with a platinum electrode in an aqueous electrolyte solution for water cleavage. Photocatalysis has significant potential for creating renewable energy resources and also for cleaning our environment. There is lots of hope from its possible applications in various fields.

2.2 PHOTOCATALYSIS

Although there is no consensus in the scientific community as to a proper definition of photocatalysis, the term can be generally used to describe a process where light is used to activate a substance—the photocatalyst—that modifies the rate of a chemical reaction without itself being involved in the chemical transformation. Thus, the main difference between a conventional thermal catalyst and photocatalyst is that the former is activated by heat whereas the latter is activated by photons of appropriate energy.

The IUPAC definition of photocatalysis is "A catalytic reaction involving light absorption by substrate." In a real sense, the term photocatalysis is used for chemical reactions occurring in the presence of light and a photocatalyst.

The principle of photocatalysis is based on the activation of a semiconductor particulate material by the action of radiation with an appropriate wavelength. Photocatalysis is used for the elimination of several pollutants (e.g., alkanes, alkenes, phenols, aromatics, pesticides) and complete mineralization of the organic

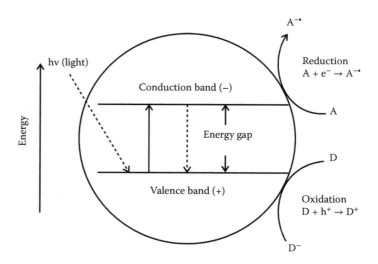

FIGURE 2.1 Mechanism of photocatalysis.

compounds. Several photocatalysts such as CdS, Fe_2O_3, ZnO, WO_3, ZnS, and so forth have been studied, but the best results have been obtained with TiO_2 P25. The field of photocatalysis has been excellently reviewed by various researchers over time. Some of the latest reviews are by Fox and Dulay (1993), Ameta et al. (1999), Chong et al. (2010), Pelaez et al. (2012), Ameta et al. (2012), Ibhadon and Fitzpatrick (2013), Schneider et al. (2014), Lang et al. (2014), and so on.

When a photocatalyst is irradiated with a light of suitable wavelength, an electron is excited to the conduction band (CB), leaving behind a positive hole in the valence band (VB). The electron in the CB can be utilized to reduce any substrate, whereas the hole in the VB can be used for oxidizing some compounds (Figure 2.1).

2.3 PHOTOCATALYTIC REACTIONS

Basically, three processes, oxidation, electron injection, and reduction, occur in photocatalytic degradation. These three steps are as follows:

1. Oxidation
   This is the common photocatalytic degradation process of organic compounds (Figure 2.2).
2. Electron injection
   This is a case of spectral sensitization, which is observed in a wet-type solar cell (Figure 2.3).
3. Oxidation and reduction
   One moiety serves as the electron acceptor that is reduced, thus suppressing recombination between electron and positive hole (Figure 2.4).

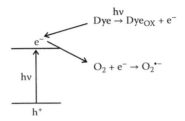

FIGURE 2.2   Oxidation.

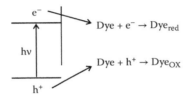

FIGURE 2.3   Electron injection.

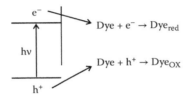

FIGURE 2.4   Redox reactions.

There are four possible combinations of semiconductor and substrate, depending upon the relative positions of bands and redox levels. These are as follows:

1. If the redox level of the substrate is lower than the CB of the semiconductor, then reduction of substrate occurs.
2. If the redox level of the substrate is higher than the VB of the semiconductor, then oxidation of substrate takes place.
3. If the redox level of the substrate is higher than the CB and lower than the VB of the semiconductor, then neither oxidation nor reduction is possible.
4. If the redox level of the substrate is lower than the CB and higher than the VB, then both reduction and oxidation of the substrate occur.

All four possibilities are presented in Figure 2.5.

2.4   VARIOUS PHOTOCATALYSTS

A number of semiconductors have been used successfully for different applications. These are either binary, tertiary, or quaternary semiconductors.

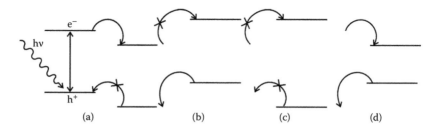

FIGURE 2.5 Various electron–hole transfer possibilities: (a) reduction, (b) oxidation, (c) no reaction, (d) redox reaction.

TABLE 2.1
Band Gap of Different Semiconductors

Binary Photocatalysts

Photocatalysts	Band Gap (eV)	Photocatalysts	Band Gap (eV)
WSe_2	1.40	WO_3	2.50–2.80
CdTe	1.49	ZnSe	2.70
CdSe	1.74	In_2O_3	2.90
Cu_2O	1.90	TiO_2	3.02–3.20
CuO	1.9	ZnO	3.20
ZnTe	2.25	GaN	3.44
GaP	2.26	SnO_2	3.50
CdS	2.40	ZnS	3.70
V_2O_5	2.70	MnO	3.60

Ternary Photocatalysts

Photocatalysts	Band Gap (eV)	Photocatalysts	Band Gap (eV)
$CuInSe_2$	1.00	$BiVO_4$	2.40
Cu_2SnS_3	1.16	$CuGaS_2$	2.53
$CuInS_2$	1.55	$InTaO_4$	2.60
$ZnFe_2O_4$	1.90	Bi_2WO_6	2.70
La_2CuO_4	2.00	$BaTiO_3$	3.00
$PbCrO_4$	2.30	$SrTiO_3$	3.40
$BiVO_4$	2.40		

Quaternary Photocatalysts

Photocatalysts	Band Gap (eV)	Photocatalysts	Band Gap (eV)
$Cu_2ZnSnSe_4$	1.00	Bi_2AlVO_7	2.06
$Cu_{1.0}Ga_xIn_{2-x}S_{3.5}$	1.43–2.42	Bi_2InTaO_7	2.81
Cu_2ZnSnS_4	1.50	$FeZn_2Cu_3O_{6.5}$	2.70
$Li_2CuMo_2O_8$	1.54–1.65		

The band gaps of different binary, tertiary, and quaternary semiconductors are given in Table 2.1.

2.5   REACTIVE OXYGEN SPECIES

Major reactive oxygen species are as follows:

- Molecules like hydrogen peroxide H_2O_2
- Ions like hypochlorite ion (OCl^-)

- Radicals like the hydroxyl radical ($^\bullet$OH)
- Superoxide anion ($O_2^{\bullet-}$), which is both an ion and a radical

A hydroxyl radical removes a hydrogen atom from the substrate. A molecule of water is eliminated and a new radical is formed, which may further degrade.

$$RH + {}^\bullet OH \rightarrow R^\bullet + H_2O \qquad (2.1)$$

The reduction of molecular oxygen (O_2) produces superoxide $O_2^{\bullet-}$, which is the precursor of most other reactive oxygen species.

$$O_2 + e^- \rightarrow O_2^{\bullet-} \qquad (2.2)$$

Dismutation of superoxide produces hydrogen peroxide. Here, one superoxide radical acts as an electron donor while another radical accepts an electron to form O_2^{2-}, which reacts with two protons to form hydrogen peroxide.

$$2H^+ + O_2^{\bullet-} + O_2^{\bullet-} \rightarrow H_2O_2 + O_2 \qquad (2.3)$$

Hydrogen peroxide may be either partially reduced to form hydroxyl radicals ($^\bullet$OH) or fully reduced to form water.

$$H_2O_2 \rightarrow {}^\bullet OH + {}^\bullet OH \qquad (2.4)$$

$$H_2O_2 + H_2O_2 \rightarrow H_2O + H_2O + O_2 \qquad (2.5)$$

The hydroxyl radical is the most reactive of these, and immediately removes electrons from any molecule present in its path, turning that molecule into a free radical, resulting in propagating a chain reaction.

The hydroxyl radical is one of the most potent oxidizing agents next to fluorine. It is too reactive to diffuse far in an environment to reach the target for oxidation (Table 2.2).

TABLE 2.2
Relative Oxidation Power of Some Oxidizing Species

Species	Relative Oxidation Power (eV)
Fluorine	2.23
Hydroxyl radical	2.06
Atomic oxygen	1.78
Hydrogen peroxide	1.31
Perhydroxyl radical	1.25
Permaganate	1.24
Chlorine dioxide	1.15
Hypochlorous acid	1.10
Hypoiodous acid	1.07
Chlorine	1.00
Bromine	0.80
Iodine	0.54

The chemical reactions of the hydroxyl radical in water can be divided into the following four types:

1. Addition

$$^{\bullet}OH + C_6H_6 \rightarrow C_6H_6(^{\bullet}OH) \xrightarrow{-H^{\bullet}} C_6H_5OH \qquad (2.6)$$

The hydroxyl radical reacts with a compound and forms a free intermediate that undergoes further reaction to form the final product.

2. Hydrogen abstraction

$$^{\bullet}OH + CH_3OH \rightarrow CH_2{}^{\bullet}OH + H_2O \rightarrow \text{Products,} \qquad (2.7)$$

where an organic free radical and water are formed.

3. Electron transfer

$$^{\bullet}OH + [Fe(CN)_6]^{4-} \rightarrow [Fe(CN)_6]^{3-} + OH^-, \qquad (2.8)$$

where ions of a higher valance state are formed or an atom free radical, if a negative ion is oxidized.

4. Radical interaction

$$^{\bullet}OH + {}^{\bullet}OH \rightarrow H_2O_2 \qquad (2.9)$$

$$^{\bullet}OH + R^{\bullet} \rightarrow ROH, \qquad (2.10)$$

where two hydroxyl radicals react with each other or the hydroxyl radical reacts with an unlike radical to form a stable product.

On irradiation, TiO_2 generates electrons in the CB (e^-) and holes in the VB (h^+). The photogenerated electron–hole pairs $(e^-\!-\!h^+)$ are trapped at the TiO_2 surface. These $e^-\!-\!h^+$ pairs recombine within 10 ns, but after 250 ns of formation, the remaining photogenerated $e^-\!-\!h^+$ pairs are trapped at the TiO_2 surface. These electrons reduce O_2 to a superoxide ion $(O_2{}^{\bullet-})$, which on further reduction produces H_2O_2. As the electron is abstracted by O_2, this step prevents recombination of the electron–hole pair.

$$TiO_2 \xrightarrow{h\nu} h_{VB}{}^+ + e_{CB}{}^- \qquad (2.11)$$

$$O_2 + TiO_2(e_{CB}{}^-) \rightarrow O_2{}^{\bullet-} + TiO_2 \qquad (2.12)$$

$$O_2{}^{\bullet-} + TiO_2(e_{CB}{}^-) + 2H^+ \rightarrow H_2O_2 \qquad (2.13)$$

When H_2O_2 reacts with superoxide ions, it may increase the rate of photodegradation. Hydroxyl ions, hydroxyl radicals, and oxygen are formed as the product, or H_2O_2 is reduced into hydroxyl ions and radicals by the CB electrons.

$$O_2{}^{\bullet-} + H_2O_2 \rightarrow {}^{\bullet}OH + {}^-OH + O_2 \qquad (2.14)$$

$$TiO_2(e_{CB}^-) + H_2O_2 \rightarrow TiO_2 + {}^{\bullet}OH + {}^-OH \qquad (2.15)$$

The positive hole in the VB reacts with hydroxyl ions to form ${}^{\bullet}OH$ radicals and oxidative decomposition of water also generates ${}^{\bullet}OH$, while the recombination of ${}^{\bullet}OH$ radicals leads to production of H_2O_2 under aerobic conditions.

$$TiO_2(h_{VB}^+) + {}^-OH \rightarrow {}^{\bullet}OH \qquad (2.16)$$

$$TiO_2(h_{VB}^+) + H_2O \rightarrow H^+ + {}^{\bullet}OH \qquad (2.17)$$

$$2{}^{\bullet}OH \rightarrow H_2O_2 \qquad (2.18)$$

Sunlight is an unlimited source of energy that can be effectively utilized for purification and conservation of the environment. One such technique is photocatalysis, which may utilize sunlight in various fields such as self-cleaning, antibacterial, antifungal, antiviral, self-sterlization, anticancer, wastewater remediation, energy generation in the form of electricity and hydrogen, carbon dioxide reduction to useful fuels, and so on. The reactive species involved in photocatalysis such as hydroxyl radical, hydroperoxide radical, superoxide anion radical, and so on are responsible for carrying out the processes involving the oxidative and/or reductive capabilities of the photocatalyst in its naive or modified forms.

REFERENCES

Ameta, S., C. Ameta, R. Vardia, J. Ameta, and Z. Ali. 1999. Photocatalysis: A frontier of photochemistry. *J. Indian Chem. Soc.* 76 (6): 281–287.

Ameta, R., S. Benjamin, A. Ameta, and S. C. Ameta. 2012. Photocatalytic degradation of organic pollutants: A review. Photocatalytic Materials and Surfaces for Environmental Cleanup-II, R. J. Tayade (Ed.). *Mater. Sci. Forum.* 734: 247–272.

Chong, M. N., Bo Jin, C. W. K. Chow, and C. Saint. 2010. Recent developments in photocatalytic water treatment technology: A review. *Water Research.* 44: 2997–3027.

Fox, M. A. and M. T. Dulay. 1993. Heterogeneous photocatalysis. *Chem. Rev.* 93 (1): 341–357.

Fujishima, A. and K. Honda. 1972. Electrochemical photolysis of water at a semiconductor electrode. *Nature.* 238 (5358): 37–38.

Ibhadon, A. O. and P. Fitzpatrick. 2013. Heterogeneous photocatalysis: Recent advances and applications. *Catalysts.* 3: 189–218.

Lang, X., X. Chen, and J. Zhao. 2014. Heterogeneous visible light photocatalysis for selective organic transformations. *Chem. Soc. Rev.* 43: 473–486.

Pelaez, M., N. T. Nolan, S. C. Pillai, M. K. Seery, P. Falaras, A. G. Kontos et al. 2012. A review on the visible light active titanium dioxide photocatalysts for environmental applications. *Appl. Catal. B: Environ.* 125: 331–349.

Schneider, J., M. Matsuoka, M. Takeuchi, J. Zhang, Y. Horiuchi, M. Anpo, and D. W. Bahnemann. 2014. Understanding TiO₂ photocatalysis: Mechanisms and materials. *Chem. Rev.* 114 (19): 9919–9986.

3 Binary Semiconductors

3.1 INTRODUCTION

Photocatalysis is a fantastic way to clean wastewater, industrial effluents, and our environment in general. One can reduce pollution in air and water by modifying and further developing this technology. It can also put a check on the spread of infections and diseases such as severe acute respiratory syndrome (SARS). This cleaner way of life would benefit everyone around the globe. A good photocatalyst should be photoactive, able to utilize visible and/or near UV light, biologically and chemically inert, photostable, inexpensive, and nontoxic in nature. The redox potential of the photogenerated valence band (VB) hole must be sufficiently positive for a semiconductor to be photochemically active so that it can generate $\cdot OH$ radicals that can subsequently oxidize the organic pollutants. The redox potential of the photogenerated conductance band electron must be sufficiently negative to be able to reduce absorbed oxygen to a superoxide. TiO_2, ZnO, WO_3, CdS, ZnS, SnO_2, WSe_2, Fe_2O_3, and so on can be used as effective photocatalysts in combating against the problem of environmental pollution.

The major advantage of photocatalysis is the fact that there is no further requirement for any secondary disposal methods as the organic contaminants are converted to carbon dioxide, water, inorganic ions, and so on. Other treatment methods such as adsorption by activated carbon and air stripping merely concentrate the chemicals present as pollutants and transfer them to the adsorbent or air. They do not convert them to nontoxic wastes as in the case of photocatalysis. Compared to other oxidation technologies, expensive oxidation methods are not required as ambient oxygen and water is used.

Photocatalysis is an important process of advanced oxidation processes (AOPs). These are widely used for the removal of recalcitrant organic constituents from industrial and municipal wastewater. The homogenous photocatalytic system is well studied and it has been reported to be a promising method for wastewater treatment, but the ions remain in the solution at the end of the process in homogeneous photocatalytic reactions. Therefore, the removal of sludge at the end of the wastewater treatment becomes necessary and increases the cost, as large amounts of chemicals and manpower is needed for this purpose. This disadvantage of homogeneous catalytic systems can be overcome by heterogeneous photocatalysis using an appropriate photocatalyst.

3.2 OXIDES

3.2.1 Titanium Dioxide

Titanium dioxide (TiO_2) is an excellent photocatalyst with wide applications in various fields. The main advantages of TiO_2 are its high chemical stability on exposure to acidic and basic conditions, nontoxic behavior, relatively low cost, and environmentally safe and high oxidizing characteristics, which make it a prospective candidate for many photocatalytic applications. TiO_2 exists in three different crystalline modifications. These are anatase, brookite, and rutile. Out of these forms, anatase exhibits the highest overall photocatalytic activity. The high oxidizing power of TiO_2 in the presence of light makes it suitable for decomposition of organic and inorganic compounds even at very low concentrations. The photocatalytic effect of TiO_2 can be used for self-cleaning surfaces, decomposing atmospheric pollutants (in water and air), self-sterilization, and so on.

The TiO_2 is both intrinsic and n-type (nonintentionally doped), due to oxygen vacancies in the TiO_2 lattice, and its properties are similar to ZnO. Of course, studies have been conducted to dope the TiO_2 and introduced a p-type conductivity in it.

Irradiation of TiO_2 with light of a suitable wavelength promotes an electron from its VB to the CB. The electron in the CB is now readily available for transfer, while the hole in VB is open for donation. Both possibilities are here: a reactant receiving the electron from TiO_2 will be reduced, while a reactant donating electron to semiconductor to occupy the hole will be oxidized.

TiO_2 possesses a large band gap value (3.2 eV), which is a disadvantage for its use in photocatalysis; hence, efforts have been made for increasing its efficiency by doping. Doping refers to the introduction of some impurities to the material for the purpose of modifying its physical properties or electrical characteristics. A dopant may increase the level of the VB edge or lower the CB level, and depending on the nature of dopant, whether it is a metal or a nonmetal; thus reduce the band gap. It improves or minimizes electron–hole recombination, so as to minimize any loss in quantum yield.

Nanosized titanium oxide powders were prepared by Bessekhouad et al. (2003) via sol–gel route. The preparation parameters were optimized using malachite green oxalate degradation. The catalyst was excellent as compared to TiO_2–P25 using two pollutants, 4-hydroxy benzoic acid and benzamide (BZ); however, their performance was found strongly dependent on the type of pollutant. The photodegradation of malachite green was studied under different pH values and amounts of TiO_2 (Chen et al. 2007). MG (99.9%) was degraded with addition of 0.5 g/L TiO_2 to solutions containing 50 mg/L of dye in 4 hours. They indicated that the N-de-methylation degradation of MG dye took place in a stepwise manner to yield mono-, di-, tri-, and tetra-N-de-methylated MG species during degradation.

Aarthi et al. (2007) evaluated the dependence of photocatalytic rate on the molecular structure of azure (A and B) and sudan (III and IV) dyes. The photocatalytic activity of combustion-synthesized TiO_2 (CS–TiO_2) was compared with that of Degussa P25. It was observed that the photodegradation rate was higher in solvents with higher polarity. The effect of pH and the presence of metal salts on degradation of azure A was also investigated.

Photocatalytic degradation of acetamiprid, a widely used pyridine-based neonicotinoid insecticide, was observed by Guzsvány et al. (2009) in UV-irradiated aqueous suspensions of O_2/TiO_2. Acetaldehyde, formic and acetic acid, and pyridine-containing intermediates (e.g., 6-chloronicotinic acid) were formed during the process. The pH changed from 5 to 2 during the photocatalytic process.

Gas-sensitive materials (both n-type and p-type) were developed from NiO_x-doped TiO_2 thin films (Wisitsoraat et al. 2009). TiO_2 gas-sensing layers were deposited over a wide range of NiO_x content (0–10 wt.%). It was reported that NiO_x content as high as 10 wt.% was needed to invert the n-type conductivity of TiO_2 into p-type conductivity. Notable gas-sensing response differences were observed between n-type and p-type NiO_x-doped TiO_2 thin film. NiO_x (p-type) doping results in enhanced response toward acetone and ethanol.

Chen et al. (2009) prepared TiO_2 photocatalyst by a surface chemical modification process with toluene 2,4-diisocyanate (TDI). As-prepared TiO_2–TDI had an absorption in the visible region because of the ligand-to-metal charge transfer (LMCT) excitation of the surface complex. This photocatalyst is stable and showed high photocatalytic performance for the degradation of 2,4-dichlorophenol. The turnover number (TON) of the photocatalyst for photodegradation of 2,4-dichlorophenol was 15.43 even after using it five times under visible light irradiation.

Rezaee et al. (2009) reported the photocatalytic degradation of reactive blue 19 (RB19) dye using TiO_2 nanofiber as the photocatalyst in an aqueous solution under UV irradiation. Nanofiber was prepared using a templating method with tetraisopropylorthotitanate as a precursor. Chemical oxygen demand (COD) measurements were also carried out. A significant decrease in the COD values was observed indicating that the photocatalytic method is quite good for the removal of RB19. It was found that photocatalytic decomposition of RB19 was most efficient in acidic medium.

Nanoglued binary titania (TiO_2)–silica (SiO_2) aerogel has been synthesized on glass substrates by Luo et al. (2009). Anatase TiO_2 aerogel was immobilized into a three-dimensional (3D) mesoporous network of the SiO_2 with the help of an about-to-gel SiO_2 sol as nanoglue. This binary aerogel exhibited high photocatalytic activity for the degradation of methylene blue (MB) under simulated solar light. It was also revealed that the hydroxyl radical was formed during the illumination of the binary TiO_2–SiO_2 aerogel, which acts as an oxidant in oxidative degradation of MB.

Mahmoodi and Arami (2009) studied the feasibility and performance of photocatalytic degradation and toxicity reduction of a textile dye (acid blue 25) at a pilot scale on immobilized titania. The effects of operational parameters such as H_2O_2, pH, and dye concentration were investigated on the photocatalytic degradation of Acid Blue 25. *Daphnia magna* bioassay was used to test the progress of toxicity during the treatment process and it was observed that the residual acute toxicity reduced during the photocatalytic degradation.

Saggioro et al. (2011) studied the photocatalytic degradation of two commercial textile azo dyes, namely reactive black 5 (RB5) and reactive red 239 (RR 239) using TiO_2 P25 Degussa as a catalyst. Photodegradation was carried out in an aqueous solution with a 125 W mercury vapor lamp. The effects of the amount of TiO_2 used, UV light irradiation time, pH initial concentration of the azo dye, and addition of different concentrations of hydrogen peroxide were investigated. It was observed that the degradation rates achieved in mono- and bi-component systems were almost identical. The rate of color lost was 77% of the initial rate even after five cycles.

A spray pyrolysis procedure was used by Stambolova et al. (2012) for preparation of nanostructured TiO_2 films. Thin films of active nanocrystalline titania were obtained from titanium isopropoxide, stabilized with acetyl acetone. The activity of this TiO_2 was tested for photocatalytic degradation of RB5 dye. The reduction of toxicity after photocatalytic treatment of RB5 was observed with TiO_2 taking mortality of *Artemia salina* as a standard. It was observed that thickness of the film, conditions of postdeposition treatment, and the type of the substrate affected the photocatalytic reaction.

Photocatalytic decolorization of brilliant green yellow (BGY), an anionic dye, has been observed in TiO_2 and ZnO aqueous dispersions under UV light irradiation (Habib et al. 2012). Adsorption was found to be prerequisite for the metal oxide-mediated photodegradation/photodecolorization of the dye. Complete decolorization of water was achieved on the UV irradiation but only 75% degradation of BGY was found. ZnO-mediated decolorization was better and faster as compared to TiO_2.

Zhang et al. (2014) successfully synthesized bicrystalline TiO_2 supported acid activated sepiolite (TiO_2/AAS) fibers by a simple method at low temperature. It was indicated that the binary mixtures of anatase and rutile exist in TiO_2/AAS composites. Photocatalytic activity of the TiO_2/AAS composites was evaluated by the degradation of gaseous formaldehyde (a main indoor air pollutant) under UV light irradiation, which was found to be excellent and superior to that of TiO_2/sepiolite (raw sepiolite) as well as pure TiO_2. This activity was attributed to the anatase–rutile mixed phase and bimodal pore structure in the TiO_2/AAS composites.

A facile synthesis of hydrogenated TiO_2 (H-TiO_2) nanobelts was reported by Tian et al. (2015). This exhibited excellent UV and visible photocatalytic activity for decomposition of methyl orange (MO) and water splitting for hydrogen production. This improved photocatalytic property was attributed to the Ti^{3+} ions and oxygen vacancies in TiO_2 nanobelts created by hydrogenation, which can enhance visible light absorption, promote charge carrier trapping, and hinder the photogenerated electron–hole recombination.

Black TiO_2 may be an excellent solution to clean polluted air and water and to produce hydrogen. Black TiO_2 had crystalline core-amorphous shell structure and was easily reduced by CaH_2 at 400°C (Zhu et al. 2015). It harvests over 80% solar absorption, while white TiO_2 harvests only 7%. It was 2.4 times faster in water decontamination and 1.7 times higher in H_2 production than pristine TiO_2. This method could provide a promising and cost-effective approach to improve the visible light absorption and performance of TiO_2.

3.2.2 ZINC OXIDE

ZnO is a wide band gap semiconductor and belongs to the II–VI semiconductor group. This semiconductor has several favorable properties, including good transparency, high electron mobility, and a wide band gap. Zinc oxide crystallizes in two main forms: hexagonal wurtzite and cubic zinc blende, out of which the wurtzite structure is most common.

The photocatalytic degradation of crystal violet (CV) dye over zinc oxide suspended in an aqueous solution has been carried out by Rao et al. (1997). The photocatalytic reaction was monitored spectrophotometrically by observing absorbance at different time intervals. The effect of various operating parameters was observed, including concentration of CV dye, pH, amount and nature of semiconductor, and light intensity. Observed similar to concentration of CV dye, pH, amount, and nature of semiconductor and light intensity.

Sunlight-mediated photocatalytic degradation of rhodamine B (RhB) dye was studied using hydrothermally prepared ZnO (Byrappa et al. 2006). Zinc chloride was used as the starting material in the hydrothermal synthesis of ZnO. The disappearance of RhB followed the first-order kinetics. The thermodynamic parameters of the photodegradation were determined similar to parameters of energy of activation, enthalpy of activation, entropy of activation, and free energy of activation. An actual textile effluent was also tried, which contained RhB as a major constituent along with other dyes and dyeing auxiliaries. The reduction in the COD of the treated effluent indicated a complete destruction of the organic molecules along with color removal.

ZnO nanostructures were synthesized by Mohajerani et al. (2009) in different shapes such as particle, rods, flower-like, and microsphere via a facile hydrothermal method. The effect of morphology on the decolorization of acid red 27 (AR 27) solution was observed under direct irradiation of sunlight. It was found that these have Wurtzite-type hexagonal structure with high crystallinity and crystallite size in the range of 67–100 nm. The decolorization of AR 27 followed the first-order kinetics.

ZnO particles with different morphologies such as rod-like, rice-like, and disc-like were successfully synthesized via the sol–gel approach by Pung et al. (2012) through adjustment of the reaction parameters such as amount of ammonia and reaction time as well as complexing agent aluminum sulfate $Al_2(SO_4)_3$. Rod-like ZnO particles were found to be the most effective in degrading the RhB solution under the illumination of the UV light. The first-order rate constants were found to be in the following order:

Rod-like ZnO particles > Rice-like ZnO particles > Disc-like ZnO particles

Ameen et al. (2013) used a simple solution method for the synthesis of ZnO flowers using zinc acetate precursor in a basic medium. It was employed for the degradation of CV dye. The average length of ZnO flower with well-defined petals was ~300 ± 50 nm. As-synthesized ZnO flowers showed efficient degradation of CV dye with a degradation rate of ~96% within 80 minutes.

Ag–N-codoped zinc oxide nanoparticles (NPs) were synthesized by Welderfael et al. (2013) through impregnation of Ag–N-doped ZnO NPs. Photocatalytic degradation of methyl red (MR) dye using Ag–N-codoped ZnO was studied under solar as well as UV irradiation. Codoped zinc oxide photocatalysts showed higher photocatalytic activity as compared to pure zinc oxide. Calcined zinc oxide showed better photocatalytic activity than commercial zinc oxide. The high catalytic activity of Ag–N-codoped zinc oxide was attributed to the lower rate of recombination of the photogenerated electrons and holes and its lower band gap energy.

Pure and nitrogen-doped ZnO nanospheres were successfully prepared by Lavand and Malghe (2015) using a microemulsion method. It was indicated that nanosized N-doped ZnO was spherical and had wurtzite phase. The incorporation of nitrogen into the oxygen site of ZnO causes lattice compression. It was observed that N doping significantly enhanced the light absorption capacity of ZnO in the visible region

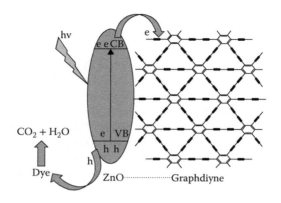

FIGURE 3.1 Graphdiyne–ZnO nanohybrids as an advanced photocatalytic material. (Adapted from Thangavel, S. et al., *J. Phys. Chem. C*, 119, 22057–22065, 2015. With permission.)

and exhibited higher photocatalytic activity as compared to that of commercial and pure ZnO NPs. As-prepared nanosized N-doped ZnO was found to be highly stable and reusable.

A novel graphdiyne–ZnO nanohybrid was prepared by Thangavel et al. (2015) using a hydrothermal method. Photocatalytic properties of as-prepared sample were evaluated on the degradation of two azo dyes (MB and RhB). The graphdiyne–ZnO nanohybrids showed superior photocatalytic properties as compared to bare ZnO NPs. The rate of degradation with graphdiyne–ZnO nanohybrids was nearly two times higher than ZnO NPs (Figure 3.1).

3.2.3 OTHERS

Bi_2O_3 NPs were prepared by means of ammonia precipitation, polyol-mediated, and microemulsion chemical methods (Peng et al. 2004). The photocatalytic oxidation reactions of benzene, toluene, and xylene were used as the model reaction to observe the photocatalytic activity of Bi_2O_3 NPs. The photocatalytic activity of Bi_2O_3 NPs prepared with the microemulsion chemical method was higher than that of the particles prepared with the polyol-mediated method. The degradation rates of the three pollutants decreased in the following sequence:

Xylene > Toluene > Benzene

A new crystal structure for nanostructured VO_2 with body-centered cubic (bcc) structure and a large optical band gap of 2.7 eV was reported by Wang et al. (2008). It showed excellent photocatalytic activity in hydrogen production. The bcc phase of VO_2 had a high quantum efficiency of 38.7% in the form of nanorods. The hydrogen production rate can be tuned by varying the incident angle of the UV light in the presence of films of the aligned VO_2 nanorods. A high rate of 800 mmol/m²/h was obtained for hydrogen production from a mixture of water and ethanol under the UV light.

Efficient removal of phenol was carried out by Gondal et al. (2009) using a UV laser induced photocatalysis process in the presence of Fe_2O_3 semiconductor catalysts. The effect of operational parameters in the removal process was also investigated with variation of laser irradiation time, laser energy, and concentration of the catalysts. Maximum phenol removal (more than 90%) was achieved in this process during 1-hour irradiation.

CeO_2 nanocrystals were synthesized by a simple precipitation method and calcination at 600°C by Pouretedal and Kadkhodaie (2010) using $(NH_4)_2Ce(NO_3)_6$ and ammonia as precursors. They studied photodegradation of MB in the presence of CeO_2 NPs under UV and sunlight irradiation. Maximum degradation was obtained with 1.0 g/L CeO_2 at pH 11 within 125 minutes. Results indicated the effect of photogenerated holes in the degradation mechanism of the dye.

Yuan and Xu (2010) prepared nanometer tin oxide (SnO_2) by constant temperature hydrolysis, microwave hydrolysis, chemical precipitation, and solid-state reaction. The photocatalytic activities of nanometer in oxides were studied using MO as a model organic pollutant. The effects of amount of photocatalysts, photocatalysts doped with different metal ions, and pH were also observed. Well-crystallized SnO_2 of 30–40 nm was obtained by constant temperature hydrolysis sintered at 800°C. The 97% decoloration of MO was achieved in 120 minutes, which shows higher photocatalytic activity of SnO_2 prepared by this method than that obtained from other methods.

Nanoporous structured tin dioxide was used by Kim et al. (2011) for photocatalytic destruction of endocrine, bisphenol A. The photocatalytic destruction of bisphenol A was enhanced by combining the nanoporous structured SnO_2 with TiO_2. It was observed from the photoluminescence curve that the recombination between electron and hole largely decreased in the TiO_2/nanoporous SnO_2 composite. Seventy-five percent decomposition of 10.0 ppm of bisphenol A was achieved after 24 hours in the presence of light and composite.

Chu et al. (2011) synthesized ultrafine SnO_2 nanocrystals via a surfactant-assisted solvothermal route in water–ethanol mixed media. Photocatalytic performance of the as-synthesized catalyst for decomposing acetaldehyde was observed at ppb level in a continuous glass-plate reactor. It was shown that SnO_2 photocatalyst of about 4 nm with a large BET surface area of 130 m^2/g exhibited the best photocatalytic oxidation properties, comparable to the properties of Degussa TiO_2 P25.

Different morphologies of Cu_2O have been synthesized by Sharma and Sharma (2012) via solution grown reduction route. Anhydrous dextrose and ascorbic acid were used as reducing agents and poly(vinyl pyrrolidone) (PVP) and sodium dodecyl sulfate (SDS) as surfactants. It was reported that Cu_2O polyhedrons possess higher (99.56%) photocatalytic activity and lower adsorption capacity (32.81%) as compared to Cu_2O NPs.

Liu et al. (2012) prepared CuO nanowires (NWs) using a Cu foil as substrate via a solution route. It was confirmed that the synthesized products were wire shaped. As-obtained NWs were well-crystalline pure CuO possessing good optical properties. About 90% of the MO was degraded in the presence of CuO NWs after 180 minutes under light.

Shao and Ma (2012) synthesized mesoporous CeO_2 NWs through a facile hydrothermal process by using a triblock copolymer F127 as the template. The surface area of the as-prepared sample was to be 273 m^2/g with pore width distribution of 6.9–13.8 nm. These mesoporous CeO_2 NWs could be used as efficient photocatalysts for organic dye degradation under UV light irradiation and were found superior as compared to the commercial photocatalyst P25 and CeO_2 powders. The NW structure facilitates the separation of CeO_2 by sedimentation so that these can be reused.

Nb_2O_5 NPs have been successfully prepared and modified by Hashemzadeh et al. (2013) via a hydrothermal process. Commercial Nb_2O_5 powder, H_2O_2, and NH_3 aqueous solutions have been used as precursors. The photocatalytic activity of modified NPs on degradation of MB and RhB dyes was examined and a higher photocatalytic efficiency for MB dye was obtained as compared to RhB.

SnO_2 NPs were synthesized by Singh and Nakate (2013) using a microwave method. SnO_2 NPs had spherical morphology with crystallite size of 35.42 nm. Synthesized NPs were used for photodegradation of methylene blue dye under UV light. These NPs were found to show 55.97% photodegradation efficiency.

Hierarchically assembled SnO_2 microflowers were prepared by Liu et al. (2013) using a facile hydrothermal process. It was found that these hierarchical nanostructures were made from two-dimensional (2D) nanosheets (50 nm thick). The as-prepared sample exhibited excellent photocatalytic performance.

ZrO_2 nanopowder was prepared by Mehrdad et al. (2013) via a sol–gel autocombustion method. The average crystalline size of ZrO_2 was determined to be 62 nm. The photocatalytic removal of nitrophenol from aqueous solution was observed in the presence of zirconia under UV light irradiation. Nitrophenol was degraded by 84%, 78%, and 66% in the presence of 0.04 g of ZrO_2 within 70 minutes using initial concentrations of 3, 5, and 10 ppm, respectively.

Cerium oxide (CeO_2) NPs have been synthesized hydrothermally by Khan et al. (2013) in the presence of urea. As-prepared CeO_2 had a well-crystalline cubic phase and NPs were optically active. CeO_2 was explored as a redox mediator for the development of a chemi-sensor for ethanol. Acridine orange (AO) was degraded to 50% in the presence of ceria in a short time. They concluded that reduction in the particle size enhanced the active surface area of the CeO_2 resulting in increase of chemical sensing of ethanol and its photocatalytic properties.

Semiconductor-based gas sensors using n-type WO_3 or p-type Co_3O_4 powder were fabricated by Akamatsu et al. (2013). They studied their gas-sensing properties toward NO_2 or NO (0.5–5 ppm in air) at 100°C or 200°C. The resistance of the WO_3-based sensor was found to increase on exposure to NO_2 and NO, while the resistance of the Co_3O_4-based sensor varied depending on the temperature and the gas species. The chemical states of the surface of WO_3 or Co_3O_4 powder on exposure to 1-ppm NO_2 and NO were also investigated.

Nanocrystalline α-Fe_2O_3 was synthesized by Jahagirdar et al. (2014) using a solution combustion method. As-formed α-Fe_2O_3 nanopowder was used as the photocatalyst for the degradation of the dye RhB under UV light. The effects of pH, amount of the photocatalyst, amount of H_2O_2, and irradiation time were observed. It was found that it acted as an efficient photocatalyst in the presence of H_2O_2. Maximum degradation of RhB was achieved only in 40 minutes with 0.8 g of the catalyst, pH 10, H_2O_2, and UV light.

Sharma et al. (2014) reported synthesis at temperature ~90°C and characteristics of well-crystalline iron oxide NPs. It was observed that the pure Fe_2O_3 NPs had a well-crystalline, rhombohedral crystal structure. They also showed highly super paramagnetic behavior. As-synthesized Fe_2O_3 NPs were used as an efficient photocatalyst for the photocatalytic degradation of MO dye. About 80% MO was degraded in the presence of Fe_2O_3 NPs within 210 minutes under UV light irradiation.

Al_2O_3 was synthesized by Tzompantzi et al. (2014) using the sol–gel method, dried at 100°C and annealed at 400°C, 500°C, 600°C, and 700°C. Al_2O_3 is a well-known insulator, but its feasibility as a catalyst was shown for the photomineralization of hazardous organic molecules. The photocatalytic activity of the sample of Al_2O_3 calcined at 400°C was higher than Degussa P25 TiO_2. It was proposed that the UV irradiation of the hydroxylated Al_2O_3 induced an effective separation of the electron–hole pairs resulting in enhancement of the photodegradation of phenolic compounds.

Liu et al. (2015a) obtained porous Fe_2O_3 nanorods using a facile chemical solution method followed by calcination. The BET surface area of the porous Fe_2O_3 nanorods was determined as 18.8 m²/g. The porous Fe_2O_3 nanorods were used as a catalyst to photodegrade RhB, MB, MO, p-nitrophenol, and eosin B (EB). As-prepared porous Fe_2O_3 nanorods exhibited higher catalytic activities as compared to the commercial Fe_2O_3 powder, due to their large surface areas and porous nanostructures. This catalyst had better stability and reusability.

A novel catalytic performance of simple-synthesized porous NiO NWs as catalyst/cocatalyst for the hydrogen evolution reaction (HER) has been reported by Shen et al. (2015). High-resolution transmission electron microscopy (HRTEM) exhibited a strong dependence of NiO NW photocatalytic and electrocatalytic HER performance on the density of exposed high-index facet (HIF) atoms. It was observed that the optimized porous NiO NWs had a long-term electrocatalytic stability of over 1 day, and 45 times higher photocatalytic hydrogen production as compared to the commercial NiO NPs.

Flower-type V_2O_5 hollow microspheres having diameters of about 700–800 nm were obtained by Liu et al. (2015b) with the assistance of carbon-sphere templates. These were used in the photodegradation of 1,2-dichlorobenzene (o-DCB) under visible light. The V_2O_5 hollow structure showed high photocatalytic activity in the degradation of gaseous o-DCB under visible light due to its strong adsorption capacity and large specific surface area. Intermediates, such as o-benzoquinone-type and organic acid species, and final degradation products (CO_2 and H_2O) were also confirmed.

Nanosized ZrO_2 powders having different structures with near-pure monoclinic, tetragonal, and cubic structures were prepared by Basahel et al. (2015) and used as catalysts for photocatalytic degradation of MO. The performance of as-synthesized ZrO_2 NPs in the photocatalytic degradation of MO was evaluated under UV light irradiation. It was found that the photocatalytic activity of the pure monoclinic ZrO_2 sample was higher than the tetragonal and cubic ZrO_2 samples. It was also revealed that monoclinic ZrO_2 NPs possessed high crystallinity and mesopores with a diameter of 100 Å. The higher activity of the monoclinic ZrO_2 sample for the photocatalytic degradation of MO was attributed to the combined effects of the presence of a small amount of oxygen-deficient zirconium oxide phase, high crystallinity, large pores, and the high density of surface hydroxyl groups.

CeO_2 NPs have strong redox ability, nontoxicity, long-term stability, and low cost. Ravishankar et al. (2015) synthesized CeO_2 NPs via a solution combustion method using ceric ammonium nitrate as an oxidizer and ethylenediaminetetraacetic acid (EDTA) as fuel at 450°C. It was observed that the particles were almost spherical, and the average size of the NPs was 42 nm. Ceria NPs exhibited photocatalytic activity against trypan blue (TB) at pH 10 under UV light irradiation. These NPs also reduced Cr(VI) to Cr(III) and showed antibacterial activity against *Pseudomonas aeruginosa*.

Sood et al. (2015) reported the synthesis of α-Bi_2O_3 nanorods and used it as an efficient sunlight-active photocatalyst for the photocatalytic degradation of RhB and 2,4,6-trichlorophenol. A simple surfactant-free sonochemical route was used for the synthesis at ambient conditions. They observed that the prepared nanorods exhibited a high purity, well-crystalline monoclinic α-Bi_2O_3 structure, and excellent optical properties. The degradation of a cationic dye (RhB), its simulated dye bath effluent, and 2,4,6-trichlorophenol under solar light irradiation were used as model systems. As-synthesized α-Bi_2O_3 nanorod catalyst exhibited excellent solar light driven photocatalysis toward RhB (97% degradation in 45 minutes) and 2,4,6-trichlorophenol (88% degradation in 180 minutes).

Metastable β-Bi_2O_3 microcrystals were prepared from $Bi(NO_3)_3$ by an aqueous crystallization strategy without calcination or other complex treatment (Lu et al. 2015). The introduction of cetyltrimethylammonium bromide (actually Br^- ions) facilitated the formation of β-Bi_2O_3 crystals. Photocatalytic activities of metastable β-Bi_2O_3 samples were evaluated in the degradation of RhB under visible light irradiation. As-prepared β-Bi_2O_3 showed excellent photocatalytic efficiency of up to 77.9% (total removal 97.2%) in 2 hours of irradiation and the sample can be used for four cycles of degradation without any loss of activity. It was concluded that a minor amount of BiOCl crystallites appearing at the surface of β-Bi_2O_3 crystals facilitated their photocatalytic performance.

3.3 SULFIDES

3.3.1 CADMIUM SULFIDE

Cadmium sulfide is yellow in color and acts as a semiconductor. It exists in nature in two different minerals, hexagonal greenockite and cubic hawleyite. Cadmium sulfide is a direct narrow band gap (2.42 eV) semiconductor.

Tristao et al. (2006) used nanometric particles of CdS to impregnate TiO_2 in order to optimize its photocatalytic properties. They used CdS/TiO_2 semiconductor composites in 1, 3, 5, and 20 mol% proportion. It was observed the CdS had hexagonal geometry and TiO_2 was in anatase form. The photocatalytic activity of the CdS/TiO_2 was investigated using UV light for the degradation of textile azo-dye, drimarene red. Better efficiency was observed for the CdS/TiO_2 5% composite as compared to other CdS/TiO_2 proportions and bare TiO_2.

CdS semiconductor photocatalyst was synthesized by Li et al. (2008) via a solvent thermal process and then platinum was added to it by photodeposition. The activities of Pt/CdS photocatalysts under visible light were evaluated in terms of hydrogen production using formic acid as the electron donor. The photoetching of CdS was monitored by measuring dissolved Cd^{2+}. The photocatalytic activity of Pt/CdS was enhanced by 126% with 1-hour photoetching treatment. However, excess photoetching was found to decrease the photocatalytic activity. Oxygen and platinum both play an important role in the photoetching process. The photocatalytic activity was enhanced through decreasing agglomeration of the CdS particles, which led to increased specific surface area, modified particle morphology, and selective removal of grain boundary defects, which acted as recombination centers of photoinduced electron–hole pairs.

Cadmium sulfide quantum dots (QDs) sensitized mesoporous TiO_2 photocatalysts were prepared by Li et al. (2009) via preplanting cadmium oxide as crystal seeds into the framework of ordered mesoporous TiO_2. Then this CdO was converted to CdS QDs through ion exchange. The presence of CdS QDs in the TiO_2 framework was responsible for its photoresponse to visible light by enhancing the photogenerated electron transfer from the inorganic sensitizer to TiO_2. This photocatalyst showed excellent photocatalytic efficiency for both the oxidation of NO gas in air and degradation of organic compounds in aqueous solution under visible light irradiation.

High-efficiency photocatalytic H_2 production was achieved by Li et al. (2011) using graphene nanosheets decorated with CdS clusters as the visible light driven photocatalysts. They used the solvothermal method, where graphene oxide (GO) served as the support and cadmium acetate as the CdS precursor. These nanosized composites achieved a high H_2 production rate of 1.12 mmol/h, which was about 4.87 times higher than pure CdS NPs, at graphene content of 1.0 wt.% and Pt 0.5 wt.% under visible light irradiation. It was attributed to the presence of graphene, which served as an electron collector and transporter to lengthen the lifetime of the photogenerated charge carriers from CdS NPs.

Nanosized cadmium sulfide containing different phase structures was synthesized by Yu et al. (2011) via complex compound thermolysis using different molar ratios of thiourea to cadmium acetate (S/Cd). Photocatalytic degradation of RhB was carried out, which showed that CdS with the cubic phase had the best photocatalytic activity due to its larger adsorption and absorbance abilities and smaller particle size of about 10–13 nm.

Cadmium sulfide is a representative material as the visible light responsive photocatalysts for hydrogen production. There has been significant progress in water splitting on CdS-based photocatalysts using solar light, especially in the presence of cocatalysts. The field of photocatalytic water splitting on CdS-based photocatalysts has been reviewed by Chen and Shangguan (2013), and includes controllable synthesis of CdS, as well as modifications with different kinds of cocatalysts, intercalated with layered nanocomposites and metal oxides, hybrids with graphenes, and so on.

Wang et al. (2015b) synthesized a novel 3D flower-like CdS via a facile template-free hydrothermal process. $Cd(NO_3)_2 \cdot 4H_2O$ and thiourea were used as precursors and L-histidine as a chelating agent. Higher photocatalytic activity was observed on the flower-like CdS photocatalyst under visible light irradiation; it was almost 13 times that of pure CdS. The imidazole ring of L-histidine captured the Cd ions from the solution, and as a result, the growth of the CdS was prevented. As-synthesized flower-like CdS with L-histidine was found to be more stable than CdS without L-histidine in hydrogen generation.

Ding et al. (2015) developed a novel heterojunction structured composite photocatalyst $CdS/Au/g-C_3N_4$ by depositing CdS/Au with a core (Au)-shell (CdS) structure on the surface of $g-C_3N_4$. Photocatalytic hydrogen production activity of this photocatalyst was evaluated under visible light irradiation using methanol as a sacrificial reagent. The activity was found to be about 125.8 times higher than $g-C_3N_4$ and was even much higher than $Pt/g-C_3N_4$. The enhancement was attributed to efficient separation of the photoexcited charges due to the anisotropic junction in the $CdS/Au/g-C_3N_4$ system.

3.3.2 ZINC SULFIDE

Zinc sulfide is a white- to yellow-colored powder. It is normally available in the more stable cubic form, also known as zinc blende or sphalerite. Its hexagonal form is known as the mineral wurtzite. Both these forms, sphalerite and wurtzite, are intrinsic and wide band gap semiconductors. The band gap of the cubic form is 3.54 eV, whereas the hexagonal form has a higher band gap of 3.91 eV. It can be either used directly as a photocatalyst or it may be modified to improve its efficiency.

Ni-doped ZnS photocatalyst ($Zn_{0.999}Ni_{0.001}S$) had 2.4 eV energy gap and showed activities for the reduction of nitrate and nitrite ions to ammonia, and dinitrogen under visible light irradiation using methanol as a reducing reagent (Hamanoi and Kudo 2002). The reduction of nitrate ions was compatible with that of water to form hydrogen. The concentration of nitrate ions and loading of a platinum cocatalyst affected the selectivity for the reduction products of nitrate ions.

Stengl et al. (2008) prepared photocatalytically active TiO_2/ZnS composites by homogeneous hydrolysis of mixture of titanium oxo-sulfate and zinc sulfate in aqueous solutions with thioacetamide. Photoactivity of the prepared TiO_2/ZnS nanocomposites was evaluated by the photocatalytic decomposition of Orange II dye in an aqueous slurry under different irradiations (255, 365, and 400 nm). The highest catalytic activity was obtained with the composite sample prepared by hydrolysis of mixture solutions of 0.63 M $TiOSO_4$ and 0.08 M $ZnSO_4 \cdot 7H_2O$.

Pouretedal et al. (2010) synthesized NPs of zinc sulfide doped with iron by controlled coprecipitation. They used these NPs in the photodegradation of Congo red (CR) as water pollutant. The effect of mole fraction of Fe^{3+} to zinc ion, pH, dosage of photocatalyst, and concentration of dye on photoreactivity of doped zinc sulfide was observed. The size of NPs was determined to be 10–40 nm. The degradation efficiency was 92% under UV and 98% under sunlight irradiation in 120 minutes and 12 hours, respectively, using 0.8 g/L of ZnS:Fe (0.5%) nanophotocatalyst, and 12 mg/L of CR at pH 6. The photoreactivity of photocatalyst was reproducible up to 90%–92% degradation in four cycles of photodegradation.

Ma et al. (2013) synthesized N-doped ZnO/ZnS photocatalysts via a simple heat treatment approach using L-cysteine as the source of N and S. Anthraquinone dye

(Reactive Brilliant Blue KNR) was used as the model contaminant to evaluate the photocatalytic activity of as-synthesized samples under sunlight illumination. N-doped ZnO/ZnS synthesized by this method showed better photocatalytic activity as compared to pure ZnO. This enhanced photocatalytic activity was attributed to N doping, ZnS/ZnO heterostructure, and covered abundant carbon species on the photocatalyst surface causing high absorption efficiency of light, efficient separation of electron–hole pairs, and quick surface reaction in doped ZnO.

ZnS–Ag nanoballs were synthesized using a chemical precipitation method at room temperature aided with the capping agent cetyl trimethyl ammonium bromide (Sivakumar et al. 2014). Photocatalytic activity of ZnS and ZnS–Ag nanoballs was evaluated by irradiating the solution of organic dye MB under visible light irradiation. The different operating parameters affecting degradation such as dye concentration, catalyst loading, and pH were studied. It was observed that the catalyst can be reused up to three cycles. It was found that ZnS–Ag nanoballs showed higher degradation efficiency as compared to ZnS.

ZnS NPs (ZnS QDs) were prepared by Al-Rasoul et al. (2014) using a simple microwave irradiation method under mild conditions. Zinc acetate and thioacetamide were used as a source of zinc and sulfur, respectively, and ethylene glycol as a solvent. As-prepared ZnS sample displayed a blue shift as compared to bulk ZnS from 310 to 345 nm. Photocatalytic degradation of MB dye catalyzed by ZnS NPs was studied under solar radiation and it was observed that photocatalytic degradation increased with increasing time exposure to solar light.

Controlling the amount of intrinsic S vacancies was achieved in ZnS spheres by Wang et al. (2015a) using a hydrothermal method. Zn and S powders were used in concentrated NaOH solution with $NaBH_4$ added as a reducing agent. Absorption spectra of ZnS was extended to the visible region because of these S vacancies. Photocatalytic activities were evaluated for H_2 production under visible light. They reported that the concentration of S vacancies in the ZnS samples can be controlled by variation in the amount of the reducing agent $NaBH_4$. As-prepared ZnS samples exhibited photocatalytic activity for H_2 production under visible light irradiation without loading any noble metal. The activity of ZnS increases steadily with increase in the concentration of S vacancies reaching an optimum value.

Apart from cadmium and zinc sulfide, various other sulfides have also been used for photocatalytic degradation of various organic pollutants particularly dyes. Some sulfides were used by Ameta et al. (2005), Sharma et al. (2011, 2013a, 2013b), Andronic et al. (2011), Yadav et al. (2012), and so on.

Some metal chalcogenides (oxide, sulfide, etc.) have been extensively used as photocatalytic materials because they have their band gaps in the desired range. They are insoluble in a wide range of pH and absorb solar radiation in the visible as well as UV range. Majority of these are low cost and eco-friendly as well, but some of them are inefficient because they cannot utilize visible radiation to the desired extent. This limitation can be overcome with slight modification of such systems like doping, sensitization, composite formation, metallization, and so on. Binary semiconductors are likely to maintain their status as effective photocatalysts in the future also, of course with certain modifications.

REFERENCES

Aarthi, T., P. Narahari, and G. Madras. 2007. Photocatalytic degradation of azure and sudan dyes using nano TiO_2. *J. Hazard. Mater.* 149 (3): 725–734.

Akamatsu, T., T. Itoh, N. Izu, and W. Shin. 2013. NO and NO_2 sensing properties of WO_3 and Co_3O_4 based gas sensors. *Sensors.* 13: 12467–12481.

Al-Rasoul, K. T., I. M. Ibrahim, I. M. Ali, and R. M. Al-Haddad. 2014. Synthesis, structure and characterization of ZnS QdS and using it in photocatalytic reaction. *Int. J. Sci. Technol. Res.* 3 (5): 213–217.

Ameen, S., M. S. Akhtar, M. Nazim, and H.-S. Shin. 2013. Rapid photocatalytic degradation of crystal violet dye over ZnO flower nanomaterials. *Mater. Lett.* 96: 228–232.

Ameta, R., A. Pandey, P. B. Punjabi, and S. C. Ameta. 2005. Modification of photocatalytic activity of antimony(III) sulphide in presence of sodium bicarbonate/carbonate. *J. Indian Chem. Soc.* 82 (9): 807–810.

Andronic, L., L. Isac, and A. Duta. 2011. Photochemical synthesis of copper sulphide/titanium oxide photocatalyst. *J. Photochem. Photobiol. A: Chem.* 221 (1): 30–37.

Basahel, S. N., T. T. Ali, M. Mokhtar, and K. Narasimharao. 2015. Influence of crystal structure of nanosized ZrO_2 on photocatalytic degradation of methyl orange. *Nanoscale Res. Lett.* 10: 73.

Bessekhouad, Y., D. Robert, and J. V. Weber. 2003. Synthesis of photocatalytic TiO_2 nanoparticles: Optimization of the preparation conditions. *J. Photochem. Photobiol. A: Chem.* 157 (1): 47–53.

Byrappa, K., A. K. Subramani, S. Ananda, K. M. L. Rai, R. Dinesh, and M. Yoshimura. 2006. Photocatalytic degradation of rhodamine B dye using hydrothermally synthesized ZnO. *Bull. Mater. Sci.* 29 (5): 433–438.

Chen, C. C., C. S. Lu, Y. C. Chung, and J. L. Jan. 2007. UV light induced photodegradation of malachite green on TiO_2 nanoparticles. *J. Hazard. Mater.* 141 (3): 520–528.

Chen, X. and W. Shangguan. 2013. Hydrogen production from water splitting on CdS-based photocatalysts using solar light. *Front. Energy.* 7 (1): 111–118.

Chen, F., W. Zou, W. Qu, and J. Zhang. 2009. Photocatalytic performance of a visible light TiO_2 photocatalyst prepared by a surface chemical modification process. *Catal. Commun.* 10: 1510–1513.

Chu, D., J. Mo, Q. Peng, Y. Zhang, Y. Wei, Z. Zhuang, and Y. Li. 2011. Enhanced photocatalytic properties of SnO_2 nanocrystals with decreased size for ppb-level acetaldehyde decomposition. 3 (2): 371–377.

Ding, X., Y. Li, J. Zhao, Y. Zhu, Y. Li, W. Deng, and C. Wang. 2015. Enhanced photocatalytic H_2 evolution over CdS/Au/g-C_3N_4 composite photocatalyst under visible-light irradiation. *APL Mater.* 3. doi:10.1063/1.4926935.

Gondal, M. A., M. N. Sayeed, Z. H. Yamani, and A. R. Al-Arfaj. 2009. Efficient removal of phenol from water using Fe_2O_3 semiconductor catalyst under UV laser irradiation. *J. Environ. Sci. Health, Part A.* 44 (5): 515–521.

Guzsvány, V. J., J. J. Csanádi, S. D. Lazić, and F. F. Gaál. 2009. Photocatalytic degradation of the insecticide acetamiprid on TiO_2 catalyst. *J. Braz. Chem. Soc.* 20 (1). doi:10.1590/S0103-50532009000100023.

Habib, A., I. M. I. Ismail, A. J. Mahmood, and R. Ullah. 2012. Photocatalytic decolorization of brilliant golden yellow in TiO_2 and ZnO suspensions. *J. Saudi Chem. Soc.* 18 (4): 423–429.

Hamanoi, O. and A. Kudo. 2002. Reduction of nitrate and nitrite ions over Ni-ZnS photocatalyst under visible light irradiation in the presence of a sacrificial reagent. *Chem. Lett.* 31: 838–839.

Hashemzadeh, F., R. Rahimi, and A. Gaffarinejad. 2013. Photocatalytic degradation of methylene blue and rhodamine B dyes by niobium oxide nanoparticles synthesized via hydrothermal method. *Int. J. Appl. Chem. Sci. Res.* 1 (7): 95–102.

Jahagirdar, A. A., M. N. Zulfiqar Ahmed, N. Donappa, H. Nagabhushana, and B. M. Nagabhushanae. 2014. Photocatalytic degradation of rhodamine B using nanocrystalline α-Fe_2O_3. *J. Mater. Environ. Sci.* 5 (5): 1426–1433.

Khan, S. B., M. Faisal, M. M. Rahman, K. Akhtar, A. M. Asiri1, A. Khan, and K. A. Alamry. 2013. Effect of particle size on the photocatalytic activity and sensing properties of CeO_2 nanoparticles. *Int. J. Electrochem. Sci.* 8: 7284–7297.

Kim, J., J. S. Lee, and M. Kang. 2011. Synthesis of nanoporous structured SnO_2 and its photocatalytic ability for bisphenol A destruction. *Bull. Korean Chem. Soc.* 32 (5): 1715–1720.

Lavand, A. B. and Y. S. Malghe. 2015. Synthesis, characterization and visible light photocatalytic activity of nitrogen-doped zinc oxide nanospheres. *J. Asian Ceram. Soc.* 3 (3): 305–310.

Li, G.-S., D.-Q. Zhang, and J. C. Yu. 2009. A new visible-light photocatalyst: CdS quantum dots embedded mesoporous TiO_2. *Environ. Sci. Technol.* 43 (18): 7079–7085.

Li, Q., B. Guo, J. Yu, J. Ran, B. Zhang, H. Yan, and J. R. Gong. 2011. Highly efficient visible-light-driven photocatalytic hydrogen production of CdS-cluster-decorated graphene nanosheets. *J. Am. Chem. Soc.* 133 (28): 10878–10884.

Li, Y., J. Du, S. Peng, and S. Li. 2008. Enhancement of photocatalytic activity of cadmium sulfide for hydrogen evolution by photoetching. *Int. J. Hydrogen Energy.* 33 (8): 2007–2013.

Liu, B., X. Li, Q. Zhao, J. Liu, S. Liu, S. Wang, and M. Tadé. 2015b. Insight into the mechanism of photocatalytic degradation of gaseous o-dichlorobenzene over flower-type V_2O_5 hollow spheres. *J. Mater. Chem. A.* 3: 15163–15170.

Liu, X., K. Chen, J.-J. Shim, and J. Huang. 2015a. Facile synthesis of porous Fe_2O_3 nanorods and their photocatalytic properties. *J. Saudi Chem. Soc.* 19 (5): 479–484.

Liu, X., Z. Li, Q. Zhang, F. Li, and T. Kong. 2012. CuO nanowires prepared via a facile solution route and their photocatalytic property. *Mater. Lett.* 72: 49–52.

Liu, Y., Y. Jiao, B. Yin, S. Zhang, F. Qu, and X. Wu. 2013. Hierarchical semiconductor oxide photocatalyst: A case of the SnO_2 microflower. *Nano-Micro Lett.* 5 (4): 234–241.

Lu, Y., Y. Zhao, J. Zhao, Y. Song, Z. Huang, F. Gao, N. Li, and Y. Li. 2015. Induced aqueous synthesis of metastable β-Bi_2O_3 microcrystals for visible-light photocatalyst study. *Cryst. Growth Des.* 15 (3): 1031–1042.

Luo, L., A.T. Cooper, and M. Fan. 2009. Preparation and application of nanoglued binary titania-silica aerogel. *J. Hazard. Mater.* 161 (1): 175–182.

Ma, H., X. Cheng, C. Ma, X. Dong, X. Zhang, M. Xue, X. Zhang, and Y. Fu. 2013. Synthesis, characterization, and photocatalytic activity of N-doped ZnO/ZnS composites. *Int. J. Photoenergy.* 2013. doi:10.1155/2013/625024.

Mahmoodi, N. M. and M. Arami. 2009. Degradation and toxicity reduction of textile wastewater using immobilized titania nanophotocatalysis. *J. Photochem. Photobiol. B: Biol.* 94 (1): 20–24.

Mehrdad, A. A., S. S. Abedini, and K. N. Assi. 2013. Photocatalytic properties of ZrO_2 nanoparticles in removal of nitrophenol from aquatic solution. *Int. J. Nano Dimens.* 3(3): 235–240.

Mohajerani, M. S., A. Lak, and A. Simchi. 2009. Effect of morphology on the solar photocatalytic behavior of ZnO nanostructures. *J. Alloys Compd.* 485 (1–2): 616–620.

Peng, D. N. G., Y.-G. Du, and Z. Xu. 2004. Effect of preparation methods of Bi_2O_3 nanoparticles on their photocatalytic activity. *Chem. Res. Chin. Univ.* 20 (6): 717–721.

Pouretedal H. R. and A. Kadkhodaie. 2010. Synthetic CeO_2 nanoparticle catalysis of methylene blue photodegradation: Kinetics and mechanism. *Chin. J. Catal.* 31 (11): 1328–1334.

Pouretedal, H. R., S. Narimany, and M. H. Keshavarz. 2010. Nanoparticles of ZnS doped with iron as photocatalyst under UV and sunlight irradiation. *Int. J. Mater. Res.* 101 (8): 1046–1051.

Pung, S.-Y., W.-P. Lee, and A. Aziz. 2012. Kinetic study of organic dye degradation using ZnO particles with different morphologies as a photocatalyst. *Int. J. Inorg. Chem.* doi:10.1155/2012/608183.

Rao, P., G. Patel, S. L. Sharma, and S. Ameta. 1997. Photocatalytic degradation of crystal violet over semiconductor zinc oxide powder suspended in aqueous solution. *Toxicol. Environ. Chem.* 60 (1–4): 155–161.

Ravishankar, T. N., T. Ramakrishnappa, G. Nagaraju, and H. Rajanaika. 2015. Synthesis and characterization of CeO_2 nanoparticles via solution combustion method for photocatalytic and antibacterial activity studies. *Chem. Open.* 4 (2): 146–154.

Rezaee, A., M. T. Ghaneian, N. Taghavinia, M. K. Aminian, and S. J. Hashemian. 2009. TiO_2 nanofibre assisted photocatalytic degradation of reactive blue 19 dye from aqueous solution. *J. Environ. Technol.* 30 (3): 233–239.

Saggioro, E. M., A. S. Oliveira, T. Pavesi, C. G. Maia, L. F. V. Ferreira, and J. C. Moreira. 2011. Use of titanium dioxide photocatalysis on the remediation of model textile wastewaters containing azo dyes. *Molecules.* 16: 10370–10386.

Shao, Y. and Y. Ma. 2012. Mesoporous CeO_2 nanowires as recycled photocatalysts. *Sci. China Chem.* 55 (7): 1303–1307.

Sharma, P., R. Kumar, S. Chauhan, D. Singh, and M. S. Chauhan. 2014. Facile growth and characterization of α-Fe_2O_3 nanoparticles for photocatalytic degradation of methyl orange. *J. Nanosci. Nanotechnol.* 14: 1–5.

Sharma, P. and S. K. Sharma. 2012. Photocatalytic degradation of cuprous oxide nanostructures under UV/visible irradiation. *Water Resour. Manage.* 26 (15): 4525–4538.

Sharma, S., R. Ameta, R. K. Malkani, and S. C. Ameta. 2011. Use of semi-conducting bismuth sulfide as a photocatalyst for degradation of rose Bengal. *Macedonian J. Chem. Chem. Eng.* 30 (2): 229–234.

Sharma, S., R. Ameta, R. K. Malkani, and S. C. Ameta. 2013a. Photocatalytic degradation of Rose Bengal using semiconducting zinc sulphide as the photocatalyst. *J. Serb. Chem. Soc.* 78 (6) 897–905.

Sharma, S., R. K. Malkani, R. Ameta, and S. C. Ameta. 2013b. Photocatalytic degradation of rose Bengal by semiconducting tin sulphide. *Poll. Res.* 32 (02): 387–391.

Shen, M., A. Han, X. Wang, Y. G. Ro, A. Kargar, Y. Lin et al. 2015. Atomic scale analysis of the enhanced electro-and photo-catalytic activity in high-index faceted porous NiO nanowires. *Sci. Rep.* 5. doi:10.1038/srep08557.

Singh, A. K. and U. T. Nakate. 2013. Microwave synthesis, characterization and photocatalytic properties of SnO_2 nanoparticles. *Adv. Nanoparticles.* 2: 66–70.

Sivakumar, P., G. K. Gaurav Kumar, P. Sivakumar, and S. Renganathan. 2014. Synthesis and characterization of ZnS-Ag nanoballs and its application in photocatalytic dye degradation under visible light. *J. Nanostruct. Chem.* 4. doi:10.1007/s40097-014-0107-0.

Sood, S., A. Umar, S. K. Mehta, and S. K. Kansal. 2015. α-Bi_2O_3 nanorods: An efficient sunlight active photocatalyst for degradation of rhodamine B and 2, 4,6-trichlorophenol. *Ceram. Int. Part A.* 41 (3): 3355–3364.

Stambolova, I., C. E. Shipochka, V. Blaskov, A. Loukanov, and S. Vassilev. 2012. Sprayed nanostructured TiO_2 films for efficient photocatalytic degradation of textile azo dye. *J. Photochem. Photobiol. B.* 5 (117): 19–26.

Stengl, V., S. Bakardjieva, N. Murafa, V. Houšková, and K. Lang. 2008. Visible-light photocatalytic activity of TiO_2/ZnS nanocomposites prepared by homogeneous hydrolysis. *Microporous Mesoporous Mater.* 110 (2–3): 370–378.

Thangavel, S., K. Krishnamoorthy, V. Krishnaswamy, N. Raju, S. J. Kim, and G. Venugopal. 2015. Graphdiyne–ZnO nanohybrids as an advanced photocatalytic material. *J. Phys. Chem. C.* 119: 22057–22065.

Tian, J., Y. Leng, H. Cui, and H. Liu. 2015. Hydrogenated TiO_2 nanobelts as highly efficient photocatalytic organic dye degradation and hydrogen evolution photocatalyst. *J. Hazard. Mater.* 299: 165–173.

Tristao, J. C., F. Magalhaes, P. Corio, M. Terezinha, and C. Sansiviero. 2006. Electronic characterization and photocatalytic properties of CdS/TiO_2 semiconductor composite. *J. Photochem. Photobiol. A: Chem.* 181: 152–157.

Tzompantzi, F., Y. Piña, A. Mantilla, O. Aguilar-Martínez, F. Galindo-Hernández, Xim Bokhimi, and A. Barrera. 2014. Hydroxylated sol–gel Al_2O_3 as photocatalyst for the degradation of phenolic compounds in presence of UV light. *Catal. Today.* 220–222: 49–55.

Wang, G., B. Huang, Z. Li, Z. Lou, Z. Wang, Y. Dai, and M.-H. Whangbo. 2015a. Synthesis and characterization of ZnS with controlled amount of S vacancies for photocatalytic H_2 production under visible light. *Sci. Rep.* 5. doi:10.1038/srep08544.

Wang, Q., J. Lian, J. Li, R. Wang, H. Huang, B. Su, and Z. Lei. 2015b. Highly efficient photocatalytic hydrogen production of flower-like cadmium sulfide decorated by histidine. *Sci. Rep.* 5. doi:10.1038/srep13593.

Wang, Y., Z. Zhang, Y. Zhu, Z. Li, R. Vajtai, L. Ci, and P. M. Ajayan. 2008. Nanostructured VO_2 photocatalysts for hydrogen production. *ACS Nano.* 2 (7): 1492–1496.

Welderfael, T., O. P. Yadav, A. M. Taddesse, and J. Kaushal. 2013. Synthesis, characterization and photocatalytic activities of Ag-N-codoped ZnO nanoparticles for degradation of methyl red. *Bull. Chem. Soc. Ethiop.* 27 (2): 221–232.

Wisitsoraat, A., A. Tuantranont, E. Comini, G. Sberveglieri, and W. Wlodarskic. 2009. Characterization of n-type and p-type semiconductor gas sensors based on NiO_x doped TiO_2 thin films. *Thin Solid Films.* 517 (8): 2775–2780.

Yadav, I., S. Nihalani, and S. Bhardwaj. 2012. Use of semi-conducting lead sulfide for degradation of azure-B: An eco-friendly process. *Der Chemica Sinica.* 3(6): 1468–1474.

Yu, Y., Y. Ding, S. Zuo, and J. Liu. 2011. Photocatalytic activity of nanosized cadmium sulfides synthesized by complex compound thermolysis. *Int. J. Photoenergy.* doi:10.1155/2011/762929.

Yuan, H. and J. Xu. 2010. Preparation, characterization and photocatalytic activity of nanometer SnO_2. *Int. J. Chem. Eng. Appl.* 1 (3): 241–246.

Zhang, G., Q. Xiong, W. Xu, and S. Guo. 2014. Synthesis of bicrystalline TiO_2 supported sepiolite fibers and their photocatalytic activity for degradation of gaseous formaldehyde. *Appl. Clay Sci.* 102: 231–237.

Zhu, G., H. Yin, C. Yang, H. Cui, Z. Wang, J. Xu, T. Lin, and F. Huang. 2015. Black titania for superior photocatalytic hydrogen production and photoelectrochemical water splitting. *ChemCatChem.* 7 (17):2614–2619.

4 Ternary Semiconductors

4.1 INTRODUCTION

Ternary chalcogenides form the most fascinating class of and exhibit a variety of structures and properties. They play an important role in many applications as they have very interesting electronic and magnetic properties. Some of the metal–oxide nanoparticles (NPs) show quite unusual optical, electronic, and magnetic properties differing from their bulk counterparts. Though the bonding in oxides is basically ionic in nature, covalent bonding, to some extent, also exists because of overlapping between p-orbitals of the filled oxide and the vacant d-orbitals of the transition metal cations. The ternary metal oxides possess multiple oxidation states, which will enable them to participate in multiple redox reactions. Ternary oxide nanostructured materials can also be used as supercapacitors (Chen et al. 2015).

4.2 STRONTIUM TITANATE

Photocatalytic (PC) activities of TiO_2 and $SrTiO_3$ (STO) photocatalysts codoped with antimony and chromium have been investigated by Kato and Kudo (2002). These are active in the visible light region and their band gap was determined as 2.2 and 2.4 eV, respectively. Antimony and chromium codoped TiO_2 evolved O_2 from an aqueous silver nitrate solution. On the other hand, STO codoped with Sb and Cr evolved H_2 from an aqueous methanol solution. The activity of codoped TiO_2 was much higher than TiO_2 doped with chromium only. Charge balance by Sb^{5+} and Cr^{3+} ions resulted in the suppression of formation of Cr^{6+} ions and oxygen defects in the lattice. These act as effectively nonradiative recombination centers between photogenerated electrons and holes.

Miyauchi et al. (2004) investigated the yellow STO powders codoped with nitrogen and lanthanum (STO:N, La) as the visible light photocatalysts. They showed that the edge of the N(2p) band is situated above the valence band (VB), which consisted of O(2p) orbitals, and the La orbitals did not exist in the band gap of STO. The optimum doping density of N and La for the visible light activity was 0.5%, and STO:N, La had activity 0.5% compared to pure STO under UV illumination.

PC activities of noble metal ion doped STO were observed by Konta et al. (2004) under visible light irradiation. Mn, Ru, Rh, and Ir are used as dopants. The doped STO possess strong absorption bands in the visible light region because of excitation from the discontinuous levels formed by the dopants to the conduction band (CB) of the STO. While Mn- and Ru-doped STO showed PC activities for O_2 evolution from an aqueous silver nitrate solution, Ru-, Rh-, and Ir-doped STO loaded with Pt cocatalysts produced H_2 from an aqueous methanol solution under visible light irradiation. The Rh-doped STO photocatalyst loaded with a Pt cocatalyst provided the quantum yield of about 5.2% at 420 nm for the H_2 evolution reaction.

Wang et al. (2005) prepared La- and N-codoped STO by a mechanochemical reaction using STO, urea, and La_2O_3 as the starting materials. La- and N-doped STO can be prepared by heating the mixture of STO and La_2O_3 under flowing NH_3 gas at 600°C. They found that the PC activity of N- and La-codoped STO was 2.6 and 2.0 times greater than that of pure STO.

PC degradation of an azo dye, C.I. reactive black 5 (RB5), has been carried out over $SiTiO_3/CaO_2$ by Song et al. (2007). The change in concentration of dye and depletion in total organic carbon content were monitored as a function of irradiation time. There was little effect of iodide ion, *tert*-butyl alcohol, fluoride ion, or persulfate ion as h_{vb}^+, $\cdot OH$, or e_{cb}^- scavengers on the decolorization, which confirmed that the decolorization of RB5 proceeded by photolysis and/or $O_2^{\cdot-}$ in the bulk solution. After decolorization, it may shift from the bulk solution to the surface of the catalysts. Here, the cleavage of naphthalene and benzene rings was attributed to the h_{vb}^+ and $\cdot OH_{ads}$, which was also verified by the effect of scavengers.

The PC and photophysical properties of pure STO as well as Fe-doped STO were investigated by Xie et al. (2008). The Fe-doped STO showed higher PC activity than pure STO in degrading RhB. It was observed that doping Fe with STO is responsible for an absorption extending up to the visible region. They attributed the shift of the absorption in the visible region partly to the metal-to-metal charge transfer (MMCT) excitation of Ti^{IV}–O–Fe^{II} linkage formed in the Fe-doped STO. This linkage was considered to be the cause of the increased degradation of RhB.

The PC degradation of C.I. direct red 23 (DR23) in aqueous solutions under UV irradiation was investigated by Song et al. (2008) using STO/CeO_2 composite as a catalyst. The STO/CeO_2 powders had more PC activity for decolorization of this dye as compared to pure STO powder under UV irradiation. The effects of catalyst dose, pH, initial concentration of dye, irradiation intensity, as well as scavenger KI were also observed. Light intensity revealed a significant positive effect on the efficiency of decolorization, while the initial dye concentration showed an adverse effect. Under the optimum conditions, complete decolorization was achieved in 60 minutes and 69% reduction in chemical oxygen demand (COD) was achieved after 240 minutes.

A sol–gel method with the aid of structure-directing surfactant was successfully used in the synthesis of STO by Puangpetch et al. (2008). The photodegradation of methyl orange (MO) by STO was observed to be affected mainly by the crystallinity, specific surface area, and pore characteristics. As-obtained mesoporous-assembled structure with a high pore uniformity of STO played the most significant role in deciding the PC activity of the STO photocatalyst.

Liu et al. (2008) synthesized STO NPs and used it for PC hydrogen production from water splitting under the UV irradiation. Cubic STO powders were prepared by three techniques: (1) the polymerized complex method, (2) solid-state reaction, and (3) the milling assistance method. The PC activity of hydrogen evolution from water splitting over STO powders by the polymerized complex method was found to be higher than that prepared by the solid-state reaction and the milling assistant method. Particle size, uniformity of components, and particle aggregation amount influence the PC activity of STO for hydrogen evolution.

Mesoporous-assembled STO nanocrystal-based photocatalysts were synthesized by Puangpetch et al. (2009) with the support of a structure-directing surfactant. The PC water splitting activity for hydrogen production was observed over this mesoporous-assembled STO nanocrystal-based photocatalysts with various hole scavengers such as methanol, ethanol, 2-propanol, d-glucose, and Na_2SO_3. The perfect mesoporous-assembled STO photocatalysts showed much higher PC activity in hydrogen production. The mesoporous assembly of nanocrystals with high pore uniformity played a major role in affecting the PC hydrogen production activity of the STO photocatalysts. The Pt cocatalyst improved the visible light harvesting ability of such photocatalyst and behaved as the active site for proton reduction. An optimum Pt loading of 0.5 wt.% on the mesoporous-assembled STO photocatalyst gave the highest PC activity, with hydrogen production rates from 50 vol.% methanol-aqueous solutions as 276 and 188 $\mu mol/h/g_{cat}$ and quantum efficiencies of 1.9% and 0.9% under UV and visible light irradiation, respectively.

Wang et al. (2009c) prepared visible light active sulfur- and nitrogen-codoped STO by high energy grinding of the mixture of STO and thiourea. A new band gap at 522 nm corresponding to 2.37 eV was formed by codoping. They reported that PC activity for nitrogen monoxide oxidation in the long wavelength range ($\lambda > 510$ nm) was enhanced. It was observed that the PC activity of nitrogen- and sulfur-codoped STO was 10.9 times greater than that of pure STO, which may be due to the formation of a new band gap that enables effective absorption of visible light.

Wang et al. (2009b) also reported a novel series of solid solution semiconductors $(AgNbO_3)_{1-x}$ $(SrTiO_3)_x$ ($0 \leq x \leq 1$) as an active photocatalyst for efficient O_2 evolution and decomposition of organic impurities. This modification revealed that the perovskite-type solid solutions $(AgNbO_3)_{1-x}$ $(SrTiO_3)_x$ are crystallized in two types, that is, an orthorhombic system ($0 \leq x < 0.9$) or a cubic system ($0.9 \leq x \leq 1$). The hybridization behaviors between Ag 4d and O 2p orbitals and between Nb 4d and Ti 3d orbitals played a crucial role in tuning the energy band structure in the mixed valent perovskites $(AgNbO_3)_{1-x}$ $(SrTiO_3)_x$, which resulted in changes of the photophysical and PC properties. It was also observed that very fine Ag particles were precipitated on the catalyst surface to construct a nanocomposite structure of Ag/ $(AgNbO_3)_{1-x}$ $(SrTiO_3)_x$, with improved PC activities.

A unique morphology of STO nanocubes precipitated on TiO_2 nanowires (NWs) was successfully obtained by Ng et al. (2010) in the form of a thin-film heterojunctioned TiO_2/STO photocatalyst via facile hydrothermal techniques. The heterostructured photocatalyst indicated the highest efficiency in PC splitting of water to produce H_2, as it was 4.9 times that of TiO_2 and 2.1 times that of STO. This enhanced PC efficiency was largely attributed to the proficient separation of photogenerated charges at heterojunctions of the two dissimilar semiconductors, as well as a negative redox potential shift at the Fermi level. STO nanocubes were precipitated on the surfaces of TiO_2 NWs, thus forming heterojunctions at interphases of the two hybridized semiconductors, which was helpful in efficient antirecombination on the photogenerated charges and facilitated interfacial electron transfer and trapping.

PC properties of STO NPs were studied by Xian et al. (2011). It was found that the selection of chelating agent had a great role on the STO synthesis. Citric acid was used as a chelating agent, which led to the synthesis of pure STO at a calcination temperature of 550°C. Use of chelating agent acetic acid prolonged the calcination temperature. The PC properties of the STO NPs have been investigated for the degradation of Congo red, RhB, MO, and MB under the UV irradiation. The sample using citric acid exhibited better PC activity than a sample prepared using acetic acid and it was assumed to be due to its relatively smaller particle size.

A composite $Sr_2TiO_4/SrTiO_3(La,Cr)$ heterojunction photocatalyst has been prepared by Jia et al. (2013) via *in situ* polymerized complex process. This catalyst had higher PC activity toward hydrogen production upon Pt cocatalyst loading than with $SrTiO_3(La,Cr)$ and $Sr_2TiO_4(La,Cr)$ in the presence of methanol sacrificial reagent. The photogenerated electrons and holes tend to migrate from $SrTiO_3(La,Cr)$ to $Sr_2TiO_4(La,Cr)$ and from $Sr_2TiO_4(La,Cr)$ to $SrTiO_3(La,Cr)$, respectively, in the composite photocatalyst. The superior PC activity of the composite heterojunction photocatalyst was due to efficient charge transfer and separation by well-defined heterojunctions formed between $SrTiO_3(La,Cr)$ and $Sr_2TiO_4(La,Cr)$, as well as preferential loading of Pt NPs on the $Sr_2TiO_4(La,Cr)$ component, and the lowest amount of Cr^{6+} in the composite photocatalyst.

Ma et al. (2015) reported that the surface plasmon (SP) enhanced UV and visible PC activities of STO after incorporating Ag NPs on its surface. An electron in the VB of STO is first excited onto the Fermi level of Ag-NP by the SP field generated on the Ag-NP, and then injected into the CB of STO from the SP band, leaving a hole at the VB of STO.

Erbium–nitrogen (Er–N) codoped STO photocatalysts have been synthesized by Xu et al. (2015) via a facile solvothermal method. It was shown that Er–N codoped STO possessed stronger absorption bands in the visible light region compared to pure STO. Er–N codoped cubic STO had higher PC activities for the degradation of MO under UV and visible light irradiations. It was found to be superior to pure STO and commercial TiO_2 (P25) powders. A Er–N codoped STO sample showed the best PC activity with a degradation rate as high as 98% after 30 minutes under visible light irradiation. The PC activity of the sample (initial molar ratios of Sr/Er/N = 1:0.015:0.1) showed no significant decrease even after five cycles, which indicated the stability of this photocatalyst under visible light irradiation.

4.3 BARIUM TITANATE

Photocatalysis utilizes near-UV or visible light to break down organic pollutants into innocuous compounds at room temperatures. Vajifdar et al. (2007) introduced the use of semiconducting optical crystals as an additive to a photocatalyst. The perovskite optical material $BaTiO_3$ (band gap of 3.7–3.8 eV) was found to increase destruction of volatile organic compound (VOC), when black light was used. This photocatalyst increased tetrachloroethylene (PERC) conversion by 12%–32%. The reaction parameters studied were space velocity, inlet concentration, and light source.

The heterojunction semiconductors $Bi_2O_3/BaTiO_3$ were prepared by Lin et al. (2007) using the milling-annealing method. PC activities of these semiconductors were evaluated by the degradation of MO and methylene blue (MB) under UV radiation. The results showed that the heterojunction semiconductors $Bi_2O_3/BaTiO_3$ exhibited better PC properties than single-phase $BaTiO_3$ or Bi_2O_3. This increased performance of $Bi_2O_3/BaTiO_3$ was attributed mainly to the electric field–driven electron–hole separation both at the interface and in the semiconductors. A strategy for designing efficient heterojunction photocatalysts was also proposed, that is, an electron-accepting semiconductor and a hole-accepting semiconductor with matching band potentials—which, respectively, possess high electron and hole conduction abilities—are tightly chemically bonded to construct the efficient heterojunction structure.

PC degradation of the herbicide pendimethalin (PM) was investigated with a $BaTiO_3/TiO_2$ UV light system in the presence of peroxide and persulphate species in aqueous medium. Devi et al. (2008) prepared NPs of $BaTiO_3$ and TiO_2 by gel to crystallite conversion method. $BaTiO_3$ exhibited comparable PC efficiency in the degradation of PM with the most widely used TiO_2 photocatalyst. The persulphate played an important role in enhancing the rate of degradation of PM. The degradation process of PM followed the first-order kinetics and is in agreement with the Langmuir–Hinshelwood model. The higher rate of degradation of PM was observed in alkaline medium at pH 11.

The PC activity of $BaTiO_3$ was also observed by Devi et al. (2011) for the degradation of chloroorganic compounds. $BaTiO_3$ exhibited better catalytic efficiency and process efficiency as compared to TiO_2. These chloroorganic compounds have at least one chlorine substituent in common, along with other functional groups such as –OH, –NH$_2$, and –NO$_2$. The effect of electron acceptors and pH on the rate of degradation was presented.

Stanca et al. (2012) reported a nanoparticulate composite TiO_2–$BaTiO_3$ film that exhibited an increased antibacterial PC activity under visible light. Although pure $BaTiO_3$, TiO_2, or their mixture do not achieve considerable PC ability under visible light, highly catalytic crystals resulted when these oxides were concurrently synthesized under controlled conditions. It was also observed that peroxidation occurred in the absence of UV light and in the presence of TiO_2–$BaTiO_3$ but not in the presence of TiO_2 alone.

Wang et al. (2013a) reported a dual chelating sol–gel method to synthesize $BaTiO_3$ NPs. Acetylacetone and citric acid were used as chelating agents. It was observed that cubic phase $BaTiO_3$ NPs of size about 19.6 nm were obtained at 600°C, and tetragonal phase at 900°C (97.1 nm). All the $BaTiO_3$ NPs showed good PC activities on the removal of humic acid under UV light irradiation. A comparison of single (acetylacetone or citric acid) and dual chelating (acetylacetone and citric acid) synthetic routes was also made, and it was demonstrated that the dual-chelating agents certainly reduced phase transformation temperature from cubic to tetragonal $BaTiO_3$.

Cui et al. (2013) used $BaTiO_3$ as a target catalyst to investigate the influence of ferroelectricity on the decolorization of rhodamine B under simulated solar light. They showed that there is a threefold increase in the decolorization rate using $BaTiO_3$ with a high tetragonal content compared to mainly cubic material. This may be due to the ferroelectricity of the tetragonal phase. The influence of ferroelectricity ensured a tightly bound layer of dye molecule that separated the photoexcited carriers due to the presence of an internal space charge layer. Both these features increased the catalytic performance of $BaTiO_3$. Complete decolorization of the dye was observed in around 45 minutes after photochemically depositing nanostructured Ag on the surface of the $BaTiO_3$.

$Bi_2O_3/BaTiO_3$ heterostructure was prepared through a solid milling and annealing process. Ren et al. (2013) observed that Bi^{3+} dissolved in the $BaTiO_3$ lattice and a chemical bond was constructed between the interface of Bi_2O_3 and $BaTiO_3$ after the annealing process. Catalytic activities of the $Bi_2O_3/BaTiO_3$ heterostructure were studied for degradation of MO dye. The band gap of the $Bi_2O_3/BaTiO_3$ heterostructure was estimated to be 3.0 eV. This heterostructure had a much higher catalytic action as compared with pure Bi_2O_3 powders. The excellent PC property of the $Bi_2O_3/BaTiO_3$ heterostructure was attributed to high mobility of species and efficient separation of photogenerated carriers driven by the photoinduced potential difference at the $Bi_2O_3/BaTiO_3$ junction interface.

$BaTiO_3$ at ZnO core-shell NPs were synthesized by Karunakaran et al. (2014) using polyethylene glycol (PEG). It was observed that the use of PEG-20k instead of PEG-4k increased the interplanar distances in ZnO of $BaTiO_3$ at ZnO core-shell NPs, while it decreased the d-spacing in sol–gel synthesized pristine ZnO NPs. The charge transfer resistances of $BaTiO_3$ at ZnO core-shell NPs were larger than perfect ZnO NPs. The PEG-assisted sol–gel prepared $BaTiO_3$ at ZnO core-shell NPs were proved better photocatalysts than the PEG-assisted sol–gel synthesized pure ZnO NPs.

Coral-like $BaTiO_3$ nanostructures were successfully synthesized by Ni et al. (2015) via a simple hydrothermal route at 150°C for 15 hours using $BaCl_2$, tetrabutyl titanate [$(C_4H_9O)_4Ti$], and NaOH as precursors without any surfactant or template. Some of the factors influencing the formation of the coral-like $BaTiO_3$ nanostructures were also investigated, such as the amount of NaOH, the source of barium ion reaction temperature, and time. As-prepared coral-like $BaTiO_3$ nanostructures had good PC activity for the degradation of MO under the irradiation of artificial sunlight. It was found that the PC activity of the coral-like $BaTiO_3$ nanostructures could be affected by the pH of the system.

Xiong et al. (2015) reported a facile hydrothermal approach to synthesize $BaTiO_3$ nanocubes of controlled sizes that were used for degradation of MB and MO. The nanocubes with a reaction time of 48 hours exhibited the maximum PC efficiency due to their narrower size distribution and better crystallinity as compared to a reaction time of 24 hours, and also a smaller particle size than a reaction time of 72 hours. $BaTiO_3$ had lower PC activity for MO than MB mainly due to the poorer absorption behavior of MO on the surface of $BaTiO_3$ nanocubes.

Yttrium-doped barium titanate (BT) nanofibers (NFs) with a significant PC effect were successfully synthesized by Shen et al. (2015) via electrospinning. A well-designed procedure was used to produce yttrium-doped BT (BYT) NFs. BT NFs and BYT NFs with pure perovskite phase showed much enhanced performance compared to BYT ceramics powders. Narrow band gap energy due to yttrium doping is among the main factors causing the novel PC effect. A direct and efficient route to obtain doped NFs was suggested, which has a wide range of potential applications in areas where complex compounds with specific surface and a special doping effect are required.

A series of $BaTiO_3$-graphene nanocomposites with different weight addition ratios of graphene oxide (GO) have been synthesized by Wang et al. (2015) by means of a facile one-pot hydrothermal approach. The PC activity of the obtained $BaTiO_3$-graphene composites for the degradation of MB was observed under visible light irradiation, and a higher PC activity with $BaTiO_3$ was observed than for pure $BaTiO_3$. They suggested a new PC mechanism in which the role of graphene in the $BaTiO_3$-graphene nanocomposites was to act as an organic dye-like photosensitizer for large band gap $BaTiO_3$. The photosensitization process of $BaTiO_3$ by graphene transformed the wide band gap $BaTiO_3$ semiconductor into a visible light photoactive material for dye degradation. This reaction could widen the applications of $BaTiO_3$-graphene nanocomposites in solar energy conversion.

$BaTiO_3$ at g–C_3N_4 composites were prepared by Xian et al. (2015) through a simple mixing-calcining method. $BaTiO_3$ NPs were uniformly assembled onto the surface of g–C_3N_4 platelets. The PC activity of as-prepared $BaTiO_3$ at g–C_3N_4 composites was evaluated by the degradation of MO under simulated sunlight irradiation, which revealed that the composites had improved PC activity as compared to bare $BaTiO_3$ and g–C_3N_4. This can be explained by the efficient separation of the photogenerated electron–hole pairs due to the migration of the carriers between g–C_3N_4 and $BaTiO_3$ and as a result, electrons and holes were increasingly available for the PC reaction. Superoxide radicals and photogenerated holes were suggested to be the main active species responsible for the dye degradation, while hydroxyl radicals played a relatively small role in these PC reactions.

The visible light active ferroelectric photocatalyst $Bi_{0.5}Na_{0.45}Li_{0.05}K_{0.5}TiO_3$–$BaTiO_3$ (BNKLBT) was synthesized by a solid-state method and its PC, photoelectrochemical, and antibacterial properties have been investigated by Kushwaha et al. (2015). The current density under the visible light was 30 $\mu A/cm$, which is three times more than that observed under dark conditions. PC activity was investigated for degradation of an organic dye (MO) and an estrogenic pollutant (estriol). High PC and photoelectrochemical activity was a result of effective separation of photogenerated charge carriers, because of the ferroelectric nature of the catalyst. The effect of different charge trapping agents on PC degradation was observed. Antimicrobial activity was investigated for *Escherichia coli* and *Aspergillus flavus*. The antibacterial action of BNKLBT was compared with that of the commercial antibiotic kanamycin (k30).

Zhang et al. (2016) prepared Au_x/$BaTiO_3$ composite thin films dispersed with Au NPs by a sol–gel spin-coating approach. The crystal structure and optical and PC properties of the films were studied with a special emphasis on the influences of x. The approximately spherical Au NPs of 20–45 nm diameter were embedded uniformly in the amorphous $BaTiO_3$ matrix. The absorption peaks appeared at the wavelength from 580 to 620 nm due to the SP resonance (SPR) effect of Au NPs, and the SPR peaks generally manifest a red shift with x increasing from 5 to 45. The Au_x/$BaTiO_3$ composite film with the optimal SPR effect ($x = 35$) exhibited the highest PC activity for the degradation of rhodamine B aqueous solution under whole spectrum sunlight irradiation.

4.4 BISMUTH VANADATE

$BiVO_4$ photocatalyst was prepared by Kudo et al. (1999) for O_2 evolution. This photocatalyst was obtained by the reaction of layered potassium vanadate powder with $Bi(NO_3)_3$ for 3 days in aqueous media at room temperature. Monoclinic and tetragonal $BiVO_4$ were obtained by changing the ratio of vanadium to bismuth in the precursors. Tetragonal $BiVO_4$ had a 2.9 eV band gap, and it is active in UV region while monoclinic $BiVO_4$ had a 2.4 eV band gap with a characteristic visible light absorption band in addition to the UV band. It was observed that the PC activity for O_2 evolution from an aqueous silver nitrate solution under the UV irradiation on the tetragonal $BiVO_4$ was similar to that of monoclinic $BiVO_4$, but monoclinic $BiVO_4$ indicated high PC activity for the O_2 evolution under visible light irradiation. The activity was further increased on calcining monoclinic $BiVO_4$ at 700–800 K.

Photocatalyst $BiVO_4$ has also been applied for the degradation of a series of linear 4-n-alkylphenols by Kohtani et al. (2003). They observed that the degradation rates increase with increase in alkyl chain length. The half-life of 4-n-nonylphenol (NP) was 18 minutes, which was about eight times less than that required for degradation of phenol. The zero-order rate law for disappearance for heptylphenol, octylphenol (OP)-, and NP was observed. It was also observed that the amount of adsorption on the $BiVO_4$ surface was much larger for longer hydrophobic alkylphenols, and therefore it was concluded that $BiVO_4$ is suitable for degradation of hydrophobic alkylphenols such as NP and OP.

Kohtani et al. (2005) also observed fine particles of silver loaded on $BiVO_4$ photocatalyst by impregnation. It improved adsorptive and PC activity of bismuth vanadate on the degradation of long-chain alkylphenols such as 4-n-NP and 4-n-OP. The Ag-loaded $BiVO_4$ powders strongly adsorbed NP and OP, which is attributed to silver oxides (Ag_2O and/or AgO) that partly covered the silver surface. This specific adsorption property vanished on reducing surface silver oxides in an H_2 stream. It was observed that degradation rates and CO_2 mineralization yields for NP were improved by the Ag loading on $BiVO_4$.

Xie et al. (2006) prepared $BiVO_4$ powder with monoclinic structure and used it as a visible light catalyst for the photooxidation of phenol and the photoreduction of $Cr(VI)$. The PC reduction of $Cr(VI)$ and PC oxidation of phenol progress more rapidly with the coexistence of phenol and $Cr(VI)$ than for the single component. It indicates a synergetic effect between these oxidation and reduction reactions. The results revealed the feasibility of using $Cr(VI)$ as the electron scavenger of the m-$BiVO_4$-mediated PC process of phenol degradation.

Pure tetragonal and monoclinic phases of $BiVO_4$ have been prepared by Zhang et al. (2008) from aqueous solutions of $Bi(NO_3)_3$ and $NaVO_3$ using a microwave-assisted method. It was observed that the highly crystalline phase was gradually converted irreversibly from tetragonal to monoclinic $BiVO_4$ with an elongated irradiation time. Variations in these phase structures led to different PC properties under the visible light.

Xu et al. (2008) prepared a series of Cu-loaded $BiVO_4$ (Cu–$BiVO_4$) catalysts by the impregnation method. It was observed that the PC activities of Cu–$BiVO_4$ catalysts for the degradation of MB were found to depend largely on the Cu content as well as the calcination temperature. The optimum Cu loading and calcination temperature were found to be 5 at.% and 300°C, respectively. The results showed that Cu (CuO in this case) was dispersed on the surface of $BiVO_4$ and the Cu–$BiVO_4$ series catalysts had significant optical absorption in the visible region ranging between 550 and 800 nm. It was found that the absorption intensity increases with the increase in Cu content. Efficient N-demethylation of MB was observed using Cu–$BiVO_4$ catalyst calcined at 300°C.

Sun et al. (2009) successfully synthesized nanoplate-stacked star-like $BiVO_4$ by a hydrothermal process. These star-like $BiVO_4$ samples exhibited a high visible light driven PC efficiency. About 91% of the MB was degraded within 25 minutes under visible light irradiation. This is much higher than for $BiVO_4$ samples prepared by other solid-state reactions and hydrothermal synthesis.

Li and Yan (2009) prepared bismuth vanadate powders by a homogeneous precipitation process with different surface dispersants. Here also, $Bi(NO_3)_3 \cdot 5H_2O$ and NH_4VO_3 were utilized as starting materials to synthesize $BiVO_4$ particles with an edge length of about 100–150 nm. Bismuth oxide was prepared by decomposing $Bi(NO_3)_3 \cdot 5H_2O$ at 600°C. The PC degradation of rhodamin B indicated that $BiVO_4$/Bi_2O_3 possessed high PC activity under visible light irradiation.

Li et al. (2009) used a facile hydrothermal process for preparing $BiVO_4$. Bi_2O_3 and NH_4VO_3 were used as starting materials. Cuboid-like, square plate-like, and flower-like $BiVO_4$ could be obtained by tailoring the pH values of the reaction suspensions in the presence of CTAB. Both pH value and CTAB played important roles in the deciding morphology of the prepared samples. The band gaps of cuboid-like, square plate-like, and flower-like $BiVO_4$ were found to be 2.39, 2.40, and 2.46 eV, respectively. The PC performance of the $BiVO_4$ was much higher than P25 for photodegradation of MO under the sunlight irradiation. High PC activities of $BiVO_4$ samples were attributed to their crystallinities and shapes.

$BiVO_4$-based photocatalysts loaded with rare earth element such as Ho, Sm, Yb, Eu, Gd, Nd, Ce, and La have been synthesized by Xu et al. (2009). The PC activities of the samples were investigated by decolorization of MB under visible light irradiation. The DRS-spectral analysis showed the following shift in the absorption edge from the UV to the visible range:

$$Ho^{3+} - BiVO_4 < Sm^{3+} - BiVO_4 < Yb^{3+} - BiVO_4 < Eu^{3+} - BiVO_4 < GD^{3+}$$

$$- BiVO_4 < Nd^{3+} - BiVO_4 < La^{3+} - BiVO_4 < Ce^{3+} - BiVO_4 < BiVO_4$$

Zhang et al. (2009a) successfully fabricated $BiVO_4$ film coated on F-doped SnO_2 (FTO) glass by a modified metal-organic decomposition (MOD) technique. It was revealed that the absorption performance of the $BiVO_4$ film was intense in the visible light region and the band gap was 2.43 eV. It was present as the monoclinic-type $BiVO_4$. The photocurrent densities of $BiVO_4$ films was found to increase with increasing film thickness from 0.23 to 1.04 μm, which may be due to enhanced amount of absorption of the incident photons. The removal rate of phenol in the photoelectrocatalytic (PEC) process by the $BiVO_4$ film electrode under visible light was 27.1 times higher than the simple PC process. It was attributed to the promoted separation of photogenerated electron–hole pairs. The $BiVO_4$ film electrode coated on FTO glass showed good stability in the PEC process.

PC active spindle-like $BiVO_4$ modified by polyaniline (PANI) was synthesized by Shang et al. (2009) via a sonochemical approach. The photocatalysts were composed of well-crystallized small NPs. Efficient PC activity in the degradation of tetraethylated rhodamine and phenol under visible light irradiation was observed by the spindle-like PANI/$BiVO_4$. A reduction in the COD was also observed. Improved PC performance for phenol oxidation was reported with the assistance of a small amount of H_2O_2. This enhanced PC activity was attributed to the synergistic effect between PANI and $BiVO_4$, which promoted the migration efficiency of photogenerated electron–hole pairs.

Zhang et al. (2009c) reported the preparation of Ag-doped $BiVO_4$ film by the photoreduction method. The PC degradation rate of the phenol on the Ag-doped $BiVO_4$ film was 1.61 times higher than PC process and 42.7 times higher than the PEC process compared to undoped $BiVO_4$ film. The transportation of the electrons from the $BiVO_4$ to Ag was driven due to the Schottky barrier formed between Ag and $BiVO_4$, which can increase the charge carrier separation, and PC performance was enhanced as a result. This ability in the PEC process could be attributed to the simultaneous movements of the photogenerated electrons to an external circuit and the photogenerated holes to the Ag particles deposited on the $BiVO_4$ film.

A highly proficient monoclinic $BiVO_4$ photocatalyst (C–BVO) was synthesized by Yin et al. (2010) using cetyltrimethylammonium bromide (CTAB). PC efficiencies were assessed by the degradation of rhodamine B under visible light irradiation. The degradation rate over the C–BVO was found to be much higher than that over the reference $BiVO_4$ prepared by an aqueous technique and solid-state reaction. The COD values of the RhB were observed after PC degradation over the C–BVO, which showed a 53% decrease in COD. It was proposed that the existence of contamination level played a major role in the high PC efficiency of the C–BVO.

4.5 ZINC FERRITE

Srinivasan et al. (2005) studied the nanocomposite heterogeneous semiconductors with suitable energy levels. TiO_2/$ZnFe_2O_4$ has been selected among different alloys because of its low band gap (1.9 eV), nontoxic nature of $ZnFe_2O_4$, and its visible light absorption characteristics resulting from its narrow band gap. It was quite interesting to note that there was a red shift of 0.25 eV in the absorption edge of the ball-milled TiO_2 sample compared to the nonball-milled TiO_2 photocatalyst.

Cao et al. (2007) prepared spinel $ZnFe_2O_4$ nanoplates embedded with Ag clusters. These nanoplates were synthesized through the coprecipitation reaction of Zn^{2+} and Fe^{3+} in the solution containing urea and oleic acid. Ag NPs were established in the nanoplates through the reduction of silver nitrate by ethylene glycol. The content and size of Ag NPs were adjusted by controlling the reflux time of silver nitrate in ethylene glycol. The irradiation with a xenon lamp in the presence of $ZnFe_2O_4$ nanocomposites resulted in the rapid decomposition of rhodamine B, and the degradation rate was further increased with the increase of Ag weight content in the composite structure.

The PC oxidation of acid dye under visible light irradiation was carried out in a recycle fluidized bed reactor. Wang et al. (2009a) employed $ZnFe_2O_4/TiO_2$-immobilized granular activated carbon ($ZnFe_2O_4/TiO_2$-GAC) as the photocatalyst. It was prepared by hydrolysis of titanium n-butoxide combined with zinc ferrite solution and calcined at different temperatures. It was observed that the apparent first-order rate constant decreased with increasing initial acid dye concentration and particle size of photocatalyst, but it increased with increasing $ZnFe_2O_4/TiO_2$-GAC loading amount.

Highly ordered $ZnFe_2O_4$ nanotube arrays (NTAs) were also successfully prepared by Li et al. (2011) via anodic aluminum oxide templates from the sol–gel solution. They observed that as-prepared samples were vertically aligned spinel $ZnFe_2O_4$ NTAs, and the nanotubes (NTs) were uniform along the axial direction. The absorption edge of $ZnFe_2O_4$ NTAs showed a blue shift unlike the $ZnFe_2O_4$ NP film. Synthesized $ZnFe_2O_4$ NTAs revealed excellent PC capability for degradation of 4-chlorophenol under visible light irradiation. The main intermediate degradation species of 4-chlorophenol were also detected.

A cubic $ZnFe_2O_4$ with spinel structure was synthesized by Sun et al. (2012) under an the ignition temperature of 573 K. The synthesized $ZnFe_2O_4$ had a sponge-like porous structure and wide absorption in the visible light region. The impurities α-Fe_2O_3 and ZnO formed in the sample increased the reducing and oxidizing ability and supported the separation of photogenerated electrons and holes. The $ZnFe_2O_4$ derived by a solution combustion technique indicated better PC activity under visible light radiation.

Magnetic $ZnFe_2O_4$ octahedra was successfully synthesized by Sun et al. (2013) via a hydrothermal method with the help of sodium oleate. Its band gap was calculated as 1.7 eV. The $ZnFe_2O_4$ nanocrystal exhibited good PC activity in the degradation of rhodamine B under the irradiation of simulated solar light. The improved PC activity can be attributed to the small size and octahedral morphology of $ZnFe_2O_4$.

Hou et al. (2013) synthesized a novel graphene supported $ZnFe_2O_4$ multiporous microbricks hybrid via a facile deposition-precipitation reaction, followed by a hydrothermal treatment. It was indicated that some intimate contact between $ZnFe_2O_4$ microbricks and graphene sheets was formed. This hybrid exhibited a much higher PC activity in PC degradation of p-chlorophenol than the pure $ZnFe_2O_4$ multiporous microbricks and $ZnFe_2O_4$ NPs. This increase could be attributed to the fast photogenerated charge separation and transfer, which is due to the high electron mobility of graphene sheets, improved light absorption, high specific surface area, and multiporous structure of the hybrid. Hydroxyl radicals were established as the main active oxygen species in this PC reaction.

$ZnFe_2O_4$-coupled TiO_2 NTAs ($ZnFe_2O_4$–TiO_2 NTAs) were synthesized by Wang et al. (2013b) using modified TiO_2 NTAs with $ZnFe_2O_4$ NPs. $ZnFe_2O_4$ NPs were deposited on the surface of TiO_2 NTs through a facile hydrothermal route. The modification of $ZnFe_2O_4$ NPs extended the photoresponse of TiO_2 NTAs to the visible light region. A 12-fold development in PC activity was achieved under visible light irradiation using $ZnFe_2O_4$–TiO_2 NTAs compared to TiO_2 NTAs. An enhanced photogenerated electron–hole separation and improved transfer of photogenerated charge carriers were considered responsible for the increased PC activity of $ZnFe_2O_4$–TiO_2 NTAs.

$ZnFe_2O_4$ NPs were synthesized by Shao et al. (2013) using a solvothermal method that served as seeding materials. Wurtzite ZnO was coated on the $ZnFe_2O_4$ particle surfaces using a chemical precipitation process. Photodegradative action of as-prepared sample was observed on MB under UV irradiation. They showed that the core-shell $ZnFe_2O_4$ at ZnO NPs exhibited higher PC activity than pure ZnO. The product showed the highest PC activity, when the molar ratio of $ZnFe_2O_4$ to ZnO was 1:10.

Magnetic composite $ZnFe_2O_4/SrFe_{12}O_{19}$ including a hard-magnetic phase ($SrFe_{12}O_{19}$) and a soft-magnetic phase ($ZnFe_2O_4$) has been prepared by Xie et al. (2013) using one-step chemical coprecipitation with a high-temperature sintering method. This composite can be easily separated, recycled, and reused after reaction. The PC activity of the composite was studied by the degradation reaction of MB under visible light irradiation. The results indicated that the degradation rate was still more than 70%, when the composite was reused four times.

4.6   OTHERS

The ternary chalcopyrite semiconductors $Cu(In/Ga)(Se/S)_2$ are currently used as an absorber layer in high efficiency thin film solar cells. Various types of I–III–VI (I = Cu, III = Ga or In, VI = S or Se) thin films ($CuGaS_2$, $CuInS_2$, and $CuInSe_2$) were prepared by Afzaal et al. (2002) from a series of organometallic precursors, $M[(S/Se)_2CNMeR]_n$ (M = Cu, In, Ga; R = alkyl) by aerosol-assisted chemical vapor deposition (AACVD). The precursors were easy to synthesize by one-pot reactions and were air stable as well.

A simple and reproducible route to ordered porous Bi_2WO_6 films with open pores was reported by Zhang et al. (2009b). The PC activity of the as-prepared porous films was evaluated using decomposition of MB under visible light irradiation. This method can be used to prepare porous complex ternary metal oxide semiconductor films with high PC activity in visible light. The porous films showed highly enhanced PC activity and higher photocurrent conversion efficiency over different applied potentials.

Ternary oxides have been used as effective photocatalysts for carrying out a number of chemical reactions. The method of preparation has a major effect on the performance of these mixed oxides catalysts. A cerium iron oxide catalyst was synthesized by Ameta et al. (2009) using a coprecipitation method and specific heating cycle. PC degradation on gentian violet dye was observed using this catalyst and the progress of the reaction was monitored spectrophotometrically. The effect of variation of different parameters such as concentration of gentian violet, pH, amount of semiconductor, and light intensity was also studied.

The fabrication and PC properties of the visible light driven $CuInSe_2/TiO_2$ heterojunction films were reported by Liao et al. (2013) using a solvothermal method. They decorated it onto self-organized anodic TiO_2 NT arrays through an electrophoretic deposition route. An increase in deposition time created an improved quantity of $CuInSe_2$ NCs loaded onto the TiO_2 NT arrays. It expanded the light absorption range of the $CuInSe_2$ NCs/TiO_2 NTs. PC degradation results revealed that the activities of the $CuInSe_2$ NCs/TiO_2 NT films were significantly improved compared to pure TiO_2 NTs.

CuInSe$_2$–ZnO nanocomposites were prepared as an efficient photocatalyst by Bagheri et al. (2014) via a solvothermal process. It was observed that crystallite size, BET surface area, and optical absorption of the samples varied with the addition of CuInSe$_2$ to ZnO. The optical band gap values of these nanocomposites were calculated to be about 3.37–2.1 eV, which means a red shift from that of pure ZnO. These red shifts support the incorporation of CuInSe$_2$ in the zinc oxide lattice. The effect of the different parameters such as pH, Congo red concentration, CuInSe$_2$ content, and irradiation sources of UV and visible light has been studied. The PC activity of as-prepared samples was improved confirming that the addition of CuInSe$_2$ was effective in improving the PC activity of ZnO.

Sheng et al. (2014) developed a high-performance hollow CuInSe$_2$ nanospheres-based photoelectrochemical cell for hydrogen evolution. Cyclic voltammetry was used to verify the stepwise CB edge and electrochemically active surface areas. A type-II-like core/shell heterojunction model was planned to reveal the charge transfer mechanism. The saturated short circuit photocurrent attained by photoelectrode under illumination of AM 1.5 (100 mW/cm^2) was 22.4 mA/cm^2.

Myung et al. (2015) converted the nanosheets of BiOCl, a V–VI–VII ternary semiconductor, to an oxygen-rich Bi$_{12}$O$_{15}$C$_{16}$ phase. The intrinsic conductivity switches from p-type to n-type in this process. The phase change was achieved using a vacuum annealing step at 500°C for 1 hour. BiOCl nanosheets alter to the Bi$_{12}$O$_{15}$C$_{16}$ phase through volatilization of BiCl$_3$, resulting in a unique superlattice-like structure. The band gap was found to decrease from 3.41 to 2.48 eV, raising the VB level.

Light-induced water splitting and carbon dioxide reduction is an existing challenge for photochemists and it requires sustainable and firm semiconductors. Huang et al. (2015) synthesized a ternary semiconductor, boron carbon nitride (B–C–N), and showed that this semiconductor can catalyze hydrogen or oxygen evolution from water and reduce carbon dioxide under visible light illumination. Such photocatalysts are sustainable, made of lightweight elements, and facilitate the novel construction of photoredox cascades to utilize solar energy for chemical conversion.

Mutika and Tubtimtae (2015) synthesized a new ternary semiconductor NP Cd$_{1-x}$In$_x$Te, as a sensitizer for solar cell devices via a one-pot mixed precursor solution. Cd$_{1-x}$In$_x$Te NPs were prepared using a chemical bath deposition process and then coated onto a TiO$_2$ photoelectrode. When the dipping cycle was increased, the energy gaps became narrower from 1.2 to 0.6 eV due to the increasing amount and the larger size of NPs. The best power conversion efficiency (η) of 0.49% under full 1 Sun illumination (100 mW/cm^2, AM 1.5G) was obtained for the seven-cycle-Cd$_{1-x}$In$_x$Te NPs with a current density of 2.64 mA/cm^2, an open-circuit voltage of 638 mV, and a fill factor of 0.29. It was suggested that the efficiency of this material can be further improved.

Different ternary chalcogenides and in particular ternary oxides have been quite commonly used as photocatalysts for different purposes. Zinc ferrite, bismuth vanadate, BT, and others have been explored for their possible use in treating wastewaters, hydrogen generation, and so on, but there are many more miles to go and a search must go on for better and efficient ternary semiconductors.

REFERENCES

Afzaal, M., D. Crouch, P. O'Brien, and J.-H. Park. 2002. New approach towards the deposition of I-III-VI thin films. *Mater. Res. Soc. Symp. Proc.* 692: 215–220.

Ameta, J., A. Kumar, R. Ameta, V. K. Sharma, and S. C. Ameta. 2009. Synthesis and characterization of $CeFeO_3$ photocatalyst used in photocatalytic bleaching of gentian violet. *J. Iranian Chem. Soc.* 6 (2): 293–299.

Bagheri, M., A. R. Mahjoub, and B. Mehri. 2014. Enhanced photocatalytic degradation of Congo red by solvothermally synthesized $CuInSe_2$-ZnO nanocomposites. *RSC Adv.* 4 (42): 21757–21764.

Cao, X., L. Gu, X. Lan, C. Zhao, D. Yao, and W. Sheng. 2007. Spinel $ZnFe_2O_4$ nanoplates embedded with Ag clusters: Preparation, characterization and photocatalytic application. *Mater. Chem. Phys.* 106 (2–3): 175–180.

Chen, Y.-C., C.-B. Siao, H.-S. Chen, K.-W. Wang, and S.-R. Chung. 2015. The application of $Zn_{0.8}Cd_{0.2}S$ nanocrystals in white light emitting diodes devices. *RSC Adv.* 5: 87667–87671.

Cui, Y., J. Briscoe, and S. Dunn. 2013. Effect of ferroelectricity on solar-light-driven photocatalytic activity of $BaTiO_3$ influence on the carrier separation and stern layer formation. *Chem. Mater.* 25 (21): 4215–4223.

Devi, G. L. and G. Krishnamurthy. 2011. TiO_2-and $BaTiO_3$-assisted photocatalytic degradation of selected chloroorganic compounds in aqueous medium. *J. Phys. Chem. A.* 115 (4): 460–469.

Devi, G., L. Naik, and G. Krishnamurthy. 2008. Photocatalytic degradation of the herbicide pendimethalin using nanoparticles of $BaTiO_3/TiO_2$ prepared by gel to crystalline conversion method. *J. Environ. Sci. Health.* 43 (7): 553–561.

Hou, Y., X. Li, Q. Zhao, and G. Chen. 2013. $ZnFe_2O_4$ multi-porous microbricks/graphene hybrid photocatalyst: Facile synthesis, improved activity and photocatalytic mechanism. *Appl. Catal. B: Environ.* 142–143: 80–88.

Huang, C., C. Chen, M. Zhang, L. Lin, X. Ye, S. Lin, M. Antonietti, and X. Wang. 2015. Carbon-doped BN nanosheets for metal-free photoredox catalysis. *Nat. Commun.* 6. doi:10.1038/ncomms8698.

Jia, Y., S. Shen, D. Wang, X. Wang, J. Shi, F. Zhang, H. Han, and C. Li. 2013. Composite $Sr_2TiO_4/SrTiO_3$(La,Cr) heterojunction based photocatalyst for hydrogen production under visible light irradiation. *J. Mater. Chem. A.* 1 (27): 7905–7912.

Karunakaran, C., P. Vinayagamoorthy, and J. Jayabharathi. 2014. Optical, electrical, and photocatalytic properties of polyethylene glycol-assisted sol-gel synthesized $BaTiO_3$@ZnO core-shell nanoparticles. *Powder Technol.* 254: 480–487.

Kato, H. and A. Kudo. 2002. Photocatalytic activities of TiO_2 and $SrTiO_3$ photocatalysts co-doped with antimony and chromium. *J. Phys. Chem. B.* 106 (19): 5029–5034.

Kohtani, S., J. Hiro, N. Yamamoto, A. Kudo, K. Tokumura, and R. Nakagaki. 2005. Adsorptive and photocatalytic properties of Ag-loaded $BiVO_4$ on the degradation of 4-n-alkylphenols under visible light irradiation. *Catal. Commun.* 6 (3): 185–189.

Kohtani, S., M. Koshiko, A. Kudo, K. Tokumura, Y. Ishigaki, A. Toriba, K. Hayakawa, and R. Nakagaki. 2003. Photodegradation of 4-alkylphenols using $BiVO_4$ photocatalyst under irradiation with visible light from a solar simulator. *Appl. Catal. B: Environ.* 46 (3): 573–586.

Konta, R., T. Ishii, H. Kato, and A. Kudo. 2004. Photocatalytic activities of noble metal ion doped $SrTiO_3$ under visible light irradiation. *J. Phys. Chem. B.* 108 (26): 8992–8995.

Kudo, A., K. Omori, and H. Kato. 1999. A novel aqueous process for preparation of crystal form-controlled and highly crystalline $BiVO_4$ powder from layered vanadates at room temperature and its photocatalytic and photophysical properties. *J. Am. Chem. Soc.* 121 (49): 11459–11467.

Kushwaha, H. S., A. Halder, D. Jain, and R. Vaish. 2015. Visible light-induced photocatalytic and antibacterial activity of Li-doped $Bi_{0.5}Na_{0.45}K_{0.5}TiO_3$–$BaTiO_3$ ferroelectric ceramics. *J. Electron. Mater.* 44 (11): 4334–4342.

Li, H., G. Liu, and X. Duan. 2009. Monoclinic $BiVO_4$ with regular morphologies: Hydrothermal synthesis, characterization and photocatalytic properties. *Mater. Chem. Phys.* 115 (1): 9–13.

Li, L. and B. Yan. 2009. $BiVO_4/Bi_2O_3$ submicrometer sphere composite: Microstructure and photocatalytic activity under visible-light irradiation. *J. Alloys Compd.* 476 (1–2): 624–628.

Li, X., Y. Hou, Q. Zhao, W. Teng, X. Hu, and G. Chen. 2011. Capability of novel $ZnFe_2O_4$ nanotube arrays for visible-light induced degradation of 4-chlorophenol. *Chemosphere.* 82 (4): 581–586.

Liao, Y., H. Zhang, Z. Zhong, L. Jia, F. Bai, J. Li, P. Zhong, H. Chen, and J. Zhang. 2013. Enhanced visible-photocatalytic activity of anodic TiO_2 nanotubes film via decoration with $CuInSe_2$ nanocrystals. *ACS Appl. Mater. Interfaces.* 5 (21): 11022–11028.

Lin, X., J. Xing, W. Wang, Z. Shan, F. Xu, and F. Huang. 2007. Photocatalytic activities of heterojunction semiconductors $Bi_2O_3/BaTiO_3$. *J. Phys. Chem. C.* 111 (49): 18288–18293.

Liu, Y., L. Xie, Y. Li, R. Yang, J. Qu, Y. Li, and X. Li. 2008. Synthesized and high photocatalytic hydrogen production of $SrTiO_3$ nanoparticles from water splitting under UV irradiation. *J. Power Sources.* 183 (2): 701–707.

Ma, L., T. Sun, H. Cai, Z.-Q. Zhou, J. Sun, and M. Lu. 2015. Enhancing photocatalysis in $SrTiO_3$ by using Ag nanoparticles: A two-step excitation model for surface plasmon-enhanced photocatalysis. *J. Chem. Phys.* 143 (8). doi:10.1063/1.4929910.

Miyauchi, M., M. Takashio, and H. Tobimatsu. 2004. Photocatalytic activity of $SrTiO_3$ codoped with nitrogen and lanthanum under visible light illumination. *Langmuir.* 20 (1): 232–236.

Mutika, S.-N. and A. Tubtimtae. 2015. One-pot synthesis of $Cd_{1-x}In_xTe$ semiconductor as a sensitizer on TiO_2 mesoporous for potential solar cell devices. *Appl. Phys. A.* 120 (2): 757–764.

Myung, Y., F. Wu, S. Banerjee, A. Stoica, H. Zhong, S. S. Lee, J. Fortner, L. Yang, and P. Banerjee. 2015. Highly conducting, n-type $Bi_{12}O_{15}C_{16}$ nanosheets with superlattice-like structure. *Chem. Mater.* 27 (22): 7710–7718.

Ng, J., S. Xu, X. Zhang, H. Y. Yang, and D. D. Sun. 2010. Hybridized nanowires and cubes: A novel architecture of a heterojunctioned $TiO_2/SrTiO_3$ thin film for efficient water splitting. *Adv. Funct. Mater.* 20 (24): 4287–4294.

Ni, Y., H. Zheng, N. Xiang, K. Yuan, and J. Hong. 2015. Simple hydrothermal synthesis and photocatalytic performance of coral-like $BaTiO_3$ nanostructures. *RSC Adv.* 5 (10): 7245–7252.

Puangpetch, T., T. Sreethawong, S. Yoshikawa, and S. Chavadej. 2008. Synthesis and photocatalytic activity in methyl orange degradation of mesoporous-assembled $SrTiO_3$ nanocrystals repaired by sol-gel method with the aid of structure-directing surfactant. *J. Mol. Catal. A: Chem.* 287 (1–2): 70–79.

Puangpetch, T., T. Sreethawong, S. Yoshikawa, and S. Chavadej. 2009. Hydrogen production from photocatalytic water splitting over mesoporous-assembled $SrTiO_3$ nanocrystal-based photocatalysts. *J. Mol. Catal. A: Chem.* 312 (1–2): 97–106.

Ren, P., H. Fan, and X. Wang. 2013. Solid-state synthesis of $Bi_2O_3/BaTiO_3$ heterostructure: Preparation and photocatalytic degradation of methyl orange. *Appl. Phys. A.* 111 (4): 1139–1145.

Shang, M., W. Wang, S. Sun, J. Ren, L. Zhou, and L. Zhang. 2009. Efficient visible light-induced photocatalytic degradation of contaminant by spindle-like $PANI/BiVO_4$. *J. Phys. Chem. C.* 113 (47): 20228–20233.

Shao, R., L. Sun, L. Tang, and Z. Chen. 2013. Preparation and characterization of magnetic core-shell $ZnFe_2O_4$@ZnO nanoparticles and their application for the photodegradation of methylene blue. *Chem. Eng. J.* 217: 185–191.

Shen, Z., Y. Wang, W. Chen, H. L. W. Chan, and L. Bing. 2015. Photocatalysis of yttrium doped $BaTiO_3$ nanofibres synthesized by electrospinning. *J. Nanomater.* doi:10.1155/2015/327130.

Sheng, P., W. Li, X. Tong, X. Wang, and Q. Cai. 2014. Development of a high performance hollow $CuInSe_2$ nanospheres-based photoelectrochemical cell for hydrogen evolution. *J. Mater. Chem. A.* 2 (44): 18974–18987.

Song, S., L. Xu, Z. He, J. Chen, X. Xiao, and B. Yan. 2007. Mechanism of the photocatalytic degradation of C.I. reactive black 5 at pH 12.0 using $SrTiO_3$/CeO_2 as the catalyst. *Environ. Sci. Technol.* 41 (16): 5846–5853.

Song, S., L. Xu, Z. He, H. Ying, J. Chen, X. Xiao, and B. Yan. 2008. Photocatalytic degradation of C. I. direct red 23 in aqueous solutions under UV irradiation using $SrTiO_3$/CeO_2 composite as the catalyst. *J. Hazard. Mater.* 152 (3): 1301–1308.

Srinivasan, S. S., N. Kislov, J. Wade, M. T. Smith, E. K. Stefanakos, and Y. Goswami. 2005. Mechanochemical synthesis, structural characterization and visible light photocatalysis of TiO_2/$ZnFe_2O_4$ nanocomposites. *Mater. Res. Soc. Symp. Proc.* 900: 307–315.

Stanca, S. E., R. Müller, M. Urban, A. Csaki, F. Froehlich, C. Krafft, J. Popp, and W. Fritzsche. 2012. Photocatalyst activation by intrinsic stimulation in TiO_2-$BaTiO_3$. *Catal. Sci. Technol.* 2 (7): 1472–1479.

Sun, S., W. Wang, L. Zhou, and H. Xu. 2009. Efficient methylene blue removal over hydrothermally synthesized starlike $BiVO_4$. *Ind. Eng. Chem. Res.* 48 (4): 1735–1739.

Sun, S., X. Yang, Y. Zhang, F. Zhang, J. Ding, J. Bao, and C. Gao. 2012. Enhanced photocatalytic activity of sponge-like $ZnFe_2O_4$ synthesized by solution combustion method. *Prog. Nat Sci. Mater. Int.* 22 (6): 639–643.

Sun, Y., W. Wang, L. Zhang, S. Sun, and E. Gao. 2013. Magnetic $ZnFe_2O_4$ octahedra: Synthesis and visible light induced photocatalytic activities. *Mater. Lett.* 98: 124–127.

Vajifdar, K. J., D. H. Chen, J. L. Gossage, K. Li, X. Ye, G. Gadiyar, and B. Ardoin. 2007. Photocatalytic oxidation of PCE and butyraldehyde over titania modified with perovskite optical crystal $BaTiO_3$. *Chem. Eng. Technol.* 30 (4): 474–480.

Wang, D., T. Kako, and J. Ye. 2009b. New series of solid-solution semiconductors $(AgNbO_3)_{1-x}(SrTiO_3)_x$ with modulated band structure and enhanced visible-light photocatalytic activity. *J. Phys. Chem. C.* 113 (9): 3785–3792.

Wang, J., H. Li, H. Li, S. Yin, and T. Sato. 2009c. Photocatalytic activity of visible light-active sulfur and nitrogen co-doped $SrTiO_3$. *Solid State Sci.* 11 (1): 182–188.

Wang, J., S. Yin, M. Komatsu, and T. Sato. 2005. Lanthanum and nitrogen co-doped $SrTiO_3$ powders as visible light sensitive photocatalyst. *J. Europ. Ceramic Soc.* 25 (13): 3207–3212.

Wang, M., L. Sun, J. Cai, P. Huang, Y. Su, and C. Lin. 2013b. A facile hydrothermal deposition of $ZnFe_2O_4$ nanoparticles on TiO_2 nanotube arrays for enhanced visible light photocatalytic activity. *J. Mater. Chem. A.* 1 (39): 12082–12087.

Wang, P., C. Fan, Y. Wang, G. Ding, and P. Yuan. 2013a. Dual chelating sol-gel synthesis of $BaTiO_3$ nanoparticles with effective photocatalytic activity for removing humic acid from water. *Mater. Res. Bull. A.* 48 (2): 869–877.

Wang, R.-C., K.-S. Fan, and J.-S. Chang. 2009a. Removal of acid dye by $ZnFe_2O_4$/TiO_2-immobilized granular activated carbon under visible light irradiation in a recycle liquid-solid fluidized bed. *J. Taiwan Inst Chem. Eng.* 40 (5): 533–540.

Wang, R.-X., Q. Zhu, W.-S. Wang, C.-M. Fan, and A.-W. Xu. 2015. $BaTiO_3$-graphene nanocomposites: Synthesis and visible light photocatalytic activity. *New J. Chem.* 39 (6): 4407–4413.

Xian, T., H. Yang, J. F. Dai, Z. Q. Wei, J. Y. Ma, and W. J. Feng. 2011. Photocatalytic properties of SrTiO$_3$ nanoparticles using polyacrylamide gel. *Mater. Lett.* 65 (21–22): 3254–3257.

Xian, T., H. Yang, L. J. Di, and J. F. Dai. 2015. Enhanced photocatalytic activity of BaTiO$_3$@g-C$_3$N$_4$ for the degradation of methyl orange under simulated sunlight irradiation. *J. Alloys Compd.* 622: 1098–1104.

Xie, B., H. Zhang, P. Cai, R. Qiu, and Y. Xiong. 2006. Simultaneous photocatalytic reduction of Cr(VI) and oxidation of phenol over monoclinic BiVO$_4$ under visible light irradiation. *Chemosphere.* 63 (6): 956–963.

Xie, T.-H., Sun, X., and J. Lin. 2008. Enhanced photocatalytic degradation of RhB driven by visible light-induced MMCT of Ti(IV)-O-Fe(II) formed in Fe-doped SrTiO$_3$. *J. Phys. Chem. C.* 112 (26): 9753–9759.

Xie, T., L. Xu, C. Liu, and Y. Wang. 2013. Magnetic composite ZnFe$_2$O$_4$/SrFe$_{12}$O$_{19}$: Preparation, characterization, and photocatalytic activity under visible light. *Appl. Surface Sci.* 273: 684–691.

Xiong, X., R. Tian, X. Lin, D. Chu, and S. Li. 2015. Formation and photocatalytic activity of BaTiO$_3$ nanocubes via hydrothermal process. *J. Nanomater.* doi:10.1155/2015/692182.

Xu, H., H. Li, C. Wu, J. Chu, Y. Yan, H. Shu, and Z. Gu. 2008. Preparation, characterization and photocatalytic properties of Cu-loaded BiVO$_4$. *J. Hazard. Mater.* 153 (1–2): 877–884.

Xu, H., C. Wu, H. Li, J. Chu, G. Sun, Y. Xu, and Y. Yan. 2009. Synthesis, characterization and photocatalytic activities of rare earth-loaded BiVO$_4$ catalysts. *Appl. Surf. Sci.* 256 (3): 597–602.

Xu, J., Y. Wei, Y. Huang, J. Wang, X. Zheng, Z. Sun, Y. Wu, X. Tao, L. fan, and J. Wu. 2015. Erbium and nitrogen co-doped SrTiO$_3$ with highly visible light photocatalytic activity and stability by solvothermal synthesis. *Mater. Res. Bull.* 70: 114–121.

Yin, W., W. Wang, L. Zhou, S. Sun, and L. Zhang. 2010. CTAB-assisted synthesis of monoclinic BiVO$_4$ photocatalyst and its highly efficient degradation of organic dye under visible-light irradiation. *J. Hazard. Mater.* 173 (1–3): 194–199.

Zhang, B. L.-W., Y.-J. Wang, H.-Y. Cheng, W.-Q. Yao, and Y.-F. Zhu. 2009b. Synthesis of porous Bi$_2$WO$_6$ thin films as efficient visible-light-active photocatalysts. *Adv. Mater.* 21 (12): 1286–1290.

Zhang, H. M., J. B. Liu, H. Wang, W. X. Zhang, and H. Yan. 2008. Rapid microwave-assisted synthesis of phase controlled BiVO$_4$ nanocrystals and research on photocatalytic properties under visible light irradiation. *J. Nanoparticle Res.* 10 (5): 767–774.

Zhang, S. W., B. P. Zhang, S. Li, X. Y. Li, and Z. C. Huang. 2016. SPR enhanced photocatalytic properties of Au-dispersed amorphous BaTiO$_3$ nanocomposite thin films. *J. Alloys Compd.* 654: 112–119.

Zhang, X., S. Chen, X. Quan, H. Zhao. 2009a. Preparation and characterization of BiVO$_4$ film electrode and investigation of its photoelectrocatalytic (PEC) ability under visible light. *Separ. Purific. Technol.* 64 (3): 309–313.

Zhang, X., Y. Zhang, X. Quan, and S. Chen. 2009c. Preparation of Ag doped BiVO$_4$ film and its enhanced photoelectrocatalytic (PEC) ability of phenol degradation under visible light. *J. Hazard. Mater.* 167 (1–3): 911–914.

5 Quaternary Semiconductors

5.1 INTRODUCTION

Rapid urbanization and transportation during the last few decades have given rise to a number of environmental problems such as wastewater supply, wastewater generation, and its collection, treatment, and disposal. Water scarcity is becoming a burning problem as water resources are being polluted at an ever-increasing pace. This problem will be more severe in the next few decades. The majority of industries do not give priority to the treatment of their effluents or to recycling the water. Photocatalytic technology utilizing abundant solar light holds great promise to tackle such challenging environmental and energy issues. In recent years, various types of photocatalysts that are active for environmental purification have been developed.

Many of the binary semiconductors like TiO_2, ZnO, and CdS and ternary semiconductors like Zn_2SnO_4 and $BaFeO_3$ have been used as a photocatalyst in wastewater treatment. Very little work has been carried out by researchers on the use of quaternary oxides and sulfides as photocatalysts compared with binary and ternary oxides. Quaternary oxides containing metals such as vanadium, iron, titanium, niobium, molybdenum, and so on, have been studied as photocatalysts for the decomposition of organic pollutants in water and air.

5.2 QUATERNARY OXIDES

In the very beginning, a number of scientists were interested in photocatalytic decomposition of H_2O into H_2 and O_2 using different quaternary oxides such as $K_2La_2Ti_3O_{10}$ (Ikeda et al. 1988), $NiOK_4Nb_6O_{17}$ powder (Kudo et al. 1988), $NiOSrTiO_3$ powder (Domen et al. 1986), and so on.

Photophysical and photocatalytic properties of $Sr_2(Ta_{1-x}Nb_x)_2O_7$ with layered perovskite structure were studied by Kato and Kudo (2001). They observed that the band gap of $Sr_2(Ta_{1-x}Nb_x)_2O_7$ decreased as the ratio of niobium to tantalum was increased. They proposed that the $Sr_2(Ta_{0.75}Nb_{0.25})_2O_7$ photocatalyst showed a relatively high activity among the solid solutions of $Sr_2(Ta_{1-x}Nb_x)_2O_7$ ($0 < x < 1$). The activities of $Sr_2(Ta_{1-x}Nb_x)_2O_7$ solid solution photocatalysts depend on the conduction band (CB) level and the efficiency of nonradiative recombination among photogenerated electron–hole pairs.

Yoshino et al. (2002) prepared $Sr_2Nb_xTa_{2-x}O_7$ (SNT, $x = 0$–2) by the Pechini-type polymerizable complex (PC) technique, which is based on polymerization between citric acid and ethylene glycol, and used it for decomposition of water. The two end compounds, $Sr_2Ta_2O_7$ ($x = 0$) and $Sr_2Nb_2O_7$ ($x = 2$), produced H_2 and O_2 in a stoichiometric ratio from H_2O under UV light irradiation without an NiO cocatalyst.

The photocatalytic activity of SNT for the water decomposition was greatly improved by loading NiO as a cocatalyst. The photocatalytic activity was decreased when Ta was replaced by Nb.

Bhavsar and Kharat (2003) synthesized a new quaternary oxide, $Li_2CuMo_2O_8$, which was prepared by the solid-state reaction between corresponding oxides and characterized by chemical analysis, X-ray diffraction, electrical conductivity, diffuse reflectance spectra, and magnetic study. They found that it is an n-type semiconducting material having a band gap of 1.54 eV and can be used as a photoanode in an electrochemical photovoltaic cell.

Kale et al. (2005) synthesized $ZnBiVO_4$ through a solid-state process and used it successfully for photodecomposition of hydrogen sulfide. Garza-Tovar et al. (2006) prepared $BiMNbO_7$ (where M = In, Al, Fe, and Sm) following the sol–gel method by gelling niobium ethoxide with bismuth acetate and In, Fe, and Sm precursors. The energy band gap values, evaluated by the UV–vis spectra of the compound, range between 1.43 and 2.24 eV. The results of kinetic half-life parameters for the photodegradation of methylene blue (MB) showed that $BiFeNbO_7$ sol–gel catalyst was more active ($t_{1/2}$ = 13 minutes) compared with TiO_2 P25 photocatalyst ($t_{1/2}$ = 45 minutes).

Chen et al. (2006) synthesized microporous solid $K_3PW_{12}O_{40}$ by precipitation of phosphotungstic acid and potassium ion followed by calcination. The photocatalyst was used for photodegradation of dye pollutants such as rhodmine B, malachite green (MG), rhodamine 6G (Rh6G), fuchsin basic (FB), and methyl violet (MV). The degradation kinetics, total organic carbon (TOC) change, degradation products, electron spin resonance (ESR) detection of active oxygen species, and the effect of radical scavengers were also investigated to understand the degradation process and the reaction pathway.

Luan et al. (2007) examined photocatalytic activity of Bi_2FeVO_7 and compared it with TiO_2 (P25) in the degradation of a model pollutant, MB. The results showed that Bi_2FeVO_7 was crystallized in the tetragonal crystal system with space group 14/mmm and the band gap was estimated to be about 2.22 eV. The reduction of the TOC and the formation of inorganic products, SO_4^{2-} and NO_3^{-}, revealed the continuous mineralization of aqueous MB during the photocatalytic process.

They also prepared Bi_2AlVO_7 using a solid-state technique and studied the structural and photocatalytic properties of Bi_2AlVO_7 and Bi_2AlTaO_7 (Luan et al. 2009). The band gaps of Bi_2AlVO_7 and Bi_2AlTaO_7 were found to be about 2.06 and 2.81 eV, respectively. The photocatalytic degradation of MB dye was observed under UV irradiation. They further found that Bi_2AlVO_7 showed higher photocatalytic activity compared with Bi_2AlTaO_7 for photocatalytic degradation of this model system.

The photocatalytic degradation of rhodamine B was also carried out using Y_2InSbO_7 or Y_2GdSbO_7 as a catalyst under visible light irradiation by Luan et al. (2011). The results showed that Y_2InSbO_7 and Y_2GdSbO_7 have higher catalytic activity compared with P25 TiO_2 or Bi_2InTaO_7 for the photocatalytic degradation of rhodamine B. Furthermore, Y_2InSbO_7 exhibited higher catalytic activity than Y_2GdSbO_7. The reduction of the TOC and the evolution of CO_2 were realized and these results indicated that there is complete mineralization of rhodamine B during this photocatalytic process. The possible pathway of the photocatalytic degradation of dye was proposed under visible light irradiation. The Y_2InSbO_7 (or Y_2GdSbO_7)/ (visible light) photocatalytic system was found to be suitable for textile industry wastewater treatment.

Wang et al. (2011) synthesized a pure phase of $Bi_2TiO_4F_2$ nanoflakes with a layered aurivillius structure by a simple hydrothermal method. The ·OH radicals produced during the photocatalytic reaction were detected by the photoluminescence (PL) technique. The photocatalytic properties of $Bi_2TiO_4F_2$ were explored by degradation of rhodamine B and phenol. The results showed that $Bi_2TiO_4F_2$ exhibited much higher photocatalytic performance than $Bi_4Ti_3O_{12}$ due to the unique layered structure and the existence of F.

Torres-Martinez et al. (2012) synthesized Sm_2FeTaO_7 by the conventional solid-state reaction and sol–gel method and used it for photocatalytic degradation of indigo carmine (IC) dye in aqueous solution. Its specific surface area and band gap values were 12 m^2/g and 2.0 eV, respectively. It was observed that the sol–gel photocatalyst was eight times more active than the solid-state photocatalyst. The photocatalytic activity was found to increase on addition of CuO as cocatalyst because CuO acts as an electron trap, thus decreasing electron–hole pair recombination rates.

Vijayasankar et al. (2013) reported an efficient Pb–Fe–Nb–O-based composite oxide photocatalyst prepared by a simple solid-state reaction method. The low band gap semiconductors—Pb_2FeNbO_6, $FeNbO_4$, and a composite (Pb_2FeNbO_6):($FeNbO_4$) system—were synthesized and also used in decolorization of MB under solar radiation. These were also found to be doubly efficient compared to $TiO_{2-x}N_x$.

Krishna et al. (2015) prepared silver- and nitrogen-doped $K_2Al_2Ti_6O_{16}$ photocatalyst by ion exchange and solid-state methods, while pure $K_2Al_2Ti_6O_{16}$ photocatalyst was prepared by the sol–gel method. The photocatalytic degradation of MB under UV light irradiation was investigated. It was observed that silver- and nitrogen-doped $K_2Al_2Ti_6O_{16}$ showed higher photocatalytic activity compared with pure $K_2Al_2Ti_6O_{16}$.

Jitta et al. (2015) prepared $KTi_{0.5}W_{1.5}O_6$ by a facile sol–gel method. Its nitrogen- and tin-doped analogs were prepared by a solid-state and ion exchange method, respectively. All compounds were characterized by X-ray power diffraction (XRD), Fourier transform infrared spectroscopy (FT-IR), Raman spectroscopy (RS), scanning electron microscope (SEM), and thermogravimetric analysis. The photocatalytic activity of compounds was reported for degradation of MB and rhodamine B. The tin-doped $KTi_{0.5}W_{1.5}O_6$ showed higher photocatalytic activity against both these dyes.

Some new fluorite-related quaternary rare earth oxides, LnY_2TaO_7 (Ln = La–Dy) and $LaLn_2RuO_7$ (Ln = Eu–Tb), have been prepared by Hinatsu and Doi (2016) and the structure and magnetic properties of these oxides were studied, while Kumawat et al. (2015) synthesized quaternary oxide, $FeZn_2Cu_3O_{6.5}$, by a ceramic technique using fine powders of Fe_2O_3, ZnO, and CuO and used it for the degradation of MG.

Quaternary oxides have wide applications in photocatalytic degradation of organic pollutants such as dye and other hazardous chemicals present in effluent, in electrochemical cells as semiconductors, and in hydrogen generation by water splitting.

5.3   QUATERNARY SULFIDES

Quaternary sulfides have been studied in addition to quaternary oxides and researchers have found various applications for them.

Wu et al. (1992) prepared new quaternary compounds $KGaSnS_4$, $KInGeS_4$, and $KGaGeS_4$ that were synthesized by the reaction of K_2S_5 with elemental Ga or In, Sn or Ge, and S at 900°C. They studied the structure of all three compounds and proposed that each structure has tetrahedra form corner-sharing chains and these chains are connected by pairs of edge-sharing tetrahedra. The layers in $KInGeS_4$ and $KGaGeS_4$ have the same structure while layers in $KGaSnS_4$ have a different structure, but all three compounds are poor conductors.

A Ti-based photocatalyst oxysulfide, $Sm_2Ti_2S_2O_5$, was studied by Ishikawa et al. (2002). $Sm_2Ti_2S_2O_5$ with a band gap of ~2 eV evolved H_2 or O_2 from aqueous solutions containing a sacrificial electron donor (Na_2S–Na_2SO_3 or methanol) or acceptor (Ag^+) under visible light (440 nm $\leq \lambda \leq$ 650 nm) irradiation, without any noticeable degradation. Therefore, they were of the opinion that this oxysulfide is a stable photocatalyst that has strong reduction and oxidation abilities under visible light irradiation. The CBs and valence bands (VBs) positions of $Sm_2Ti_2S_2O_5$ were also determined by electrochemical measurements.

A solid solution photocatalyst $(AgIn)_xZn_{2(1-x)}S_2$ with a wide band gap was synthesized by Tsuji et al. (2004) and its photocatalytic activities for H_2 evolution from aqueous solutions under visible light irradiation were studied. They further proved that its photocatalytic activity was increased on loading Pt cocatalysts. Pt (3 wt%) loaded $(AgIn)_{0.22}Zn_{1.56}S_2$ with a 2.3 eV band gap showed the highest activity for H_2 evolution. It was further indicated from SEM and transmission electron microscope (TEM) observations that the solid solutions partially had nanostep structures on their surfaces. The Pt cocatalysts were selectively photodeposited on the edge of these surface nanosteps. It was suggested that the specific surface nanostructure was effective for the suppression of recombination between photogenerated electrons and holes and also for the separation of H_2 evolution sites from oxidation reaction sites.

A novel $Cd_xCu_yZn_{1-x-y}S$ photocatalyst was synthesized using a coprecipitation method by Liu et al. (2008). The results of UV–vis diffuse reflectance spectra showed that the absorption edge of $Cd_{0.1}Cu_{0.01}Zn_{0.89}S$ was at 560 nm with a red shift compared with that of $Cd_{0.1}Zn_{0.9}S$ and $Cu_{0.01}Zn_{0.99}S$. The $Cd_{0.1}Cu_{0.01}Zn_{0.89}S$ photocatalyst showed a high activity for hydrogen evolution under visible light irradiation ($\lambda \geq$ 430 nm) even without cocatalysts such as Pt.

A new wurtzite phase Cu_2ZnSnS_4 was discovered and the corresponding nanocrystals have been successfully synthesized by Lu et al. (2011). It showed photoelectric response that demonstrated its possible application in photovoltaic devices. Noncrystalline Cu_2ZnSnS_4 (CZTS) nanoparticles (NPs) were prepared using a hydrothermal method by Hsiung et al. (2013). CZTS NPs having different Cu/(Zn + Sn) ratios (0.77, 0.89, 1.02) were separately prepared. The morphology, composition, and crystal structure of as-prepared NPs and the vulcanized films were investigated. They proved that as the temperature and time of vulcanization exceeded 400°C and 60 minutes, respectively, then the Cu-rich phase (Cu_3SnS_4) started dominating.

Iron-containing diamond-like materials such as Ag_2FeSiS_4, Li_2FeSnS_4, and Li_2FeGeS_4 were synthesized by Brunetta et al. (2013) via high temperature, solid-state synthesis. They found that the as-synthesized quaternary compounds have the wurtz–kesterite structure that crystallizes in the noncentrosymmetric space group Pn. The visible light photocatalytic ability of the Cu_2ZnSnS_4 NPs for degradation of MB was studied by Zhou et al. (2014). It was characterized for its crystal structure, surface morphology, and microstructure. Yu et al. (2014) observed that Cu_2ZnSnS_4–Pt and Cu_2ZnSnS_4–Au heterostructured NPs have excellent photocatalytic properties for degradation of rhodamine B as well as hydrogen generation by water splitting.

5.4 QUATERNARY SELENIDES

This is another type of group, similar to quaternary oxide and sulfides, which has drawn the attention of many scientists. Some work has been already done in this direction and a lot has to be done in coming decades.

Ramanathan et al. (2005) presented results on the growth and characterization of $CuInGaSe_2$ (CIGS) thin-film solar cells by a three-stage process. They observed that the conversion efficiency of solar cells made from such absorbers changes from 19.3% to 18.4%, when the band gap values are 1.15 and 1.21 eV, respectively. They further related these improvements to material and device properties and suggested that it might be possible to produce a 20% efficient solar cell by further optimization.

Chen and Lee (2009) synthesized quaternary selenides $Sn_2Pb_5Bi_4Se_{13}$ and $Sn_{8.65}Pb_{0.35}Bi_4Se_{15}$ from the elements in sealed silica tubes. Their crystal structures were determined by a single crystal and powder x-ray diffraction. Both compounds were found to crystallize in a monoclinic shape. They measured electrical conductivity of quaternary selenides and found that these materials are semiconductors with narrow band gaps. $Sn_2Pb_5Bi_4Se_{13}$ belongs to the n-type category, whereas $Sn_{8.65}Pb_{0.35}Bi_4Se_{15}$ is a p-type semiconductor.

The quaternary-alloyed $Zn_xCd_{1-x}S_ySe_{1-y}$ arrays were prepared by Liu et al. (2010) and its photoelectrochemical properties were evaluated by scanning electrochemical microscopy using an optical fiber tip attached to a Xenon lamp as the excitation source. The spot with a precursor composition $Zn_{0.3}Cd_{0.7}S_{0.8}Se_{0.2}$ (elemental ratio, 1:2.12:1.75:0.81) showed the highest photocurrent under 150 W Xenon lamp irradiation. The difference of the photocurrent onset indicated that addition of Zn raises the CB position of CdS_ySe_{1-y}.

Kush and Deka (2015) developed anisotropic quaternary $Cu_2ZnSnSe_4$ (CZTSe) semiconductor NPs of kesterite using a single pot colloidal synthesis route. The initially grown CZTSe NPs were approximately 10 nm in size, which aggregated and joined into three-dimensional anisotropic morphology. The photoresponse property and photocatalytic activity of the CZTSe NPs have also been investigated. Thin films of CZTSe NP ink displayed five times the photocurrent under illumination compared with a dark state. As a first case of photocatalytic application of CZTSe NPs, this anisotropic CZTSe sample demonstrated effective degradation of polluting dyes and proved to be a novel semiconductor for wastewater treatment under indoor visible light illumination, with active recyclability of course.

A new wide band gap quaternary selenide $Cu_2MgSnSe_4$ has been studied by Kumar et al. (2015). They showed good thermoelectric properties for this quaternary selenide. The electrical and thermal transport properties of Cu- and In-doped $Cu_2MgSnSe_4$ in the temperature range of 300–700K were also studied. They proved that on substitution of In^{3+} for Sn^{4+} and Cu^{2+} for Mg^{2+}, the thermoelectric efficiency was further improved.

Considerable research attention has been given to quaternary oxide, sulfides, and selenides because of their exciting photoelectrochemical and photocatalytic properties, so that more efficient solar or photoelectric cells can be prepared as well as effective and successful treatment of pollutants. A lot of work had already been done on binary and ternary semiconductors, but quaternary semiconductors also have a wide variety of applications from photodegradation of organic water pollutants, to generation of electricity in photoelectric cells, or photosplitting of water. Therefore substantial and sincere efforts have to be made in order to make these more useful to mankind.

REFERENCES

Bhavsar, R. S. and R. B. Kharat. 2003. Characterization of newly synthesized quaternary oxide Li–Cu–Mo–O for photoelectric devices. *Mater. Chem. Phys.* 80 (1): 143–149.

Brunetta, C. D., J. A. Brant, K. A. Rosmus, K. M. Henline, E. Karey, J. H. MacNeil, and J. A. Aitken. 2013. The impact of three new quaternary sulfides on the current predictive tools for structure and composition of diamond-like materials. *J. Alloys Compd.* 574:495–503.

Chen, C., Q. Wang, P. Lei, W. Song, W. Ma, and J. Zhao. 2006. Photodegradation of dye pollutants catalyzed by porous $K_3PW_{12}O_{40}$ under visible irradiation. *Environ. Sci. Technol.* 40 (12): 3965–3970.

Chen, K. B. and C. S. Lee. 2009. Synthesis and characterization of quaternary selenides $Sn_2Pb_5Bi_4Se_{13}$ and $Sn_{8.65}Pb_{0.35}Bi_4Se_{15}$. *Solid State Sci.* 11 (9): 1666–1672.

Domen, K., A. Kudo, and T. Onishi. 1986. Mechanism of photocatalytic decomposition of water into H_2 and O_2 over $NiOSrTiO_3$, *J. Catal.* 102 (1): 92–98.

Garza-Tovar, L. L., L. M. Torres-Martinez, D. B. Rodriguez, R. Gomez, and G. D. Angel. 2006. Photocatalytic degradation of methylene blue on Bi_2MNbO_7 (M = Al, Fe, In, Sm,) sol-gel catalysts, *J. Mol. Catal. A: Chem.* 247 (1–2):283–290.

Hinatsu, Y. and Y. Doi. 2016. Structures and magnetic properties of new fluorite-related quaternary rare earth oxides LnY_2TaO_7 and $LaLn_2RuO_7$ (Ln = rare earths). *J. Solid State Chem.* 233:37–43.

Ikeda, S., M. Hara, J. N. Kondo, K. Domen, H. Takahashi, T. Okubo, and M. Kakihana. 1988. Preparation of $K_2La_2Ti_3O_{10}$ by polymerized complex method and photocatalytic decomposition of water. *Chem. Mater.* 10 (1): 72–77.

Ishikawa, A., T. Takata, J. N. Kondo, M. Hara, H. Kobayashi, and K. Domen. 2002. Oxysulfide $Sm_2Ti_2S_2O_5$ as a stable photocatalyst for water oxidation and reduction under visible light irradiation ($\lambda \leq 650$ nm). *J. Am. Chem. Soc.* 124 (45): 13547–13553.

Jitta, R. R., R. Guje, N. K. Veldurthi, S. Prathapuram, R. Velchuri, and V. Muga. 2015. Preparation, characterization and photocatalytic studies of N, Sn-doped defect pyrochlore oxide $KTi_{0.5}W_{1.5}O_6$. *J. Alloys Compd.* 618:815–823.

Kale, B., J. Baeg, J. S. Yoo, S. M. Lee, C. W. Lee, S. Moon, and H. Chang. 2005. Synthesis of a novel photocatalyst, $ZnBiVO_4$, for the photodecomposition of H_2S. *Canad. J. Chem.* 83 (6–7):527–532.

Kato, H. and A. Kudo. 2001. Energy structure and photocatalytic activity for water splitting of Sr_2 $(Ta_{1-x}Nb_x)_2O_7$ solid solution. *J. Photochem. Photobiol. A. Chem.* 145 (1–2):129–133.

Krishna, S. R., P. Shrujana, S. Palla, K. Sreenu, R. Velchuri, and M. Vithal. 2015. Preparation, characterization and photocatalytic studies of $K_2Al_2Ti_6O_{16}$, $K_{2-x}Ag_xAl_2Ti_6O_{16}$ and $K_2Al_2Ti_6O_{16-x}Ny$. *Mater. Res. Express.* 2. doi: 10.1088/2053-1591/2/3/035008.

Kudo, A., A. Tanaka, K. Domen, K. I. Maruya, K. I. Aika, and T. Onishi. 1988. Photocatalytic decomposition of water over $NiOK_4Nb_6O_{17}$ catalyst. *J. Catal.* 111 (1): 67–76.

Kumar, V. P., E. Guilmeau, B. Raveau, V. Caignaert, and U. V. Varadaraju. 2015. A new wide gap thermoelectric quaternary selenide $Cu_2MgSnSe_4$. *J. Appl. Phys.* 118: 155101.

Kumawat, P., M. Joshi, R. Ameta, and S. C. Ameta. 2015. Synthesis and use of quaternary oxide $FeZn_2Cu_3O_{6.5}$ as a photocatalyst. *Ind. J. Sci. Res. Technol.* 3 (4): 72–77.

Kush, P. and S. Deka. 2015. Anisotropic kesterite $Cu_2ZnSnSe_4$ colloidal nanoparticles: Photoelectrical and photocatalytic properties. *Mater. Chem. Phys.* 162:608–616.

Liu, G., C. Liu, and A. J. Bard. 2010. Rapid synthesis and screening of $Zn_xCd_{1-x}S_ySe_{1-y}$ photocatalysts by scanning electrochemical microscopy. *J. Phys. Chem. C.* 114 (49): 20997–21002.

Liu, G., L. Zhao, L. Ma, and L. Guo. 2008. Photocatalytic H_2 evolution under visible light irradiation on a novel $Cd_xCu_yZn_{1-x-y}S$ catalyst. *Catal. Commun.* 9 (1): 126–130.

Lu, X., Z. Zhuang, Q. Peng, and Y. Li. 2011. Wurtzite Cu_2ZnSnS_4 nanocrystals: A novel quaternary semiconductor. *Chem. Commun.* 47 (11): 3141–3143.

Luan, J., H. Cai, X. Hao, J. Zhang, G. Luan, X. Wu, and Z. Zou. 2007. Structural characterization and photocatalytic properties of novel Bi_2FeVO_7. *Res. Chem. Intermed.* 33 (6): 487–500.

Luan, J., M. Li, K. Ma, Y. Li, and Z. Zou. 2011. Photocatalytic activity of novel Y_2InSbO_7 and Y_2GdSbO_7 nanocatalysts for degradation of environmental pollutant rhodamine B under visible light irradiation. *Chem. Eng. J.* 167 (1): 162–171.

Luan, J., W. Zhao, J. Feng, H. Cai, Z. Zheng, B. Pan, X. Wub, Z. Zouc, and Y. Li. 2009. Structural, photophysical and photocatalytic properties of novel Bi_2AlVO_7. *J. Hazard. Mater.* 164:781–789.

Ramanathan, K., G. Teeter, J. C. Keane, and R. Noufi. 2005. Properties of high-efficiency $CuInGaSe_2$ thin film solar cells. *Thin Solid Films.* 480–481:499–502.

Torres-Martinez, L. M., M. A. Ruiz-Gomez, M. Z. Figueroa-Torres, I. Juarez-Ramirez, and E. Moctezuma. 2012. Sm_2FeTaO_7 photocatalyst for degradation of indigo carmine dye under solar light irradiation. *Int. J. Photoenergy.* 2012:7. doi: org/10.1155/2012/939608.

Tsuji, I., H. Kato, H. Kobayashi, and A. Kudo. 2004. Photocatalytic H_2 evolution reaction from aqueous solutions over band structure-controlled $(AgIn)_xZn_{2(1-x)}S_2$ solid solution photocatalysts with visible-light response and their surface nanostructures. *J. Am. Chem. Soc.* 126 (41): 13406–13413.

Vijayasankar, K., N. Y. Hebalkar, H. G. Kim, and P. H. Borse. 2013. Controlled band energetics in Pb-Fe-Nb-O metal oxide composite system to fabricate efficient visible light photocatalyst. *J. Ceram. Proc. Res.* 14 (4): 557–562.

Wang, S., B. Huang, Z. Wang, Y. Liu, X. Qin, X. Zhang, and Y. Dai. 2011. A new photocatalyst: $Bi_2TiO_4F_2$ nanoflakes synthesized by a hydrothermal method. *Dalton Trans.* 40 (47): 12670–12675.

Wu, P., Y. J. Lu, and J. A. Ibers. 1992. Synthesis and structures of the quaternary sulfides $KGaSnS_4$, $KInGeS_4$ and $KGaGeS_4$. *J. Solid State Chem.* 97 (2): 383–390.

Yoshino, M., M. Kakihana, W. S. Cho, H. Kato, and A. Kudo. 2002. Polymerizable complex synthesis of pure $Sr_2Nb_xTa_{2-x}O_7$ solid solution with high photocatalytic activities for water decomposition into H_2 and O_2. *Chem. Mater.* 14 (8): 3369–3376.

Yu, X., A. Shavel, X. An, Z. Luo, M. Ibanez, and A. Cabot. 2014. Cu_2ZnSnS_4-Pt and Cu_2ZnSnS_4-Au heterostructured nanoprticles for photocatalytic water splitting and pollutant degradation. *J. Am. Chem. Soc.* 136 (26): 9236–9239.

Zhou, Z., P. Zhang, Y. Lin, E. Ashalley, H. Ji, J. Wu, H. Li, and Z. Wang. 2014. Microwave fabrication of Cu_2ZnSnS_4 nanoparticles and its visible light photocatalytic properties. *Nanoscale Res. Lett.* 9 (1): 477. doi:10.1186/1556-276X-9-477.

6 Metallization

6.1 INTRODUCTION

Semiconductor-based heterostructures have the ability to modify the characteristics of materials and hence they find wide applications in biomedicine, photocatalysis, and nanodevices. Among these heterostructures, metal/semiconductor is one of the most extensively studied because of its excellent catalytic activity.

The photocatalytic reaction occurring in the semiconductor is a redox reaction. When the semiconductor is irradiated, its irradiated surface acts as a sink for electrons (or holes) according to the direction of band bending. The other charge carrier moves under the influence of electric field into the bulk of the semiconductor or to its surface that receives incident radiation of lowest intensity. Hence, efficiency of the photocatalytic activity of the semiconductor depends on the charge carrier (electrons and holes) separation. The photocatalytic efficiency of a semiconductor can be further enhanced by metallization.

Photocatalysts made by coating semiconductor powders on metals are gaining much interest worldwide. When a metallized semiconductor is irradiated with light energy greater than the band gap, it generates electrons and holes that have a tendency to recombine. The presence of metal prevents this process by trapping the photoexcited electrons and uses it to perform a reduction reaction. In the same manner, holes can be used to carry out an oxidation reaction. Thus, metal deposition on a semiconductor improves its efficiency considerably. The first ever use of a metallized semiconductor for water cleavage was reported by Bulatov and Khidekel (1976), who used platinized TiO_2 in 1N H_2SO_4 for hydrogen and oxygen production. Thereafter, metallization of semiconductors with noble metals has gained considerable attention throughout the world, particularly in the field of heterogeneous photocatalysis (Hermann et al. 1997; Kamat 2002).

The photocatalytic activity of metallized semiconductors is governed by the following factors.

6.1.1 NATURE OF METAL

The selection of metal to be loaded depends on the electron affinity of the metal. The higher the electron affinity, greater the ability of the metal to trap the photoexcited electron. The electron affinities of some metals suitable for metal loading are provided in Table 6.1. Besides the electron affinity, the metal must have a suitable work function so that it can make favorable contact with the semiconductor.

Efficient hydrogen evolution from water has been reported by Kalyanasundaram et al. (1981) by loading CdS powder with Pt and a marked increase in the photocatalytic activity of CdS was observed. Li and Li (2002) have reported that 0.75%

TABLE 6.1
Electron Affinity of Some Metals

Metal	Electron Affinity (eV)
Palladium (Pd)	0.562
Ruthenium (Ru)	1.050
Rhodium (Rh)	1.137
Silver (Ag)	1.304
Platinum (Pt)	2.125
Gold (Au)	2.308

Pt–TiO$_2$ showed better photocatalytic performance in methyl blue (MB) and methyl orange (MO) degradation. Maicu et al. (2011) also reported enhancement in the photooxidation of phenol upon loading TiO$_2$ with platinum.

6.1.2 NATURE OF THE SEMICONDUCTOR

The lower edge of the conduction band (CB) of a semiconductor is the measure of the reducing strength of a photoexcited electron, whereas the upper edge of the valence band (VB) is a measure of the oxidizing power of the hole. Depending on the oxidation and reduction reactions of the semiconductors in terms of splitting of water, semiconductors have been classified (Gratzel 1983) as follows:

- Those having strong oxidation and reduction power in terms of evolution of both hydrogen and oxygen, such as TiO$_2$, CdS, SrTiO$_3$, and so on
- Those having strong reducing power that can reduce water, such as CdSe, CdTe, and so on
- Those having strong oxidation power, such as WO$_3$, Fe$_2$O$_3$, Bi$_2$O$_3$, and so on
- Those having weak reducing and oxidizing powers and hence neither hydrogen nor oxygen is evolved

6.2 MECHANISM

The following are the two main factors that are responsible for higher activity of noble metals:

- Schottky barrier formation
- Surface plasmon resonance (SPR)

The Schottky junction results from the contact of the noble metal and the semiconductor. This junction creates an internal electric field close to the interface of metal/semiconductor, which forces the electrons and holes to move in different directions (Zhu et al. 2012). The ohmic contact between the metal and the semiconductor causes the electron to move from the CB of the semiconductor to the metal and as a result electron–hole recombination is suppressed, which facilitates the transfer of holes to the surface of semiconductor. Hence, the photocatalytic activity of metallized semiconductor increases in terms of relatively longer electron–hole pair separation under UV irradiation.

The photocatalytic activity under visible light is enhanced due to SPR (Tung 2001). Surface plasmons are the oscillations that can propagate at the interface of the metal and the semiconductor. Oscillations of the metal electrons are induced by the light waves. Metal electron density is polarized to one side and it oscillates in resonance with the light frequency as the light wave passes through the metal. Here, photogenerated electrons have higher negative potential than the CB of the semiconductor, thus causing electron transfer from excited metal to the CB of the semiconductor. This results in the rapid formation of electron–hole pairs in the semiconductor. The photocatalytic activity in the visible range is increased due to the rise of the local electric field under visible light irradiation. As the energy of the UV irradiation is much higher than the SPR, the enhancement in the photocatalytic activity under the UV light is not related to the surface plasmon resonance of the noble metal, but only under the visible light.

6.3 TITANIUM DIOXIDE

Semiconductor material, such as TiO_2, is widely used in heterogeneous photocatalytic decomposition of organic compounds and finds important applications in the treatment of contaminated water and air streams (Carp et al. 2004; Fujishima et al. 2008). Due to its high photosensitivity, nontoxicity, stability, and commercial availability, TiO_2 scores highest among the large band gap semiconductors as a photocatalyst. In spite of these desirable characteristics, it has also some limitations such as that the band gap of TiO_2 (~3.23 eV for the anatase form and ~3.02 eV for the rutile form) corresponds to the adsorption band between 380 and 410 nm, which means it does not absorb the major part of solar light as it has only ~5% UV radiation.

Hence, it is unsuitable for photocatalytic processes depending on the sunlight. Furthermore, the efficiency of this photocatalyst is limited by the fast recombination rate of charge carriers produced along with a slow transfer rate of electrons to oxygen. The surface deposition of noble metals has been employed to overcome these limitations. TiO_2 materials modified by the surface deposition of Pt, Au, and Pd have been reported to exhibit enhanced photocatalytic activity. Three commercially available TiO_2 catalysts, namely Degussa P25, Sachtleben Hombikat UV100, and Millennium TiONA PC50, were platinized by a photochemical impregnation method in two ratios of platinum deposits (0.5 wt.% and 1 wt.%) (Hufschmidt et al. 2002). The physical characterization of the newly synthesized catalysts was carried out by measurements of the Brunauer–Emmett–Teller (BET) surface area, the light absorption properties, and the adsorption of the model compounds. The photocatalytic activities of these samples were determined using three different model compounds: ethylenediaminetetraacetic acid (EDTA), 4-chlorophenol (4-CP), and dichloroacetic acid (DCA). In the case of EDTA, its disappearance was studied, while total mineralization was measured for 4-CP and DCA.

In all these cases, the photocatalytical activity was found to increase by increasing amounts of Pt, for example, the photonic efficiency for DCA degradation increased from 12.2% for pure Hombikat UV100 to 32.1% for Hombikat UV100/0.5 wt.% Pt and to 42.7% for Hombikat UV100/1 wt.% Pt. Promising results were also achieved for the total mineralization of 4-CP. The photonic efficiency was increased from 0.82% using unmodified PC50 to 1.14% with PC50/0.5 wt.% Pt. Similar results were obtained with the other synthesized catalyst samples and for the model compound EDTA.

The commercially available TiO_2 photocatalyst (Degussa P25) was modified with nanosized gold particles by the photoreduction method at four different pH values of the medium (Iliev et al. 2007). A remarkable influence of the pH on the particle size of Au was observed. The size of the gold nanoparticles (NPs) on the TiO_2 surface decreases with increase in the pH of the medium. The degradation of oxalic acid has been studied in aqueous solution photocatalyzed by band gap irradiated TiO_2, modified with nanosized gold particles. The photocatalytic activity of modified TiO_2 was found to increase with the decrease in the size of the gold NPs on the surface of the photocatalytic material. The maximal value of the photocatalytic activity (twice that of the semiconducting support) was registered in the case of gold photoreduction at pH 7.

The adsorption properties of the catalysts, as well as the size of the noble metal NPs on the surface of the support, influence the efficiency of the photocatalytic process. The reaction rate of photocatalytic degradation of the oxalic acid followed a zero kinetic order according to the Langmuir–Hinshelwood model. An increase in the quantum yield of the photodestruction reaction of the model pollutant was due to the formation of Schottky barriers on the metal–semiconductor interface that served as efficient electron traps, preventing electron–hole recombination.

Li and Li (2002) found that 0.75% Pt–TiO_2 would achieve the best photocatalytic degradation of MB and MO and that the Pt–TiO_2 catalyst can be sensitized by the visible light. The existence of Pt^0, Pt^{2+}, and Pt^{4+} species on the surface of Pt–TiO_2 and Ti^{3+} in its lattice was also confirmed by them. Liu et al. (2004) stated that the interaction of Pt impurity and the crystal surface of TiO_2 might form the Schottky barrier at the interface. Vamathevan et al. (2004) suggested that this barrier expedites electron capture and consequently hinders the recombination rate between electrons and holes. Li et al. (2007) reported the synthesis of metallized TiO_2. The chemical vapor deposition (CVD) technique was used to deposit TiO_2 on the surface of Pyrex glass tubes. Then reduced $AgNO_3$ and $H_2PtCl_6 \cdot 6H_2O$ were deposited on the surface of TiO_2 by the photoreduction deposition method. The Taguchi orthogonal array method was used to perform the experiments on metal ion concentration, light intensity, and duration of illumination. The results suggested that metal ion concentration had a great influence on the preparation of Ag/TiO_2 photocatalysts.

Silver ion concentration of 1.5 mM had maximum contribution to the conversion of salicylic acid, while the effect of illumination time was insignificant. However, the illumination time had a huge effect on the preparation of Pt/TiO_2 catalyst. The maximum contribution was obtained with 8-hour illumination time, whereas platinum ion concentrations showed no visible influence. The results of X-ray power diffraction (XRD) and scanning electron microscopy coupled with energy dispersive X-ray (SEM–EDX) analysis implied that the photoreduction deposition method deposited silver and platinum on the surface of the TiO_2 catalyst and the modified TiO_2 photocatalyst retained its anatase form.

Pt, Au, and Pd deposited TiO_2 have been prepared and characterized by Sakthivel et al. (2004). The photocatalytic activity of the doped catalysts was ascertained by the photooxidation of a leather dye, acid green 16, in aqueous solution illuminated with a low-pressure mercury lamp (~254 nm). The effect of metal contents on the photocatalytic activity was also investigated. The highest photonic efficiency was observed with metal deposition level of less than 1 wt.%.

The photocatalytic production of hydrogen from liquid ethanol, a renewable bio-fuel, over Rh/TiO_2, Pd/TiO_2, and Pt/TiO_2 has been studied (Yang et al. 2006). TiO_2 shows negligible production of molecular hydrogen in the absence of the metal. The addition of Pd or Pt dramatically increased the production of hydrogen and a quantum yield of about 10% was reached at 350 K. On the contrary, the Rh-doped TiO_2 was far less active. The low activity of Rh compared to that of Pd and Pt was not due to poor dispersion or less available Rh sites on the surface. Transmission electron microscopy (TEM) showed most particles with a size less than 10 nm for all three catalysts. XPS results showed that while the state of Pd and Pt particles in the as-prepared catalysts was mostly metallic, Rh was composed of a nonnegligible contribution of Rh cations. The extent of reaction for a series of alcohols was also studied on Pt/TiO_2. It was found that the reaction was governed by the solvation of the alcohol. The production of molecular hydrogen over Pt/TiO_2 showed the following trend:

$$Methanol \approx Ethanol > Propanol \approx Isopropanol > n\text{-}Butanol$$

Zou et al. (2007) have reported an efficient method of Pt/TiO_2 synthesis. This method is known as the cold plasma method. The plasma method comprises impregnation, plasma treatment, calcinations, and reduction. The sol–gel method was used to prepare TiO_2 powders. TiO_2 powders were impregnated with H_2PtCl_6 solution and dried under ambient conditions for 24 hours. These were then treated with glow discharge plasma for 30 minutes. The treated powders were calcined in air at 300°C for 2 hours and finally reduced with hydrogen for another 2 hours. Characterization by high-resolution TEM showed that a greatly distorted metal–support interface was formed on the plasma-prepared catalyst. This interface between the metal atoms and support lattices allowed close contact, which clearly indicated the enhanced interaction. This interface was expected to enable electron transfer during a photocatalytic reaction and remained the major cause for the high activity of the plasma-prepared photocatalysts.

Pt/TiO_2 catalyst synthesized by the cold plasma method greatly improved the production of hydrogen from water/alcohols. A quantity of 33.0-mmol hydrogen was produced from water/methanol mixture in 7.5 hours over Pt/TiO_2 prepared by the plasma method, but only 14.4 mmol over Pt/TiO_2 was prepared by the impregnation method. The water/ethanol mixture also showed similar results. The amount of H_2 produced over Pt/TiO_2–P (P for the plasma method) is 1.4 times higher than that over Pt/TiO_2–C (C for the impregnation method). It is thus clear that the photocatalyst prepared by the plasma method is more active than that prepared by the impregnation method.

Monometallic 2.0 wt.% Pd/TiO_2 catalysts were prepared by Tapin et al. (2013) from the impregnation and deposition–precipitation (DP) method. Several palladium salt precursors such as $Pd(NH_3)_4Cl_2$, $Pd(NH_3)_4(NO_3)_2$, $PdCl_2$, K_2PdCl_4, and $Pd(C_5H_7O_2)_2$ were used for the impregnation technique (IMP). In this case, impregnation was carried out using acetone as solvent. An aqueous solution was prepared with a fixed pH value (1 or 11, controlled by addition of 32 wt.% chlorhydric acid or 28 wt.% ammonia, respectively) in order to favor ionic interactions (of $PdCl_4^{2-}$ or $Pd(NH_3)_4^{2+}$ complex, respectively) with the support during impregnation. After the impregnation step, the solvent was evaporated and the catalysts were further dried overnight in an oven at 120°C. The supported catalysts were calcined under an artificial air flow (80% N_2 + 20% O_2, 3.6 L/h) at 300°C or 400°C for 4 hours. Finally, these were reduced for 4 hours in flowing H_2 (3.6 L/h) at 300°C or 400°C.

K_2PdCl_4 was exclusively used for the DP method. The support was slurried with water, and an appropriate amount of this precursor salt was added to the suspension. Then pH was adjusted and maintained at 11 by addition of solid KOH. The suspension was refluxed for 1 hour. The mixture was cooled, filtered, washed, dried, and reduced with H_2 flow (3.6 L/h) at 300°C for 3 hours and finally passivized in 1% O_2/N_2 (1.8 L/h, 30 minutes). The structural and textural properties of the catalytic systems were fully characterized by several techniques.

The catalytic performances were further estimated for the hydrogenation of an aqueous solution of succinic acid performed in a batch reactor at a temperature of 160°C and total pressure of 150 bar. The results showed that all the Pd catalysts were very selective to produce γ-butyrolactone, the first hydrogenated product. However, the rate of succinic acid conversion is a function of both Pd dispersion and the preparation method. The DP method allowed one to obtain the highest performing 2 wt.% Pd/TiO_2 samples during succinic acid hydrogenation in terms of activity and stability.

Photocatalytic activity of titanium dioxide was modified by Badawy et al. (2014) through doping with silver metal at room temperature. Photocatalytic activities of as-prepared samples under simulated sunlight were evaluated for the degradation of five pharmaceutical compounds that are commonly present in hospital wastewater. These samples showed very high efficiency for photodegradation. Maximum photodegradation rate of the simulated hospital wastewater was obtained using 1000 ppm of the material with 0.1% Ag/TiO_2 (weight ratio) calcined at 300°C and pH of 5.

Noble metal deposition onto the TiO_2 surface was achieved by Kmetykó et al. (2014) via *in situ* chemical reduction (CRIS) or by mixing chemically reduced Pt NP containing sols to the aqueous suspensions of the photocatalysts (sol-impregnated samples, CRSIM). Aeroxide P25 TiO_2 photocatalyst was deposited with differently sized Pt NPs having identical platinum content (1 wt.%). Fine and low-scale control of the size was obtained by taking different concentrations of trisodium citrate. $NaBH_4$ was used as a reducing reagent. The photocatalytic activity of the samples was observed in the presence of oxalic acid (50 mM) as a sacrificial hole scavenger component. The H_2 evolution rates were found to be strongly dependent on the Pt particle size, as well as the irradiation time.

Cybula et al. (2014) prepared TiO_2 (rutile) loaded with Au/Pd NPs using a water-in-oil microemulsion system of water/AOT/cyclohexane followed by calcination. The effect of calcination temperature (from 350°C to 700°C) on the structure of Au/Pd NPs and the photocatalytic properties of Au/Pd–TiO_2 was investigated for two model reactions. Toluene was irradiated over Au/Pd–TiO_2 using light-emitting diodes (LEDs with 415 nm). The sample 0.5 mol% Pd/TiO_2 exhibited the highest activity under visible light irradiation in gas and aqueous phase reaction among all the photocatalysts calcined at 350°C. The Au/Pd–TiO$_2$ sample showed the highest photocatalytic activity in degrading phenol under visible light, which is 14 times higher than that obtained with the sample calcined at 450°C.

A novel Pd reduction method has been developed by Wang et al. (2015b) for preparation of Pd-supported TiO$_2$ catalysts, in which the Pd^{2+} ions were partly reduced and homogeneously loaded on the surface of 8 nm TiO$_2$ with the aid of preproduced surface Ti^{3+} ions. This facile preparation is calcination free and inexpensive. The excellent catalysts can produce 33 mmol/g of hydrogen per hour under UV–visible light irradiation, and degrade about 10% MO in 1 hour under visible light irradiation, which is much higher than that of unloaded TiO$_2$ or Pd/TiO$_2$ prepared by conventional high-temperature calcination methods. As-prepared resulting Pd-supported TiO$_2$ nanocatalysts were also effective for phenol degradation under visible light.

The coexistence of Pd^0 and Pd^{2+} on the surface of TiO$_2$ NPs, which is closely related to the calcination-free synthetic process, is primarily responsible for its superior photocatalytic activity. The final product 0.53 Pd/TiO$_2$ has been observed as a visible light driven photocatalyst with a high specific surface area (203 m^2/g) and it produced the highest amount of hydrogen, that is, 32.7 and 98.11 mmol/g/h after 3 hours, which is about 22 times higher than the SD-TiO$_2$ (self-doped TiO$_2$) catalyst that produced only 4.38 mmol/g/h. It has also been found that 0.53 Pd/TiO$_2$ irradiated under visible light for 60 and 120 minutes resulted in 100% MO and 80% phenol degradation, respectively.

Fine metal NPs (2–3 nm; Au, Pt, and alloyed Au–Pt) with a narrow size distribution were deposited on active TiO$_2$ through a facile chemical reduction method by Wang et al. (2015a). Compared to the bare TiO$_2$, a remarkable enhancement of up to 10-fold for the photocatalytic hydrogen evolution was achieved on the alloyed nanocomposites. Two electronic properties contributed to the promoted photocatalytic activity: (1) stronger metal–support interaction between the alloyed structures and TiO$_2$ and (2) higher electron population on the Au–Pt/TiO$_2$ photocatalysts in comparison to bare TiO$_2$. An improved charge separation over TiO$_2$ using Au–Pt NPs was clearly evidenced by the significant increase of photocurrent responses obtained from the photoelectrochemical measurements. The surface-adsorbed methanol was first oxidized to formaldehyde via a two-electron oxidation pathway, followed by spontaneous hydrolysis and methanolysis to methanediol and methoxymethanol, rather than methyl formate and formic acid as reported earlier in gaseous CH$_3$OH photocatalysis. The *in situ* monitoring also revealed that deposition of metal NPs would not alter the reaction pathways while making the reaction faster compared to the bare TiO$_2$.

The heterostructure of Ag-doped TiO$_2$ has been fabricated by Zhang et al. (2016) through the facile synthesis of the electrospinning technique and hydrothermal reaction. The concentrations of reactants play a significant role during the hydrothermal process, in controlling the size and load of Ag NPs on the surface of the TiO$_2$ fibers. It was observed that surface modification by Ag rendered visible light–induced photocatalytic activity to the TiO$_2$ fibers. Rhodamine B was employed as a pollutant to evaluate the photocatalytic activity of the Ag/TiO$_2$ samples. The heterojunction structure of Ag/TiO$_2$ exhibited higher degradation efficiency than the pure TiO$_2$ fibers (Figure 6.1).

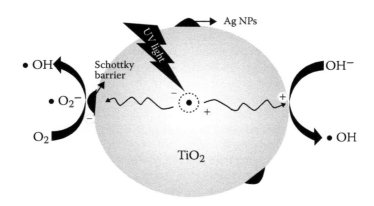

FIGURE 6.1 Schottky barrier in Ag NPs–TiO$_2$ system. (Adapted from Zhang, F. et al., *RSC Adv.*, 6, 1844–1850, 2016. With permission.)

6.4 ZINC OXIDE

Among semiconducting materials, ZnO offers a vital opportunity in providing electronic, photonic, and spin-based functionality (spintronics) because of its wide band gap (3.37 eV) and large excitation binding energy of ~60 meV. This makes it suitable in potential applications such as ceramics, piezoelectric transducers, optical coatings, and display devices (Bahadur and Rao 1995; Lee et al. 2001; Pan et al. 2001). Furthermore, ZnO finds UV screening applications as well because of its high chemical stability and low toxicity. These semiconductors are well known and are of substantial interest as they have significant applications in the field of photocatalysis (Dionysiou et al. 2000).

Many studies have reported that the photocatalytic activity of metal–oxide semiconductors can be improved by doping semiconductors with cations (Sibu et al. 2002), anions (Padmanabhan et al. 2007), metals (Subramanian et al. 2001), and nonmetals (Asahi et al. 2001; Pillai et al. 2007). The photocatalytic activity of ZnO has been widely explored and reported (Gouvea et al. 2000; Hariharan, 2006). ZnO has some advantages over TiO$_2$, mainly the fact that ZnO is quite strongly luminescent. This promotes a study of the recombination of electron–hole pairs and hence it is an appropriate probe in the study of highly active photocatalysts. The presence of organic molecules in its immediate vicinity (Kamat et al. 2002) can also be sensed due to the emission properties of ZnO. Moreover, ZnO is relatively cheaper than TiO$_2$.

The presence of noble metals on its surface has greatly influenced the photocatalytic activity of ZnO both in the UV and visible regions of light. Liqiang et al. (2004) synthesized ZnO NP photocatalysts by depositing Pd on its surface with a photoreduction method. The results showed that the content of crystal lattice oxygen on the surface of ZnO NP decreased, whereas adsorbed oxygen increased after deposition of a considerable amount of palladium indicating that Pd was mainly deposited on the crystal lattice oxygen. Additionally, the activity of ZnO NPs in the gas phase photocatalytic oxidation of n-C$_7$H$_{16}$ was also greatly improved by depositing an appropriate amount of Pd. Thus, it was concluded that an appropriate amount of Pd on its surface increased the surface content of adsorbed oxygen and hence the efficiency of photocatalysis.

Georgekutty et al. (2008) reported an efficient synthesis of Ag–ZnO and studied the role of Ag in enhancing the photocatalytic activity of ZnO. ZnO powder was prepared by the sol–gel method. Zinc acetate (10.98 g, 50 mM) was dissolved in ethanol (500 mL) at 60°C and stirred for 30 minutes. Then oxalic acid (12.55 g, 100 mM) dissolved in ethanol (200 mL) at 60°C was slowly added to the warm ethanolic solution of zinc acetate. This mixture was stirred for 2 hours. This was followed by drying the thick white colloidal semigel formed at 80°C overnight. The dried xerogel was further calcined at different temperatures (300°C–1000°C) for 2 hours to form ZnO powder.

Silver-modified ZnO was synthesized by dissolving various concentrations (1, 3, 5, and 10 mol%) of silver nitrate in ethanol and then adding it to the zinc acetate–oxalic acid solution while stirring. This was dried and calcined at different temperatures for 2 hours. A nonaqueous sol–gel process adopted helped Ag particles to disperse well in the ZnO matrix. Such a distribution was observed to favor the tuning of structural features for achieving better photocatalytic activity. It was found to be three times better than the commercial photocatalyst, Degussa P25.

Silver can perform as an amphoteric dopant according to Equation (6.1), where Ag(Zn) is the silver occupied in the Zn site and Ag(I) is the Ag in the interstitial site.

$$ZnO + Ag_2O \longrightarrow Ag(Zn) + Ag(I) + ZnO + 1/2\ O_2, T > 350 \qquad (6.1)$$

The silver particles preferentially choose to segregate around the ZnO grain boundaries because of the difference in ionic radii between Ag^+ (1.22 Å) and Zn^{2+} (0.72 Å). It was observed from XRD that metallic Ag particles are formed only at 400°C and the incorporation of Ag into the ZnO matrix did not make considerable change in the crystalline growth of nano-ZnO compared to unmodified ZnO. However, the incorporation of 3 mol% Ag at 400°C was found to inhibit the temperature-dependent crystal growth of ZnO. Therefore, 3 mol% can be assumed as the optimum concentration of Ag particles required for the effective homogeneous distribution in the ZnO matrix. The photocatalytic activity of all the samples was determined by analyzing the degradation of rhodamine 6G in the presence of the powdered suspensions. Silver modification caused the material to show significant improvement in photocatalytic activity. The concentration (3 mol%) of Ag on ZnO showed a four times higher rate of degradation of dye than that with unmodified ZnO. The result showed that silver has a significant role to play in the trapping of electrons in such materials, and these materials can be efficiently used as a photocatalyst for both environmental purification and energy production processes.

The role of gold NPs supported on ZnO in its photocatalytic activity for dye degradation was investigated by Kim et al. (2013). Gold NPs supported on ZnO (Au–ZnO) were prepared using a simple coprecipitation method. The photocatalytic degradation of methylene blue was highly enhanced by these Au–ZnO nanocatalysts having gold particle size ranging from 2 to 7 nm, with an average size of 3.8 nm. It was found that the recombination rate of the photoexited electron–hole pairs was significantly reduced by the addition of Au NPs because photoelectrons from the ZnO CB could quickly and easily transfer to the Au NPs surface, which allowed efficient charge carrier separation. Consequently, the photocatalytic efficiency for methylene blue degradation was found to be highly enhanced.

Hydroxyapatite-modified Pt–ZnO NPs (Pt–ZnO–HAP) were synthesized by Mohamed and Aazam (2013) via a template-ultrasonic assisted method. The photocatalytic activities were evaluated by decomposition of benzene under visible light irradiation. The results showed that the coupled system (Pt–ZnO–HAP) indicated a maximum photocatalytic activity and photochemical stability under visible light irradiation compared to all the other catalysts. The enhanced photocatalytic activity of the Pt–ZnO–HAP hybrids could be attributed to its strong absorption in the visible light region, low recombination rate of the electron–hole pair, and large BET-specific surface area.

Sun et al. (2015) prepared an Au–ZnO hybrid nanostructure array on a transparent conductive oxide (TCO) substrate via annealing an Au-coated ZnO nanorod array in a nitrogen atmosphere. As-prepared Au–ZnO hybrid nanostructure showed better photocatalytic activity over ZnO nanorods in degradation of MO. It was found that 8-mL MO aqueous solution (2.5 ppm) could be totally degraded with four sheets of Au–ZnO hybrid nanostructure (12×12 mm^2) within 4 hours.

The enhancement of the photocatalytic activity of ZnO catalyst by doping with Pt was reported by Shojaei and Golriz (2015) in the reaction of nitrate reduction in the presence of formic acid. It was found that a suitable amount (1 wt.%) of the Pt dopant effectively increased the photocatalytic activity of ZnO. The Pt particles doped on the ZnO surface behaved as sites where electrons could accumulate. Better separation of electrons and holes on the modified ZnO surface allowed more efficient channeling of the charge carriers into useful reduction and oxidation reactions rather than recombination reactions. They demonstrated that more than 70% of the nitrate was degraded in aqueous Pt/ZnO suspension within 60 minutes. The reduction of nitrate released nitrogen gas.

Senthilraja et al. (2015) have successfully synthesized Sn–Au–ZnO by the precipitation–decomposition method. The catalyst Sn–Au–ZnO has been reported to be more efficient than bare ZnO and other commercial catalysts at pH 11 for the mineralization of acid red 18 dye under UV-A light. The mineralization of dye has been confirmed by chemical oxygen demand (COD) measurements and the catalyst was found to be reusable.

Methyl *tert*-butyl ether (MTBE) is a commonly used fuel oxygenate present in gasoline to eliminate the use of leaded gasoline and to improve the octane quality. MTBE appeared as an environmental and human health threat because of its nonbiodegradable nature and carcinogenic potential. It was degraded using zinc oxide as a photocatalyst in the presence of visible light (Seddigi et al. 2015). A mixture of zinc oxide and MTBE solution was irradiated with visible light and about 99% photocatalytic degradation was observed. The photoactivity of 1% Pd-doped ZnO was tested under similar conditions to understand the effect of Pd doping on ZnO.

Pd-doped ZnO photocatalysts were prepared by Güy et al. (2016) using microwave irradiation, UV irradiation, and borohydride reduction methods. The optical band gap value was calculated as 3.25 eV. Pd/ZnO prepared by the borohydride reduction method exhibited higher photocatalytic activity than others in photocatalytic degradation of Congo red in aqueous solutions under UV light. The kinetics of photodecomposition of Congo red and the identification of photoproducts were investigated by using liquid chromatography–mass spectrometry (LC–MS).

Synthesis of silver–zinc oxide (Ag–ZnO) nanostructures with a plant extract mediated hydrothermal method was reported by Patil et al. (2016). The eco-friendly plant extract of *Azadirachta indica* (Neem) was used as a reducing agent. Hierarchical ZnO nanostructures were found decorated with 10–50 nm diameter Ag NPs. The Ag NPs acted as electron acceptors, inhibiting electron–hole recombination. The photocatalytic activity of the Ag–ZnO nanostructures was observed in the degradation of aqueous methylene blue dye under the natural sunlight. Such a plant extract–mediated synthetic route could also be applied to the synthesis of other Ag-semiconductor oxide nanostructures.

6.5 CADMIUM SULFIDE

Considering the band gap magnitude and the position of band edges, CdS is quite suitable for photocatalytic water splitting but it is ineffective in the absence of appropriate electron donors. Hence, CdS-based systems can be efficiently applied to evolution of hydrogen from water containing electron donors (e.g., wastewaters). The photocatalytic efficiency of CdS is substantially influenced by various factors such as crystallinity (Matsumura et al. 1985), surface area (Reber and Rusek 1986), surface etching (Buhler et al. 1984; Jin et al. 1989) pH, and related properties, for example, flat band potential and surface charge (Guindo et al. 1996; Matsumura et al. 1984; Watanabe and Honda 1974).

Cleavage of water in visible light has been reported by Kalyanasundaram et al. (1981) in CdS dispersion loaded with Pt and RuO_2. The deposition of RuO_2 on the particle surface greatly sped up the transfer of holes from the semiconductor VB to the aqueous solution, thus inhibiting photocorrosion.

CdS particles are usually platinized by physical mixing with Pt particles or by photoplatinization. The effects of platinization have been reported to be quite diverse. Reber and Rusek (1986) reported that platinized CdS showed an increase in hydrogen evolution rate of 300 mL/h at 1.5 wt.% loading of platinum. While photoplatinization is likely to make better contact at the Pt/CdS interface than physical mixing, the former was reportedly much less active than the latter for the photocatalytic degradation of lactic acid (Harada et al. 1985).

The photochemical reduction of Pt ion (Pt^{4+}) is hindered by surface chemistry of CdS (Li et al. 1992) as

$$CdS + H_2O \longrightarrow Cd(OH)^+ + SH^- \tag{6.2}$$

$$PtCl_6^{2-}(ad) + 2\ e^- \longrightarrow PtCl_4^{2-}(ad) + 2\ Cl^- \tag{6.3}$$

$$PtCl_4^{2-}(ad) + SH^- \longrightarrow PtS(s) + H^+ + 4\ Cl^- \tag{6.4}$$

$$PtCl_4^{2-}(ad) + 2\ OH^-(ad) \longrightarrow Pt(OH)_2(s) + 4\ Cl^- \tag{6.5}$$

In acidic medium, PtS was deposited on the surface of CdS while at higher pH, $Pt(OH)_2$ was deposited. The presence of undesirable Pt species lower the efficiency and thus heat treatment becomes imperative to convert PtS or $Pt(OH)_2$ to metallic Pt (Pt^0).

The photoefficiency of CdS can be improved by hybridization with metal oxides such as TiO_2. The physical mixing of CdS with platinized TiO_2 has been shown to enhance hydrogen production (Sabate et al. 1990). Park et al. (2008) studied CdS/Pt–TiO_2 and CdS/TiO_2 catalysts and found that CdS/Pt–TiO_2 produced hydrogen at the millimolar level, which is far more efficient than CdS/TiO_2 hybrid photocatalysts. Electron transfer from CdS to TiO_2 increased due to direct contact of CdS with TiO_2, and further Pt deposition on the TiO_2 surface maximized the overall efficiency of hydrogen production.

Among the semiconductor materials, TiO_2 has attracted the greatest attention due to its chemical and photochemical stability, low cost, and biological inertness. But TiO_2 has a major disadvantage in that it can be activated only by photons in the UV region. This is because of its large band gap. CdS overcomes this limitation as it has a suitable band gap (2.4 eV) and a sufficiently negative CB potential (–0.7 V vs. NHE). However, CdS when used as a photocatalyst for water splitting undergoes photocorrosion (Ni et al. 2007). The decomposition of CdS by photogenerated holes takes place as

$$CdS + 2\,H^+ \longrightarrow Cd^{2+} + S \tag{6.6}$$

The presence of sacrificial electron donors in solution, such as sulfide and/or sulfite ions (Koca and Sahin 2002; Li et al. 2007; Tambwekar et al. 1999) could be used to suppress the anodic decomposition of CdS. The sulfide ions react with photogenerated holes to form sulfur. The sulfur so produced is dissolved by sulfite ions into thiosulfate ions. Thus, the unfavorable effect induced by deposition of sulfur onto CdS is inhibited (De et al. 1995).

$$S^{2-} + 2\,H^+ \longrightarrow S \tag{6.7}$$

$$S + SO_3^{2-} \longrightarrow S_2O_3^{2-} \tag{6.8}$$

Due to thermodynamic and kinetic reasons, the photooxidation of sulfide in solution is favored over photooxidation of sulfide anions present in semiconductor lattice (CdS/ZnS) (Sabate et al. 1990). Consequently, the photocorrosion of photocatalyst was prevented and hydrogen production occurred if sacrificial hole scavengers were present in the solution. A continuous supply of sulfide and sulfite ions was required by the reaction system to maintain the photocatalytic reaction. Since these chemicals are expensive, their use is inappropriate for any practical applications. An alternative method for improving the stability and photocatalytic activity of CdS is dispersion of noble metals such as platinum on the surface of the photocatalyst (Sathish et al. 2006).

There is a need to develop visible light–responsive photocatalysts with a high quantum efficiency (QE) to convert solar energy efficiently into chemical energy by artificial photosynthesis. Yan et al. (2009) reported that a photocatalyst, Pt–PdS/CdS, could achieve a QE up to of 93% in photocatalytic H_2 production in the presence of sacrificial reagents under visible light irradiation. This system is very stable under photocatalytic reaction conditions. The high quantum efficiency was achieved by loading 0.30 wt.% of Pt and 0.13 wt.% of PdS as cocatalysts on CdS.

Dasakalki et al. (2010) synthesized powdered Pt/CdS/TiO_2 photocatalysts containing variable amounts of CdS (0%–100%) by precipitation of CdS NPs on TiO_2 (Degussa P25) followed by platinum deposition (0.5 wt.%). The photocatalysts were used to oxidize inorganic (S^{2-} and SO_3^{2-}) and organic (ethanol) sacrificial agents/water pollutants. It was possible to produce hydrogen efficiently (20% quantum efficiency at 470 nm) using virtual solar light and photocatalytic consumption of inorganic and organic compounds. They also showed that deposition of platinum on powdered CdS/TiO_2 semiconductors was responsible for complete oxidation of inorganic compounds and efficient hydrogen production, whereas degradation of organic compounds occur efficiently by spatially separated Pt and CdS/TiO_2 on transparent conductive fluorine-doped SnO_2 electrodes.

Strataki et al. (2010) have reported another example of visible light photocatalytic hydrogen production from ethanol–water mixtures using a Pt–CdS–TiO_2 photocatalyst. Conductive glass slides (electrodes) bearing a fluorine-doped tin oxide (FTO) layer were deposited by nanocrystalline titania films, made of Degussa P25. Two-thirds of the electrode region was covered by titania film. The remaining area of the electrode was coated by platinum using a solution casting method. Further, the nanocrystalline titania was deposited on CdS. This system acts as a photoelectrochemical cell with anode and cathode and is used for treating water–ethanol mixtures photocatalytically to produce hydrogen under irradiation by visible light.

A modified photoetching process was found to improve the photocatalytic activity of Pt/CdS photocatalyst significantly for solar hydrogen production (Yao et al. 2013). The photoetching process was conducted by dispersing Pt/CdS photocatalyst particles in an aqueous ammonium sulfite solution under a vacuum-degassed condition. The rate of hydrogen production via visible light photooxidation of aqueous $(NH_4)_2SO_3$ solution over the photochemically treated Pt/CdS was about 130 times higher than that of untreated Pt/CdS samples. Photochemically deposited Pt/CdS photocatalyst in an aqueous ammonium sulfite solution has shown a 100% increase in hydrogen production rate over photocatalysts traditionally photoplatinized in an aqueous glacial acetic acid solution.

A hydroxyl anion/radical redox couple has been used by Simon et al. (2014) to efficiently relay the hole from the semiconductor to the scavenger leading to a significant increase in the H_2 generation rate with Ni-decorated CdS nanorods, avoiding the use of expensive noble metal cocatalysts. The apparent quantum yield and the formation rate exceeded 53% and 63 mmol/g/h, respectively. The fast hole transfer conferred long-term photostability to the system and it also opened new pathways to improve the oxidation side of water splitting.

Park et al. (2015) synthesized CdS nanorods by a facile and rapid microwave-assisted method, and Au dots were decorated on the surface by a reduction method. It was observed that addition of the Au dots hindered the recombination of electron–hole pairs and enhanced the photocatalytic activity under visible light.

The photocatalytic activity of the Au/CdS nanorods was evaluated by the photodegradation of methylene blue under visible light irradiation (455 nm LED lamp). The photocatalytic efficiency of Au/CdS nanorods was compared to the conventional TiO_2 (P25) and CdS prepared by an autoclave method. An improvement of the Au/CdS system was attributed to the reduced band gap energy (2.31 eV).

CdS- and Ag-doped CdS (Ag/CdS) NPs were synthesized by Fard et al. (2016) via an ultrasonic-assisted sol–gel method. The calculated band gaps of CdS and Ag/CdS were found to be 2.62 and 2.46 eV, respectively. Photocatalytic degradation of direct red 264 azo dye was investigated with CdS and Ag/CdS under UV-C and visible light irradiation.

Co-doped ZnO nanorods (Co–ZnO NRs) were synthesized by the hydrothermal method by Chouhan et al. (2016) using cationic surfactant cetyltrimethylammonium bromide (CTAB). Successful loading of the nanosized sensitizer CdS onto the Co–ZnO NRs' surface resulted in the band gap reduction of the CdS/Co–ZnO NR sample (e.g., 2.25 eV). Gradual modification in pristine ZnO NRs enhanced the photocatalytic activity. Hetero-assembly of 1.5% Pt/CdS/Co–ZnO NRs exhibited excellent photocatalytic responses in terms of quantum efficiency (1.98%) and hydrogen generation capacity (67.20 mmol H_2 per gram) under one sunlight (1.5AM G) exposure.

Photocatalysis has emerged as a useful technique for wastewater treatment and production of hydrogen by splitting of water that will be used as an energy source (fuel) in the future. Various photocatalysts have been synthesized for this purpose. In this context, metallization of photocatalyst has served as a boon in this field, as it can considerably enhance the efficiency of photocatalysts. Some new and efficient metallized semiconductor photocatalysts may be prepared, which might play a vital role in solar cells, artificial photosynthesis (photocatalytic water splitting, CO_2 photoreduction) and environmental clean-up. The scope and applications of metallized photocatalysts should be further explored to a greater extent.

REFERENCES

Asahi, R., T. Morikawa, T. Ohwaki, K. Aoki, and Y. Taga. 2001. Visible-light photocatalysis in nitrogen-doped titanium oxides. *Science*. 293: 269–271.

Badawy, M. I., E. M. R. Souaya, T. A. Gad-Allah, M. S. Abdel-Wahed, and M. Ulbricht. 2014. Fabrication of Ag/TiO_2 photocatalyst for the treatment of simulated hospital wastewater under sunlight. *Environ. Prog. Sustain. Energy*. 33 (3):886–894.

Bahadur, L. and T. N. Rao. 1995. Photoelectrochemical investigations on particulate ZnO thin film electrodes in non-aqueous solvents. *J. Photochem. Photobiol. A. Chem.* 91: 233–240.

Buhler, N., K. Meier, and J. F. Reber. 1984. Photochemical hydrogen production with cadmium sulfide suspensions. *J. Phys. Chem.* 88: 3261–3268.

Bulatov, A. V. and M. I. Khidekel. 1976. Decomposition of water under the effect of UV irradiation in the presence of platinized titanium dioxide. *Izv. Akad. Nauk SSRR SerKhim.* 8: 1902–1903.

Carp, O., C. L. Huisman, and A. Reller. 2004. Photo-induced reactivity of titanium dioxide. Prog. *Solid State Chem.* 32: 33–177.

Chouhan, N., R. Ameta, R. K. Meena, N. Mandawat, and R. Ghildiyal. 2016. Visible light harvesting Pt/CdS/Co-doped ZnO nanorods molecular device for hydrogen generation. *Int. J. Hydrogen Energy.* 41 (4): 2298–2306.

Cybula, A., G. Nowaczyk, M. Jarek, and A. Zaleska. 2014. Preparation and characterization of Au/Pd modified-TiO$_2$ photocatalysts for phenol and toluene degradation under visible light—The effect of calcination temperature. *J. Nanomater.* 2014. doi: 10.1155/2014/918607.

Dasakalki, V., M. Antoniadou, G. Puma, D. Kondarides, and P. Lianos. 2010. Solar light-responsive Pt/CdS/TiO$_2$ photocatalysts for hydrogen production and simultaneous degradation of inorganic or organic sacrificial agents in wastewater. *Environ. Sci. Technol.* 44: 7200–7205.

De, G. C., A. M. Roy, and S. S. Bhattacharya. 1995. Photocatalytic production of hydrogen and concomitant cleavage of industrial waste hydrogen sulphide. *Int. J. Hydrogen Energy.* 20: 127–131.

Dionysiou, D. D., M. T Suidan, E. Bekou, I. Baudin, and M. J. Laiine. 2000. Effect of ionic strength and hydrogen peroxide on the photocatalytic degradation of 4-chlorobenzoic acid in water. *Appl. Catal. B.* 26: 153–171.

Fard, N. E., R. Fazaeli, and R. Ghiasi. 2016. Band gap energies and photocatalytic properties of CdS and Ag/CdS nanoparticles for azo dye degradation. *Chem. Eng. Technol.* 39 (1): 149–157.

Fujishima, A., X. Zhang, and D. A. Tryk. 2008. TiO$_2$ photocatalysis and related surface phenomena. *Surf. Sci. Rep.* 63: 515–582.

Georgekutty, R., M. K. Seery, and S. C. Pillai. 2008. A highly efficient Ag-ZnO photocatalyst: Synthesis, properties, and mechanism. *J. Phys. Chem. C.* 112: 13563–13570.

Gouvea, C. A. K., F. Wypych, S. G. Moraes, N. Duran, and P. Peralta-Zamora. 2000. Semiconductor-assisted photodegradation of lignin, dye, and craft effluent by Ag-doped ZnO. *Chemosphere.* 40: 427–432.

Gratzel, M. 1983. *Energy Resources Through Photochemistry and Catalysis.* New York, NY: Academic Press.

Guindo, M. C., L. Zurita, J. D. G. Duran, and A. V. Delgado. 1996. Electrokinetic behavior of spherical colloidal particles of cadmium sulfide. *Mater. Chem. Phys.* 44: 51–55.

Güy, N., S. Çakar, and M. Özacar. 2016. Comparison of palladium/zinc oxide photocatalysts prepared by different palladium doping methods for Congo red degradation. *J. Colloid Interface Sci.* 466: 128–137.

Harada, H., T. Sakata, and T. Ueda. 1985. Effect of semiconductor on photocatalytic decomposition of lactic acid. *J. Am. Chem. Soc.* 107: 1773–1774.

Hariharan, C. 2006. Photocatalytic degradation of organic contaminants in water by ZnO nanoparticles. *Appl. Catal. A.* 304: 55–61.

Hermann, J. M., H. Tahiri, H. Y. Ait-Ichou, G. Lossaletta, A. R. Gonzalez-Elipe, and A. Fernandez. 1997. Characterization and photocatalytic activity in aqueous medium of TiO$_2$ and Ag-TiO$_2$ coatings on quartz. *Appl. Catal. B Environ.* 13: 219–228.

Hufschmidt, D., D. Bahnemann, J. J. Testa, C. A. Emilio, and M. I. Litter. 2002. Enhancement of the photocatalytic activity of various TiO$_2$ materials by platinisation. *J. Photochem. Photobiol. A. Chem.* 148: 223–231.

Iliev, V., D. Tomova, L. Bilyarska, and G. Tyuliev. 2007. Influence of the size of gold nanoparticles deposited on TiO$_2$ upon the photocatalytic destruction of oxalic acid. *J. Mol. Catal. A: Chem.* 263: 32–38.

Jin, Z., Q. Li, L. Feng, Z. Chen, X. Zheng, and C. Xi. 1989. Investigation of the functions of CdS surface composite layer and Pt on treated Pt/CdS for photocatalytic dehydrogenation of aqueous alcohol solutions. *J. Mol. Catal.* 50: 315–333.

Kalyanasundaram, K., E. Borgarello, and M. Gráutzel. 1981. Visible light induced water cleavage in CdS dispersions loaded with Pt and RuO$_2$, hole scavenging by RuO$_2$. *Helv. Chim. Acta.* 64: 362–366.

Kamat, P. V. 2002. Photophysical, photochemical and photocatalytic aspects of metal nanoparticles. *J. Phys. Chem. B.* 106: 7729–7744.

Kamat, P. V., R. Huehn, and R. Nicolaescu. 2002. A "sense and shoot" approach for photocatalytic degradation of organic contaminants in water. *J. Phys. Chem. B.* 106: 788–794.

Kim, K. J., P. B. Kreider, C. H. Chang, C. M. Park, and H. G. Ahn, 2013. Visible-light-sensitive nanoscale Au-ZnO photocatalysts. *J. Nanopart. Res.* 15: 1–11.

Kmetykó, Á., K. Mogyorósi, V. Gerse, Z. Kónya, P. Pusztai, A. Dombi, and K. Hernádi. 2014. Photocatalytic H_2 production using Pt-TiO$_2$ in the presence of oxalic acid: Influence of the noble metal size and the carrier gas flow rate. *Materials.* 7: 7022–7038.

Koca, A. and M. Sahin. 2002. Photocatalytic hydrogen production by direct sun light from sulfide/sulfite solution. *Int. J. Hydrogen Energy.* 27: 363–367.

Lee, J. B., H. J. Lee, S. H. Seo, and J. S. Park. 2001. Characterization of undoped and Cu-doped ZnO films for surface acoustic wave applications. *Thin Solid Films.* 398–399: 641–646.

Li, C. H., Y. H. Hsieh, W. T. Chiu, C. Liu, and C. L. Kao. 2007. Preparation and characterization of Ag/TiO$_2$ and Pt/TiO$_2$. *J. Environ. Eng. Manage.* 17: 163–167.

Li, F. B. and X. Z. Li. 2002. The enhancement of photodegradation efficiency using Pt-TiO$_2$ catalyst. *Chemosphere.* 48(10): 1103–1111.

Li, Q., Z. Chen, X. Zheng, and Z. Jin. 1992. Study of photoreduction of hexachloroplatinate(2−) on cadmium sulphide. *J. Phys. Chem.* 96 (14): 5959–5962.

Liqiang, J., W. Baiqi, X. Baifu, L. Shudan, S. Keying, C. Weimin, and F. Honggang. 2004. Investigations on the surface modification of ZnO nanoparticle photocatalyst by depositing Pd. *J. Solid State Chem.* 177: 4221–4227.

Liu, S. X., Z. P. Qu, X. W. Han, and C. L. Sun. 2004. A mechanism for enhanced photocatalytic activity of silver-loaded titanium dioxide. *Catal. Today.* 93: 877–884.

Maicu, M., M. C. Hidalgo, G. Colon, and G. J. Navio. 2011. Comparative study of the photodeposition of Pt, Au and Pd on pre-sulphated TiO$_2$ for the photocatalytic decomposition of phenol. *Photochem. Photobiol. A Chem.* 217: 275–283.

Matsumura, M., S. Furukawa, Y. Saho, and H. Tsubomura. 1985. Cadmium sulfide photocatalyzed hydrogen production from aqueous solutions of sulfite: Effect of crystal structure and preparation method of the catalyst. *J. Phys. Chem.* 8: 1327–1329.

Matsumura, M., M. Hiramoto, T. Iehara, and H. Tsubomura. 1984. Photocatalytic and photoelectrochemical reactions of aqueous solutions of formic acid, formaldehyde, and methanol on platinized CdS powder and at a CdS electrode. *J. Phys. Chem.* 88 (2): 248–250.

Mohamed, R. M. and E. Aazam. 2013. Synthesis and characterization of Pt–ZnO-hydroxyapatite nanoparticles for photocatalytic degradation of benzene under visible light. *Desalin. Water Treat.* 51 (31–33): 6082–6090.

Ni, M., M. K. H. Leung, D. Y. C. Leung, and K. Sumathy. 2007. A review and recent developments in photocatalytic water-splitting using TiO$_2$ for hydrogen production. *Renew. Sustain. Energy Rev.* 11: 401–425.

Padmanabhan, S. C., S. C. Pillai, J. Colreavy, S. Balakrishnan, D. E. McCormack, T. S. Perova, Y. Gun'ko, S.> J. Hinder, and J. M. Kelly. 2007. A simple sol–gel processing for the development of high-temperature stable photoactive anatase titania. *Chem. Mater.* 19: 4474–4481.

Pan, Z. W., Z. R. Dai, and Z. L. Wang. 2001. Nanobelts of semiconducting oxides. *Science.* 291: 1947–1949.

Park, H., W. Choi, and M. R. Hoffmann. 2008. Effects of the preparation method of the ternary CdS/TiO$_2$/Pt hybrid photocatalysts on visible light-induced hydrogen production. *J. Mater. Chem.* 18: 2379–2385.

Park, J., S. Park, R. Selvaraj, and Y. Kim. 2015. Microwave-assisted synthesis of Au/CdS nanorods for a visible-light responsive photocatalyst. *RSC Adv.* 5: 52737–52742.

Patil, S. S., M. G. Mali, M. S. Tamboli, D. R. Patil, M. V. Kulkarni, H. Yoon et al. 2016. Green approach for hierarchical nanostructured Ag-ZnO and their photocatalytic performance under sunlight. *Catal. Today.* 260: 126–134.

Pillai, S. C., P. Periyat, R. George, D. McCormack, M. K. Seery, M. K., H. Hayden, J. Colreavy, D. Corr, and S. J. Hinder. 2007. Synthesis of high-temperature stable anatase TiO_2 photocatalyst. *J. Phys. Chem. C.* 111: 1605–1611.

Reber, F. and M. Rusek. 1986. Photochemical hydrogen production with platinized suspensions of cadmium sulfide and cadmium zinc sulfide modified by silver sulfide. *J. Phys. Chem.* 90: 824–834.

Sabate, J., S. Cerveramarch, R. Simarro, and J. Gimenez, 1990. A comparative study of semiconductor photocatalysts for hydrogen production by visible light using different sacrificial substrates in aqueous media. *Int. J. Hydrogen Energy.* 15: 115–124.

Sakthivel, S., S. M. Shankar, M. Palanichamy, B. Arabindoo, D. W. Bahnemann, and V. Murugesan. 2004. Enhancement of photocatalytic activity by metal deposition: Characterisation and photonic efficiency of Pt, Au and Pd deposited on TiO_2 catalyst. *Water Res.* 38: 3001–3008.

Sathish, M., B. Viswanathan, and R. P. Viswanath. 2006. Alternate synthetic strategy for the preparation of CdS nanoparticles and its exploitation for water splitting. *Int. J. Hydrogen Energy.* 31: 891–898.

Seddigi, Z. S., S. A. Ahmed, A. Bumajdad, E. Y. Danish, A. M. Shawky, M. A. Gondal, and M. Soylak. 2015. The efficient photocatalytic degradation of methyl tert-butyl ether under Pd/ZnO and visible light irradiation. *Photochem. Photobiol.* 91(2): 265–271.

Senthilraja, S., B. Krishnakumar, B. Subhash, A. J. F. N. Sobral, and M. Swaminathan. 2015. Sn loaded Au-ZnO photocatalyst for the degradation of AR 18 dye under UV-A light. *J. Int. Eng. Chem.* 2643: 1–8.

Shojaei, A. F. and F. Golriz. 2015. High photocatalytic activity in nitrate reduction by using Pt/ZnO nanoparticles in the presence of formic acid as hole scavenger. *Bulg. Chem. Commun.* 47: 509–514.

Sibu, C. P., S. R. Kumar, P. Mukundan, and K. G. K. Warrier. 2002. Structural modifications and associated properties of lanthanum oxide doped sol-gel nanosized titanium oxide. *Chem. Mater.* 14: 2876–2881.

Simon, T., N. Bouchonville, M. J. Berr, A. Vaneski, A. Adrović, David Volbers et al. 2014. Redox shuttle mechanism enhances photocatalytic H_2 generation on Ni-decorated CdS nanorods. *Nat Mater.* 13: 1013–1018.

Strataki, N., M. Antoniadou, V. Dracopoulos, and P. Liano. 2010. Visible-light photocatalytic hydrogen production from ethanol–water mixtures using a Pt–CdS–TiO_2 photocatalyst. *Catal. Today.* 151: 53–57.

Subramanian, V., E. Wolf, and P. V. Kamat. 2001. Semiconductor-metal composite nanostructures. To what extent metal nanoparticles (Au, Pt, Ir) improve the photocatalytic activity of TiO_2 films? *J. Phys. Chem. B.* 105: 11439–11446.

Sun, Y., L. Jiang, T. Zeng, J. Wei, L. Liu, Y. Jin, Z. Jiao, and X. Sun. 2015. Synthesis of Au–ZnO hybrid nanostructure arrays and their enhanced photocatalytic activity. *New J. Chem.* 39: 2943–2948.

Tambwekar, S. V., D. Venugopal, and M. Subrahmanyan. 1999. H2 production of (CdS-ZnS)-TiO_2 supported photocatalytic system. *Int. J. Hydrogen Energy.* 24: 957–963.

Tapin, B., F. Epron, C. Especel, B. K. Ly, C. Pinel, and M. Besson, 2013. Study of monometallic Pd/TiO_2 catalysts for the hydrogenation of succinic acid in aqueous phase. *ACS Catal.* 3: 2327–2335.

Tung, R. T. 2001. Recent advances in Schottky barrier concepts. *Mater. Sci. Eng. R.* 35: 1–138.

Vamathevan, V., R. Amal, D. Beydoum, G. Low, and S. McEvoy. 2004. Silver metallization of titania particles: Effects on photoactivity for the oxidation of organics. *Chem. Eng. J.* 98: 127–349.

Wang, F., Y. Jiang, D. J. Lawes, G. E. Ball, C. Zhou, Z. Liu, and R. Amal. 2015a. Analysis of the promoted activity and molecular mechanism of hydrogen production over fine Au–Pt alloyed TiO_2 photocatalysts. *ACS Catal.* 5: 3924–3931.

Wang, Y. F., L. P. Li, X.-S Huang, and G.-S. Li. 2015b. A new Ti^{3+}-assisted synthesis of Pd-supported TiO$_2$ nanomaterial with enhanced photocatalytic activity for hydrogen generation and methyl orange degradation. *Chinese J. Struct. Chem.* 34: 1203–1216.

Watanabe, A. F. and K.-I. Honda. 1974. Potential variation at the semiconductor-electrolyte interface through a change in pH of the solution. *Chem. Lett.* 3 (8): 897–900.

Yan, H., J. Yang, G. Ma, G. Wu, X. Zong, Z. Lei, J. Shi, and C. Li. 2009. Visible-light-driven hydrogen production with extremely high quantum efficiency on Pt–PdS/CdS photocatalyst. *J. Catal.* 266: 165–168.

Yang, Y. Z., C. H. Chang, and H. Idriss. 2006. Photo-catalytic production of hydrogen form ethanol over M/TiO$_2$ catalysts (M = Pd, Pt or Rh). *Appl. Catal. B.* 67: 217–222.

Yao, W., X. Song, C. Huang, Q. Xu, and Q. Wu. 2013. Enhancing solar hydrogen production via modified photochemical treatment of Pt/CdS photocatalyst. *Catal. Today.* 199: 42–47.

Zhang, F., Z. Cheng, L. Cui, T. Duan, A. Anan, C. Zhang, and L. Kang. 2016. Controllable synthesis of Ag@TiO$_2$ heterostructures with enhanced photocatalytic activities under UV and visible excitation. *RSC Adv.* 6: 1844–1850.

Zhu, S., S. Liang, Q. Gu, L. Xie, J. Wang, Z. Ding, and P. Liu. 2012. Effect of Au supported TiO$_2$ with dominant exposed {0 0 1} facets on the visible-light photocatalytic activity. *Appl. Catal. B.* 119–120: 146–155.

Zou, J. J., H. He, L. Cui, and H. Y. Du, 2007. Highly efficient Pt/TiO$_2$ photocatalyst for hydrogen generation prepared by a cold plasma method. *Int. J. Hydrogen Energy.* 37: 1762–1770.

7 Doping

7.1 INTRODUCTION

The solar spectrum is constituted of nearly 7% or even less of ultraviolet (UV) light, while the rest is visible light and infrared (IR) radiation. Therefore, the harvesting and utilization of sunlight from the UV–vis to near-infrared (NIR) regions and preferably the full solar light spectrum in the photocatalysis process is gaining increasing popularity (Sang et al. 2015) and has attracted the extensive attention of researchers (Baruah et al. 2012). Metal chalcogenides, in general, and metal oxides, in particular, are the most investigated photocatalysts in the contemporary material sciences. But their large band gap is one of the major drawbacks in their widespread use, which increases sensitivity of the metal oxides in the UV part of the solar spectrum and not in the visible and/or IR range. Therefore, an active research area these days is to synthesize a narrow band gap semiconductor that absorbs longer wavelengths of the solar spectrum. A semiconductor can be used successfully and efficiently for environmental remediation such as degradation and decontamination of organic pollutant on a large scale if it can harness solar energy through an electron transfer. There are many shortcomings such as a wide band gap, colorless metal oxide, high recombination rate, and so on, which restrict the wide usage of these photocatalysts. Therefore, a search is being made to find some amicable solution to these problems. Many methods have been tried from time to time to overcome these issues. One of the approaches that has attracted the attention of material scientists is doping of a semiconductor with metal and nonmetals (Nah et al. 2010).

Doping is a process of adding a very small amount of a foreign substance (impurity) to a very pure semiconductor (one dopant atom per 1.0×10^4 to 1.0×10^8 atoms) (Ali et al. 2012). Doping of a semiconductor is an important approach in band gap engineering as it modifies some important properties of a semiconductor such as structural, morphological, electrical, and optical properties that influence the absorbance of light, redox potential, charge-carrier mobility, and so on (Rehman et al. 2009; Liu et al. 2010). Basically doping leads to a bathochromic shift, which means a decrease in band gap or addition of intraband gap state, enabling a semiconductor to harness more photons from the visible light of solar insolation.

7.2 EFFECTS OF DOPING

A number of properties of a semiconductor are likely to be affected by the process of doping. The major ones are as follows:

- Narrowing of the band gap
- Addition of impurity energy level

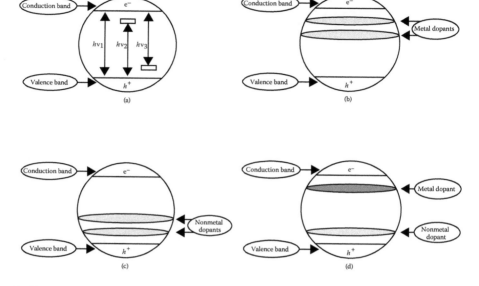

FIGURE 7.1 Band structure of (a) bare photocatalyst and with (b) metal dopant, (c) non-metal dopant, and (d) metal–nonmetal codopant. hv_1, hv_2, and hv_3 represent the band gap for pure, metal-doped, and nonmetal-doped semiconductors, respectively.

- Formation of oxygen vacancies
- Optical properties
- Crystallinity
- Surface morphology, surface area, porosity, and wettability
- Restricting some transformation to different forms, for example, anatase to rutile in TiO_2

Metal and nonmetal dopants provide new energy levels to reduce the band gap of a photocatalyst. It is due to the creation of new bands below the conduction band (CB) or above the valence band (VB) by a metal or nonmetal, respectively (Figure 7.1). Thus, the doping process leads to the formation of new energy levels (reduction of band gap) and as a consequence, less energy (hv) is required for excitation of electrons from the VB. It also improves trapping of electrons and avoids electron–hole recombination during irradiation. Thus, efficiency of a photocatalyst increases on adding a metal as an impurity because of a decrease in charge-carrier recombination (Zaleska 2008).

7.3 METAL DOPING

Most of the metal oxides have a limited range of light absorption and an inefficient charge separation, which leads to a high recombination rate with concomitant diminishing of their photocatalytic activity and limitation of future applications. Therefore, introducing a metal ion as a dopant leads to development of new visible light–induced photocatalysts with improved physicochemical properties. Thus, metal dopants improve the morphology, electronic and magnetic properties, and photocatalytic performance of photocatalytic semiconductors. A large number of studies have been carried out on various metal dopants such as vanadium (Klosek and Raftery 2001), chromium (Chang et al. 2014), cobalt (Suriye et al. 2005; Hsieh et al. 2009), copper (Chen et al. 2009), iron (Janes et al. 2004; Deng et al. 2009), manganese (Zhang et al. 2006), zinc (Xu et al. 2005), palladium (Yao et al. 2011), silver (Hsu and Chang 2014), and so on, to enhance activity of metal oxides under the sunlight for different applications (Paul et al. 2010; Zhang et al. 2012; Neamtu and Volmer 2014). The performance of a photocatalyst increases due to shifting the absorption spectra to a lower energy region and limiting the recombination rate of the photogenerated electron and hole pair (Kandula and Jeevanandam 2015).

Metal ions get incorporated in the semiconductor lattice on metal doping, which effectively enhances the photocatalytic performance of metal oxides (photocatalyst) either by broadening the absorption range in the visible region of the solar spectrum or by modifying the redox potential of the photoexcited species. In titania, modification in the presence of a dopant does not always exhibit positive results, and sometimes it may lead to adverse results. Mostly doping of metal ions increases the activity of a photocatalyst, but none of them shows stable activity after a certain period of time because of the instability of the dopant against photocorrosion (Hoffmann et al. 1995; Choi et al. 1994; Lin et al. 1999; Apno 2000; Pal et al. 2001; Beydoun et al. 2002; Karvinen et al. 2003).

7.3.1 TITANIUM DIOXIDE

In some cases, substituting metal oxide with a transition metal or doping of metal has also shown a detrimental effect on photocatalytic activity. Sivalingam et al. (2003) studied the adverse effect of doping of transition metal on titania prepared by the solution combustion method, but this inhibition effect was not observed with Pt-impregnated TiO_2.

A dip coating method was used to develop Cu-, Fe-, and Al-doped TiO_2. It was noticed that Cu was more effective as a dopant in increasing photocatalytic activity of semiconductor TiO_2 compared with Fe and Al (Maeda and Yamada 2007). Doping led to the narrowing of the band gap, but in the presence of Fe dopant in the semiconductor recombination of photogenerated electron–hole pairs took place quickly, whereas the charge separation of the photogenerated electron–hole pairs occurred effectively in the Cu-doped TiO_2 film. Al-doped semiconductor showed much less photocatalytic activity. The higher efficiency of Cr-doped TiO_2 was reported for photodecomposition of gaseous acetaldehyde by Fan et al. (2008).

Xin et al. (2008) varied content of Cu dopant in TiO_2 and found that use of 0.06 mol% Cu avoided the recombination of electron–hole pairs because of an abundant electronic trap. When dopant content was more than 0.06 mol%, recombination took place because of the excessive oxygen vacancies and Cu species. A decrease in the photocatalytic activity of the photocatalyst was observed at very high Cu concentration due to excessive covering of p-type Cu_2O on the surface of TiO_2.

Carvalho et al. (2010) studied the effect of the dopant content and depth profile of Cu atoms on the photocatalytic and surface properties of TiO_2 films. They used TiO_2-coated glass substrate on which a 5-nm-thick Cu layer was deposited. The film was annealed at 100°C and 400°C for 1 second. They showed that Cu content increased. The increasing red shift and absorption of the UV–visible spectra exhibited change of the surface properties. Methylene blue (MB) was taken as a model system in this case.

The effect of Ag dopant content from 0.1 to 1.0 mol% on composites of three metal oxides ($TiO_2/SnO_2/SiO_2$ nanocomposite) has been investigated by Yaithongkum et al. (2011). As-prepared sample transformed into anatase phase on calcination at 500°C. Its enhanced antifungal behavior (*Penicillium expansum*) under the UV light was observed with 1.0 mol% Ag dopant in composite.

Ashkarran (2011) prepared Ag-TiO_2 by combining the following two methods:

- Bottom-up
- Top-down

The sample with 0.15 g Ag exhibited highest efficiency for the inactivation of *Escherichia coli* bacteria under visible light irradiation. In this doped sample, noble metal Ag acts as an electron trap.

Kumaresan et al. (2011) observed that changes in the oxidation state of the dopant also affects the photocatalytic performance. They used different percentages (i.e., 0.5, 1.0, 2.0, and 3.0 wt.%) of Zr^{4+}, La^{3+}, and Ce^{3+} dopants to prepare a series of doped mesoporous TiO_2 by a simple, yet efficient method, that is, the sol–gel method. The characterization of the sample showed an isomorphic substitution of Zr^{4+} ion into the lattice of TiO_2, and the surface binding nature of La^{3+} and Ce^{3+} ions on mesoporous TiO_2. Mesoporous TiO_2 doped with 1 wt.% Ce^{3+} exhibited higher activity for photodegradation of alachlor as a pollutant than mesoporous TiO_2 doped with pure and other metal ions because of the change in its oxidation state.

Siriwong et al. (2012) prepared metal-doped metal oxides and used them for mineralization of methanol, sucrose, glucose, oxalic acid, and formic acid under UV–visible light illumination in a pyrex spiral photoreactor. They doped TiO_2 and CeO_2 with Fe separately by using modifying sol–gel/impregnation and homogeneous precipitation/impregnation methods, respectively. Results led to the conclusion that doped metal oxides could improve the photocatalytic activity of the pure metal oxides.

Various Ag-modified TiO_2 such as Ag/TiO_2, Ag(I)–TiO_2, and Ag/Ag(I)–TiO_2 have been prepared by impregnation, and calcined at 450°C. Their activity was examined by the photocatalytic decomposition of methyl orange and phenol solution under UV and visible light. The isolated energy level of Ag 4d contributed to the visible light absorption, while the surface metallic Ag promoted the effective separation of photogenerated electrons and holes in the Ag/Ag(I)–TiO_2 nanoparticles (NPs) under visible light irradiation. As a result, Ag/Ag(I)–TiO_2 exhibited higher visible light photocatalytic activity than the one component of Ag-modified TiO_2 (i.e., Ag(I)–TiO_2 and Ag/TiO_2) under UV light irradiation. The doping energy level of Ag(I) ions in the band gap of TiO_2 behaves as a recombination center for photogenerated electrons and holes, which resulted in lower photocatalytic performance of Ag-doped TiO_2 (such as $Ag/Ag(I)–TiO_2$ and $Ag(I)–TiO_2$) than the corresponding undoped photocatalysts (such as Ag/TiO_2 and TiO_2) (Liu et al. 2012a).

Behnajady et al. (2013) studied the effect of the doping procedure on photocatalytic activity. They prepared Cu-doped TiO_2 by two different methods: (1) doping during synthesis (DDS) via a hydrolysis method and (2) doping on the provided TiO_2 NPs (DOP) by the impregnation method. Photocatalytic activity was examined by photodegradation of C.I. acid red 27 as a model contaminant. The results indicated that samples doped by the DDS procedure showed higher photocatalytic efficiency than the samples doped by the DOP procedure. TiO_2 was doped with different contents of Cu from 0.5 to 5.0 mol%. TiO_2 doped with 2.0 mol% Cu showed increased photocatalytic activity compared with bare TiO_2 in mineralization of oxalic acid and formic acid under visible light irradiation.

The combination of sol–gel and hydrothermal methods was used to prepare lanthanum-doped TiO_2 nanotubes ($La^{3+}–TNTs$) by Cheng et al. (2013). Gaseous ethylbenzene (EB) was used to test the activity of $La^{3+}–TNTs$ under 254 nm UV light. As-prepared photocatalyst exhibited enhanced EB conversion compared with undoped TiO_2. They also reported that the photocatalytic activity of $La^{3+}–TNTs$ was affected by the initial EB concentrations and relative humidity.

The doping process does not always increase the rate of the reaction; in some reactions it may lead to a decrease in rate. Munusamy et al. (2013) reported a decrease in the rate of degradation of brilliant green using Zn and Cu as dopants in TiO_2. They observed higher degradation, that is, 99% by pure TiO_2, whereas 87% and 46% degradation was observed in Zn- and Cu-doped TiO_2.

The photocatalytic activity and antimicrobial activity of $Fe–TiO_2$ were also studied by Stoyanova et al. (2013). The titanium tetrachloride, benzyl alcohol, and iron(III) nitrate were used in the nonhydrolytic sol–gel method for preparation of pure and iron-doped TiO_2. The average particle size of about 12–15 nm was calculated by X-ray power diffraction (XRD). The activity of pure and Fe-modified titanium dioxide samples was tested by photooxidation of reactive black 5 and photodisinfection of *E. coli* under the UV–visible light.

Cu-doped mesoporous TiO_2 NPs were synthesized by the hydrothermal method at relatively low temperatures (Wang et al. 2014). The XRD results revealed that the NPs prepared by this method were approximately 20 nm in size, which aggregated together to shapes of approximately 1100 nm and led to a porous aggregate structure. It was observed that methyl orange degraded very fast in the presence of Cu-doped mesoporous TiO_2 due to the formation of stable Cu(I) and the mesoporous structure.

Doping increases the surface area and pore volume, and provides good crystallinity, strong visible light absorption, and effective charge separation of photogenerated electron–hole pairs in the catalyst. Not only this, but it also inhibits phase transition from anatase to rutile in the case of TiO_2. Zhan et al. (2014) observed the same characteristic properties in mesoporous sulfated rare earth ions (Nd^{3+}, La^{3+}, Y^{3+}) doped TiO_2 at fumed SiO_2 photocatalysts in which

P123 ($EO_{20}PO_{70}EO_{20}$) was used as a template. The sample was prepared by the sol–gel method and its photocatalytic activity was evaluated by degradation of methyl orange. Results revealed that rare earth metal doped samples were more efficient and effective than undoped samples and Degussa P25.

The degradation of 1.1.1-trichloroethane, trichloroethene, and tetrachloroethene under UV irradiation was carried out by Cu-doped titania, undoped titania, and Degussa P25 (Ndong et al. 2014). The Cu-doped sample was synthesized using tetrabutyl titanate, hydrofluoric acid, and cupric nitrate through hydrothermal solution. As-prepared photocatalyst had anatase form and sheet structure. Cu-doped TiO_2 showed good stability and can be reused up to five cycles.

Harikishore et al. (2014) used the sol–gel method for preparation of nanocrystalline pure TiO_2 and 5 mol% silver-doped TiO_2 (Ag–TiO_2) powders. Addition of Ag reduced the band gap of TiO_2 from 3.1 to 2.9 eV. The photocatalytic activity was evaluated by degradation of MB. Complete inhibition in growth of *E. coli* was observed within 24 hours. They also observed the highest efficiency when TiO_2 was annealed at 500°C compared with as-synthesized TiO_2; however, efficiency decreased with further rise in temperature. The average particle size was reported to be around 6–15 nm.

Highly stable, low-cost, hierarchical structure, high density, and efficient secondary Ag NPs grown on primary TiO_2 fibers have been fabricated by a combination of electrospinning and hydrothermal processes. The photocatalytic activity was monitored by degradation of rhodamine B under UV light illumination (Zhang et al. 2015). Suwarnkar et al. (2014) reported 99.5% photodegradation efficiency of methyl orange by doping of Ag in TiO_2 matrix using light of 365 nm wavelength. Ag-doped TiO_2 with different Ag contents were synthesized by a microwave-assisted method and characterized.

Tsuruoka et al. (2015) studied photocatalytic decomposition activity and water adsorption capability of Ti-doped hydroxyapatite (Ti–HAp) and anatase-type TiO_2 powders under UV irradiation. Anatase-type TiO_2 showed photocatalytic wettability conversion to a hydrophilic state, but Ti–HAp does not exhibit similar properties. This proves that Ti–HAp exhibited a different trend than anatase.

Pham et al. (2015) carried out photooxidation of MB in water by thin films of Cu–TiO_2 reduced graphene oxide (rGO) on quartz substrates. The thin films were fabricated by spraying a sol of copper metal–doped titanium dioxide combined with rGO. The Cu-doped TiO_2/rGO film photocatalysts showed better efficiency in the photodegradation of MB than undoped TiO_2/rGO film. Doping of copper increased hydrophilicity of the materials, and decreased the band gap.

Jose et al. (2015) prepared nanotubes of undoped and silver-doped anatase titania. They used the hydrothermal method with slight modification, using acetic acid modified sol–gel process treated nanocrystalline undoped and Ag-doped anatase TiO_2 as precursors. The effect of MB dye concentration and pH on the reaction rate was also studied. When the Ag/Ti ratio was varied from 0.01 to 0.05, maximum dye adsorption capacity of 39 mg/g was observed by the 0.01 Ag/Ti ratio, while dye adsorption capacity of undoped nanotubes of anatase was only 32 mg/g at the initial solution pH of ~10.

7.3.2 ZINC OXIDE

The photoelectrochemical production of hydrogen was observed by Ullah and Dutta (2007) using Mn^{2+} and Cu^{2+} dopants in ZnO. It was observed that manganese-doped ZnO ($ZnO:Mn^{2+}$) absorbed more visible light compared with the copper-doped ZnO ($ZnO:Cu^{2+}$), when exposed to tungsten lamp irradiation. These samples exhibited a significant enhancement in the optical absorption when compared with bare ZnO.

Mohan et al. (2012) studied photocatalytic activity of Cu-doped ZnO nanorods with different Cu concentrations on degradation of resazurin dye. It was synthesized via the vapor transport method. The needle-like shape of undoped ZnO and rod-like shape of Cu-doped ZnO samples with an average diameter and length of 60–90 nm and 1.5–3 μm, respectively, were confirmed by the field emission-scanning electron microscopy (FE–SEM) images. The rate constant was equal to 10.17×10^{-2} per minute with 15% Cu-doped ZnO, which was almost double that of the pure ZnO. This degradation followed psuedo-first-order kinetics. Higher efficiency of Cu-doped ZnO was due to intrinsic oxygen vacancies, because of high surface to volume ratio in nanorods, and extrinsic defects due to Cu doping.

A series of Al-doped ZnO (AZO) were prepared by the combustion method by Ahmad et al. (2013). As-prepared samples were calcined at 500°C for 3 hours. Dopant concentration was varied from 0.5 to 6.0 mol% and it was found that the optical band gap energy for the AZO nanopowders was in the range of 3.12–3.21 eV up to 4.0 mol%, which further decreased with increasing Al dopant. Their efficiency was observed by degradation of methyl orange at a wavelength of 420 nm. A sample with 4.0 mol% Al showed a maximum rate of dye decomposition and showed five times higher efficiency than pure ZnO. The enhanced photocatalytic activity was observed due to extended visible light absorption, reduced electron–hole pair recombination, and increased adsorptivity of MO dye molecule on the surface of AZO nanopowders.

Cu–ZnO and Ag–ZnO nanorods were synthesized by the precipitation method. The effect of Cu and Ag dopant on the optical property, that is, narrowing of the band gap, was studied by Rahimi et al. (2013). The doped photocatalysts were tested for their tendency to remove MB in aqueous solution under UV–visible light.

Saleh and Djaja (2014a) used Fe-doped wurtzite ZnO NPs for degradation of two dyes: methyl orange and MB. The prepared sample with different dopant contents was characterized. The electron spin resonance (ESR) analysis showed the presence of Fe^{2+} and Fe^{3+} valence states of iron. The concentration of these states had a major influence on the magnetization property. It was reported that on increasing dopant contents, the number of spins arises due to an increase in Fe^{2+} ions and spins. The catalysts with the highest number of spins due to Fe^{2+} ions showed optimum photocatalytic performance for the degradation of both dyes.

Phuruangrat et al. (2014) prepared and characterized single crystalline flower-like ZnO and Eu-doped ZnO structures. These were synthesized by the sonochemical method, and structures, morphologies, and photocatalytic activities were examined by XRD, SEM, transmission electron microscopy, Raman spectroscopy, X-ray photoelectron spectroscopy, and UV–visible absorption spectroscopy. As-prepared samples were used for removal of MB under UV illumination. Doped photocatalyst proved to be an excellent photocatalyst for dye degradation from wastewater.

Saleh and Djaja (2014b) prepared Co- and Mn-doped ZnO NPs with wurtzite structure by a coprecipitation process. They studied the effect of dopant contents on the different properties of ZnO particles such as structural and optical properties, spin resonance, and photocatalytic activity under UV irradiation. Results showed that the degradation efficiency of Mn-doped ZnO was higher than Co-doped ZnO.

Zinc oxide was modified by doping with manganese (Abdollahi et al. 2011), cobalt (Kuriakose et al. 2014), iron, and nickel (Liu et al. 2014) by precipitation and wet chemical methods for various applications. The percentage of palladium in the range of 0.5%–1.5% was used to prepare a series of Pd-doped ceria–ZnO (PdCeO$_{2-x}$–ZnO) (Seddigi et al. 2014). Degradation of methyl *tert*-butyl ether (MTBE) was highest with the ceria–ZnO catalyst doped with 1% Pd. This also indicates that photocatalytic activity of metal oxide depends on the amount of dopant.

7.3.3 Tin Oxide

The band gap depends on the amount of the metal dopant and thus the activity of semiconductor is also affected by concentration of dopants. Ray and Podder (2009) showed that optical transmission of Cu–SnO$_2$ increased up to 79% for 200 nm thickness of film, which has only 71% pure SnO$_2$. This increase was observed only from 1% to 4% of Cu doping, but the activity was found to decrease on further increase of dopant concentration. They also observed that band gap shifted to lower energies and then increased with further increase of dopant concentrations. Thus, it may be concluded that amount of dopant in a semiconductor also has a great influence on the activity of a semiconductor.

A thin film of Cu–SnO$_2$ was prepared from an aqueous solution of tin chloride pentahydrate on ultrasonically cleaned glass substrates at a temperature of 350°C by spray pyrolysis (Patil et al. 2013). The crystallite size with pyramid-type nanostructures was found to increase with an increase in Cu content in the SnO$_2$ films. Gas-sensing characteristics of this photocatalyst were studied on different gases such as carbon monoxide, ammonia, H$_2$S, and ethanol. The films with 3% Cu content showed high response and excellent selectivity for H$_2$S compared with other gases at room temperature.

Surface-modified Ag-doped SnO$_2$ NPs and Ag-SnO$_2$ modified with curcumin (Cur–Ag–SnO$_2$) have been synthesized by Vignesh et al. (2013). They used the following two routes for sample preparation:

- Precipitation method
- Chemical impregnation process

The bare SnO$_2$ and Ag–SnO$_2$ showed lower degradation efficiency for rose Bengal (RB) than Cur–Ag–SnO$_2$. Surface-modified samples also showed a red shift in the visible region and effective electron–hole separation. The antifungal activity of the photocatalyst and the reusability of Cur–Ag–SnO$_2$ were also tested.

Folic acid (FA) biosensor was synthesized with Cu-doped SnO$_2$ NPs using a simple microwave irradiation method by Lavanya et al. (2014). It was found that as the dopant was increased from 10 to 20 wt.%, the crystalline size of NPs decreased. Thus, Cu-doped NPs of SnO_2 (higher wt.%) proved to be useful for the estimation of FA content in pharmaceutical samples. The biosensor showed lowest detection amount (0.024 nM) of FA over a wide FA concentration range of 1.0×10^{-10} to 6.7×10^{-5} M at a physiological pH of 7.0.

Feng et al. (2015) utilized 3D transition metals such as Cr, Mn, Fe, Ni, and so on, as dopants in tin oxide and observed changes in its magnetic, electronic, and optical properties. Activity of SnO_2 was enhanced significantly in the visible light region, which makes it very useful for the design of solar cells, photoelectronic devices, and as a photocatalyst.

Ran et al. (2015) prepared hollow-structured SnO_2 with an adjustable Ti doping content using SiO_2 microspheres as hard templates via an improved Stober method. The comparative study of pure SnO_2 hollow spherical sample, and Ti-doped SnO_2 with a doping content of 20 mol% exhibited 92% and 54% photocatalytic degradation of MB within 3 hours. The homogeneous doping of Ti into the lattice of SnO_2 avoided the condition of electron–hole pair recombination and also expanded the range of usable excitation light to the visible region. In addition, the highly crystalline state, larger surface area, and large pore size of Ti-doped SnO_2 were also directly related with photocatalytic activity of the Ti-doped SnO_2 samples.

7.3.4 Others

Several decades ago, hydrogen was produced from carbohydrates, formed from water and carbon dioxide, in plants using UV light, which is only a small portion of the solar spectrum. Therefore, efforts have been made to create photocatalysts by the process of doping, which are capable of using visible light. It has been observed that doping has a major influence on surface, optical, gas sensing, and crystalline properties of metal oxides, but only up to a certain concentration of dopant. Zou et al. (2001) prepared nickel-doped indium tantalum oxide ($In_{(1-x)}Ni_x TaO_4$) with x equal to zero to 0.2 and used it as a photocatalyst for hydrogen and oxygen production by water splitting with 0.66% quantum yield. The increase in surface area and modification of the surface site enhanced the process of water splitting.

The impact of different transition metals such as Fe, Co, Ni, Cu, and Zn as dopants on WO_3 at various concentrations was studied by Hameed et al. (2004). The photocatalytic activity of WO_3 was evaluated for splitting of water into hydrogen and oxygen under the UV laser irradiation. The role of the configuration of 3D-orbitals of the doped transition metals in enhancing or hindering the production of hydrogen and oxygen was also reported.

Rajabi et al. (2013) used a chemical precipitation method for synthesizing pure and Fe^{3+} ion-doped ZnS quantum dots. They used 2-mercaptoethanol as a capping agent. The XRD patterns showed that the doped NPs were crystalline, with a cubic zinc blende structure. The doped sample showed a higher decolorization rate than pure ZnS.

Complete decolorization of four dyes, namely methylene blue, malachite green, methyl orange, and methyl red, was achieved under UV light in less than 2 hours using Fe ion-doped polyaniline on indium tin oxide (ITO)–coated glass substrate as a photocatalyst (Haspulat et al. 2013). The Fe doping led to an increase in the surface roughness and wettability of the produced polyaniline films, which favored photocatalytic activity in water-based solutions.

Pure CeO_2 NPs and Fe-doped CeO_2 NPs were prepared by flame spray pyrolysis by varying the Fe dopant concentrations. Average sizes of 6.39 and 5.94 nm were observed, respectively. It was found that band gap of doped semiconductor decreases from 3.18 to 2.90 eV. Fe-doped CeO_2 NPs were responsible for an increased degradation of the formic and oxalic acids (Channei et al. 2013).

Platinum-doped ZrO_2–SiO_2 mixed oxides showed an increase in photodegradation of cyanide under illumination of visible light because of increased specific surface area (Kadi and Mohamed 2013). Cu-doped ZnS quantum dots were fabricated in aqueous solution by Labiadh et al. (2014) using 3-mercaptopropionic acid (MPA). Enhanced photocatalytic activities of TiO_2/Cu:ZnS NPs were reported as compared with pure TiO_2 NPs or undoped TiO_2/ZnS NPs in the oxidation of salicylic acid aqueous solutions under UV light irradiation.

Satheesh et al. (2014) fabricated transition metal (M = Cu, Ni, and Co) doped iron oxide (Fe_2O_3) NPs via a simple coprecipitation technique. The photocatalytic activity was tested by the degradation of acid red 27 dye under visible light irradiation. It was observed that the photocatalytic activity of Cu–Fe_2O_3 was more than that of Fe_2O_3, Ni–Fe_2O_3, and Co–Fe_2O_3, and the photocatalyst can be reused four times without any remarkable loss of its activity.

Sr^{2+} cations were used as dopant by An and Onishi (2015) in perovskite-type sodium tantalate ($NaTaO_3$) to make $NaTaO_3$–$SrSr_{1/3}Ta_{2/3}O_3$ solid solutions through solid-state or hydrothermal reactions. It was concluded that alkaline earth metal dopant restricted the recombination in $NaTaO_3$ photocatalysts. Wang et al. (2015) prepared Au-doped Cu_2SnSe_3 hetero-nanostructure and bare Cu_2SnSe_3 by a seed-mediated growth method and made a comparative study.

Some metal-doped photocatalysts along with their applications are provided in Table 7.1.

Expensive techniques were required for incorporating a metal ion as a dopant in titanium dioxide (Lui et al. 2005; Wong et al. 2006). The metallic cations in TiO_2 have been identified as the main cause for the partial blockage of surface sites available for photocatalytic activity (Xiao et al. 2006). Not only this, it also increases the carrier-recombination centers, and subsequently reduces the photocatalytic performance of metal oxides. Aluminum (Al^{3+}), chromium (Cr^{3+}), and gallium (Ga^{3+}) dopants (p-type dopants) have the ability to create acceptor centers. These acceptor centers trap electrons generated by photon absorption and become negatively charged. Then positive holes get attracted and recombine with the electrons. In the same manner, niobium (Nb^{5+}), tantalum (Ta^{5+}), and antimony (Sb^{5+}) dopants (n-type dopants) create donor centers and trap photogenerated holes and become positively charged. Then they react with electrons and thus behave as recombination centers. Electron–hole pair separation in noble metal–doped titanium dioxide has been ascribed to the difference in Fermi level of noble metals and that of TiO_2. Yet, once the metal center becomes negatively charged, holes will be attracted and they recombine with electrons. This is especially obvious

TABLE 7.1

Metal-Doped Photocatalysts and Their Applications

Photocatalyst	Metal Dopant	Applications	References
TiO_2	C activated and W doped	Photodegradation of rhodamine B	Li et al. (2012b)
TiO_2	Cu and Zn	Degradation of methyl orange	Khairy et al. (2014)
TiO_2	Sm^{3+} ion	Degradation of methyl blue	Xiao et al. (2008)
Bi_2O_3	Ag-modified Ti-doped	Degradation of crystal violet	Zhang et al. (2014a)
MgO	Ca^{2+}	Wide applications such as heterogeneous catalysis, optoelectronics, and so on	Stankic et al. (2005)
Y_2WO_6	La^{3+}	Photoluminescense	Ding et al. (2015)
$ZnWO_4$	Eu^{3+}	Luminescent properties	Dai et al. (2007)
WO_3	Mo	Degradation of rhodamine B	Li et al. (2015)
Y_2WO_6	Ln^{3+} = Dy, Eu, and Sm	DSSC	Huang et al. (2014b)
Lu_2WO_6	Eu^{3+}, Pr^{3+}	Luminescence properties	Zhang et al. (2008)
$PbWO_4$	La^{3+}	Electronic structures	Chen et al. (2007)
$CaMoO_4$ and $CaWO_4$	Dy^{3+}	Fluorescent lamps and display panels	Sharma et al. (2013)
$ZnWO_4$	La	Phase, morphologies and optical properties	Arin et al. (2014)
$SrWO_4$	Eu^{3+}	Potential red emitting phosphors for white LEDs	Maheshwary et al. (2014)
Lu_6WO_{12} and Lu_6MoO_{12}	Eu^{3+} ions	Photoluminescence property	Li et al. (2011)
$CaMoO_4{:}Eu^{3+}$	Gd^{3+} ions	Photoluminescence property	Singh et al. (2014)

for highly loaded samples, where the metal content is more than 5 wt.% (Burda et al. 2003).

Besides all positive effects, metal doping has many drawbacks also. TiO_2 photocatalysts doped with metals have been known to suffer from thermal instability. It has been shown that the desired band gap narrowing can be obtained by using an anionic nonmetal as the dopant rather than metallic action (Xu et al. 2009).

7.4   NONMETAL DOPING

On the basis of spin-restricted local density approximation calculation, Asahi et al. (2001) investigated the substitutional doping of N for O and interaction of N 2p state of nitrogen dopant with O 2p state in anatase TiO_2 because of their very close energy levels. Thus, nitrogen doping led to narrowing of the band gap and also increases photocatalytic activity of a semiconductor in visible light. There are three different main opinions regarding the modification mechanism of TiO_2 doped with nonmetals. These are as follows:

- Band gap narrowing
- Impurity energy levels
- Oxygen vacancies

An additional benefit of doping is the increase in electron trapping, which inhibits electron–hole recombination during irradiation and results in enhanced photoactivity. A nonmetal dopant can react with oxides of a photocatalyst in the following three manners:

- The dopant can hybridize with the oxide of the photocatalyst
- The oxygen site gets substituted by the dopant
- Addition of dopant in the oxygen-deficient site acts as a blocker for reoxidation

Irie et al. (2003) reported addition of impurity in energy levels above the VB when doping titanium dioxide with nitrogen. These levels are formed due to substitution of the oxygen site by a nitrogen atom. Irradiation with UV light excited electrons in both the VB and the impurity energy level, but irradiation with visible light excited electrons present only in impurity level. Zhao and Liu (2008) discussed some modifications in the mechanism of activity of N-doped TiO_2. The experimental results showed that TiO_2 doped with substitutional nitrogen had shallow acceptor states above the valence state. On the contrary, TiO_2 doped with interstitial nitrogen has isolated impurity states in the middle of the band. The oxygen-deficient sites formed in the grain boundaries were essential to emerge visible activity, and N-doped TiO_2 in a part of the oxygen-deficient sites were important as blockers of the redox reaction (Ihara et al. 2003).

The surface chemistry of a photocatalyst gets affected by surface defects. Diebold (2003) investigated different defects in bare TiO_2. An intriguing surface 1×2 reconstruction on an N-doped single crystal rutile with (110) surface was reported by Batzill et al. (2006). C-, N-, and S-doped TiO_2 showed red shift of the absorption edge of TiO_2, because of the formation of oxygen vacancies and the color centers (Sakthivel and Kisch 2003).

Nonmetal (carbon) led to formation of oxygen vacancy state because of the formation of Ti^{3+} species between the VBs and CBs in the TiO_2 band structure (Li et al. 2005). Anpo (2004) explained the red shift of the optical absorption edge and formation of oxygen vacancy. He proposed that on N doping, the N 2p orbitals get localized above the O 2p VBs and the excitation from the occupied high energy states to the CB resulted in the optical absorption edge shift to the lower energy of the visible light region.

Nonmetal doping not only improves photocatalytic activity of a photocatalyst, but also affects its morphology. The improved morphology and photocatalytic efficiency of TiO_2 was observed on adding nonmetal dopants such as C, N, and S (Chen and Mao 2007). The presence of nonmetal anions increased the percentage of the anatase phase, affected the crystallinity of the semiconductor, and increased the specific surface area (Yu et al. 2009). Yu et al. (2002) prepared anatase and brookite phase of F^--doped TiO_2 by hydrolysis of titanium tetraisopropoxide in a solution of NH_4F-H_2O and tested its activity for oxidation of acetone in air. They observed that F^- doping suppressed the formation of the brookite phase and improved the crystallinity of the anatase form of titania. Moreover, fluoride ions also prevented transition from the anatase to rutile phase. Phosphorous was used as a nonmetal dopant by Raj et al. (2009) to enhance the thermal stability of titanium dioxide through formation of titanyl phosphate. High specific surface area of 154 m^2/g and crystallite size of 8.6 nm of nano-doped TiO_2 were successfully produced at a P/Ti molar ratio of 0.14.

Tang et al. (2006) doped CeO_2/TiO_2 mixed oxides with boron and observed photocatalytic performance of catalyst by degradation of acid red B dye under UV irradiation. Bettinelli et al. (2007) used boron as a nonmetal dopant to modify TiO_2. The reactivity was studied by photooxidation of MB under visible light.

The change of calcination temperature led to transformation from anatase phase to rutile phase. Nitrogen-doped TiO_2 NPs prepared from the sol–gel method using titanium(IV) tetraisopropoxide with 25% ammonia solution exhibited change in its phase, when it was treated at different temperatures from 300°C to 600°C (Bangkedphol et al. 2010). The XRD results showed that the sample was amorphous at 300°C, but at 400°C, it was transformed into anatase phase and then transformed to the rutile phase at 600°C. N-doped TiO_2, undoped TiO_2, and commercial TiO_2 showed 28%, 14.8%, and 18% degradation of tributyltin (TBT), respectively, in 3 hours under natural light.

Nitrogen-doped TiO_2 and its applications in the areas of energy conversion and environmental cleanup were reviewed by Zhang et al. (2010). The effect of various nitrogen precursors such as triethylamine, hydrazine hydrate, ethylenediamine, ammonium hydroxide, and urea on activity of TiO_2 nanocrystalline powders prepared by the sol–gel method was reported by Hu et al. (2011a). The photocatalytic activity was evaluated by decomposition of methyl orange dye.

The pyrogenation of the mixture of urea and In_2TiO_5 was used to prepare N-doped In_2TiO_5, modified by carbon nitride composite (NICN) via a polymerizable complex method. The prepared samples were characterized by different techniques. The XRD results showed that nitrogen dopant does not change the crystal structure of In_2TiO_5, and precursor sintered at 1000°C was pure. The wavelength shifts from 410 to 450 nm with increasing dopant content revealed significant narrowing of the band gap. The complete decomposition of rhodamine B within 20 minutes under visible light and its reusablity indicated that NICN has a stable structure and durable photocatalytic activity (Liu et al. 2011).

Different precursors such as NH_3 plasma, N_2 plasma, and annealing in flowing NH_3 were used for nitridation to synthesize N-doped TiO_2. The samples prepared with different nitrogen sources were examined by carrying out the degradation of an aqueous solution of a reactive dyestuff, MB, under visible light (Hu et al. 2011b). The results showed that the photocatalytic efficiency and stability of TiO_2 prepared by NH_3 plasma was much higher than that of the samples prepared by other nitridation procedures. The nitrogen-doped photocatalyst showed higher activity due to increased lattice-nitrogen content and decreased adsorbed NH_3 on the catalyst surface. The lattice nitrogen stability of N-doped TiO_2 samples was improved after HCl solution washing.

Takeuchi et al. (2011) studied photodecomposition of methanol to carbon dioxide and water by N-doped WO_3. N–WO_3 photocatalyst was prepared by thermal decomposition of an ammonium paratungstate $[(NH_4)_{10}W_{12}O_{41} \cdot 5H_2O]$ containing NH_4^+ ions as a nitrogen source.

Zhang et al. (2011) prepared a sandwich-structured photocatalyst using a combination of nonmetal doping and plasmonic metal decoration of TiO_2 nanocrystals, which exhibited potential application in the elimination of various organic compounds under UV, visible light, and direct sunlight.

Temperature influences the band gap of a semiconductor on doping it with a nonmetal (Nolan et al. 2012). Nitrogen-doped titanium dioxide was synthesized by the sol–gel method using 1,3-diaminopropane as a nitrogen source. The sample was annealed at 500°C, 600°C, and 700°C and the percentages of rutile observed by XRD were 0%, 46%, and 94%, respectively. At higher temperatures, nitrogen remained in the lattice of titania as indicated by the XPS. Sample annealed at 500°C showed maximum degradation rate of 4-chlorophenol under solar irradiation and MB under 60 W house bulb, whereas samples annealed at higher temperature exhibited lower efficiency.

Triantis et al. (2012) carried out the degradation of microcystin-LR (MC–LR), one of the most common and more toxic water soluble cyanotoxin compounds released by cyanobacteria blooms, under UV-A, solar, and visible light. They observed that under UV-A, Degussa P25 TiO_2 showed higher degradation (99%) than N-doped TiO_2 (96%). Under the sunlight, both samples exhibited the same efficiency. Doped TiO_2 displayed remarkable efficiency under visible light, whereas commercial TiO_2 has not shown any response. This means source of light is also one of the factors that affects the performance of pure and doped metal oxide.

Yu et al. (2012) investigated the role of ultrasound in doping of F in TiO_2. A sonochemical technique was used to prepare F-doped square-shaped TiO_2 nanocrystals with varied F contents. High photocatalytic activity for degradation of phenol was observed with doped TiO_2, which was attributed to the fact that F doping increased the surface hydroxyl groups over TiO_2 and effectively reduced the recombination rate of photogenerated electron–hole pairs, which will produce more ·OH radicals to decompose phenol molecules.

Carbon-doped zinc oxide nanostructures were designed using vitamin C, which resulted in red shift in the absorption band. As a result, photocatalytic activities of C-doped ZnO nanostructures were found to be much better than the activities of pure ZnO nanostructures under visible light of wavelength > 420 nm (Cho et al. 2010).

The relation between ferromagnetism and intrinsic defects of C-doped ZnO thin films was investigated by Subramanian et al. (2012). The mediation of ferromagnetic interaction and the existence of hybridization between Zn and C, respectively, affect oxygen- and zinc-related defects in C-doped ZnO. Zhang et al. (2014b) prepared carbon-doped zinc oxide without using any precursor. They used $Zn(OAc)_2 \cdot 2H_2O$ as a source of both carbon and zinc. Rate of photodegradation of rhodamine B in aqueous solutions at room temperature with near-UV light irradiation in the presence of C–ZnO increased because of more photons being absorbed and reduced electron–hole pair recombination.

Lee et al. (2013) prepared nitrogen-doped three-dimensional polycrystalline anatase TiO_2 photocatalysts (N-3D TiO_2) at temperature less than 90°C via a modified hydrothermal process under ultrasound irradiation and visible light. It was observed that as-prepared photocatalyst retained its initial decolorization rate (91.8%) even after 15 cycles. The efficiency of N-3D TiO_2 (N-3D TiO_2; [k] = 1.435 per hour) was 26.1 times higher than that of 3D TiO_2 ([k] = 0.055 per hour). N-3D TiO_2 showed strong antimicrobial properties against both Gram-negative $E.$ $coli$ and Gram-positive $Staphylococcus$ $aureus$, and therefore, it has several promising applications such as highly efficient water/air treatment, inactivation of pathogenic microorganisms, and solar energy conversion.

Yin et al. (2013) carried out the photocatalytic degradation of MB and O_2 evolution from water splitting using C-doped $BiVO_4$ as a photocatalyst. It was fabricated by the sol–gel method with fine hierarchical structures templated from $Papilioparis$ butterfly wings. The photocatalytic activity of this photocatalyst was much higher, that is, 16 times and 6.3 times than pure semiconductor for O_2 evolution and dye degradation, respectively.

The percentage of nonmetal dopant affects pore volume and structure also. A series of boron-doped Bi_2WO_6 was prepared with different amounts of boron atoms, that is, 0.1%, 0.5%, 1.0%, 5.0%, and 10%, using a hydrothermal method. The photodegradation of rhodamine B under simulated solar light was investigated by Fu et al. (2013). They showed that total pore volume increased only up to 0.5% of dopant and then it decreased on further increase in dopant concentration. Thus, 0.5% B/Bi_2WO_6 displayed stronger adsorption capacity to RhB and also trapped electrons. As a consequence, higher photodegradation of dye was observed with rate constant 8.8 times that of pure Bi_2WO_6.

Well-positioned band alignments were observed for Se- and I-doped β-Ga_2O_3. They also doped $SrTiO_3$ surface with the same dopants (Guo et al. 2015a). The dopants with smaller atomic size such as C, N, and F substituted the O atom in the TiO_2-terminated surface, whereas the larger atomic size dopants such as P, S, Cl, Se, and Br replaced O in the SrO-terminated surface. The discrete midgap state was observed using C, Si, and P dopants. Thus, due to the appearance of surface O 2p states, the band gaps were approximately 2.60 eV in the pure TiO_2-terminated surface and 3.4 eV in the bulk $SrTiO_3$.

Enhanced efficiency of semiconductor using various nonmetal dopants such as carbon, nitrogen, and boron has also been reported by a number of researchers (Solanki et al. 2015; Rajkumar and Singh 2015). Zhang et al. (2013) studied some nonmetal-doped photocatalysts and explained principles of density-functional calculation for the electric properties. They synthesized boron-, carbon-, nitrogen-, fluorine-, phosphorus-, and sulfur-doped $SrTiO_3$.

Mohamed et al. (2015) fabricated N-doped TiO_2 nanorod-assembled microspheres. The XRD results showed the presence of sample in an anatase–rutile mixed phase, while SEM, TEM, and AFM images showed the formation of TiO_2 microspheres as TiO_2 nanorods or rice-like nanorods. The XPS study indicated the incorporation of nitrogen as a dopant in TiO_2 with binding energies of 396.8, 397.5, 398.7, and 399.8 eV. The photocatalytic activity of the as-prepared TiO_2 resulted in excellent photodegradation of hazardous water pollutants such as MB under the UV and visible light irradiation.

Guo et al. (2015b) studied the effect of various nonmetals, that is, C, N, F, Si, P, S, Cl, Se, Br, and I on the performance of β-Ga_2O_3 (4.5 eV) in both photooxidation and photoreduction of water. It has been proved to be a promising photocatalyst for water splitting in the visible region. Their results showed that the doping was energetically favored under Ga-rich growth conditions with respect to O-rich growth conditions. The substitution of the threefold coordinated O atom with a nonmetal element was much easier than the fourfold coordinated O atom. The dopants C, Si, and P exhibited similar band gaps to that of semiconductor along with the presence of discrete midgap states, which resulted in an adverse effect on the photocatalytic properties. On the other hand, other dopants such as N, S, Cl, Se, Br, and I showed enhanced photocatalytic redox ability.

Some nonmetal-doped photocatalysts along with their applications are provided in Table 7.2.

TABLE 7.2
Nonmetal-Doped Photocatalysts and Their Applications

Photocatalyst	Nonmetal dopant	Applications	References
Graphitic (C_3N_4)	S	Removal of phenol	Liu et al. (2010)
WO_3	S	Water splitting	Li et al. (2012a)
WO_3 and In_2O_3	C	Photoelectrochemical	Sun et al. (2009)
$WO_3 \cdot 0.33H_2O$	C	Degradation of rhodamine B	Yi et al. (2016)
TiO_2	P	Photodegradation of bisphenol A	Kuo et al. (2014)
TiO_2	C-modified N-doped	Degradation of methyl orange	Wang et al. (2012)
ZnO	C	Degradation of malachite green	Lavand and Malghe (2015)
ZnO	C	Degradation of rhodamine B	Haibo et al. (2013)
ZnO	C	Degradation of MB	Zhang et al. (2015)
ZnO	C	Water splitting (IPCE value of 95%)	Lin et al. (2012)

7.5 CODOPING

The combination of different donor and acceptor dopants leads to the narrowing of the band gap that results in the bathochromic shift (red shift). Thus, due to the synergistic effect of dopants, codoping shifts the absorption edge successfully from the UV region to visible light region, that is, it helps in broadening of the absorption band. The process of adding donor–acceptor dopants is known as codoping. It helps to resolve some problems such as the solubility limit, carrier recombination, low carrier mobility, and nonresponse to the visible light in a host material (Yan et al. 2013).

Some of the adverse features of photocatalysts such as wide band gap, being colorless (or light colored) in most cases, high recombination rate, and so on, are responsible for their lower photoactivity in the visible region of solar spectrum. As a major portion of solar spectrum consists of the visible region, it is of utmost importance to modify the photocatalyst so that it could be used in the visible region along with the UV region of solar spectrum. Doping of a semiconductor has proved to be an effective way to overcome this problem of the bare photocatalyst. This doping process has a major influence on certain properties such as structural, morphological, electrical, and optical properties of a semiconductor. Thus, doping of various metal oxides and mixed oxides can be used as one of the major strategies to reduce the large band gap of semiconductor materials and make them effective in the visible light range. Modified photocatalysts have wide applications in the field of environmental remediation such as pollutant degradation, solar fuel generation (Marschall and Wang 2014), decolorization, removal of synthetic dyes (Khataee and Kasiri 2010; Kirupavasam and Raj 2012, Munusamy et al. 2013; Julkapli et al. 2014), degradation of gaseous acetaldehyde (Asahi et al. 2001), acetone (Singkammo et al. 2015), gas sensors such as H_2 (Liewhiran et al. 2009), and so on.

7.5.1 METAL AND METAL

ZnO was doped with Co and Al using a pulsed laser deposition method to get $Zn_{0.895}Co_{0.100}Al_{0.005}O$ photocatalyst. As-prepared sample showed the ferromagnetic nature due to the Al interstitial defects and their hybridization with Co substitutional dopants (Chang et al. 2009). Wang et al. (2011) selected a series of Er^{3+}/Yb^{3+} for codoping Sb_2O_3–WO_3–Li_2O glasses, which showed intense green up-conversion fluorescence, which was a two-photon adsorption process near 524 and 544 nm under excitation at 980 nm.

Thirupathi and Smirniotis (2011) synthesized codoped titanium dioxide, where Mn was combined with different metals (M´) (where M´ = Cr, Fe, Co, Ni, Cu, Zn, Ce, and Zr). As-prepared Mn/M' TiO_2 photocatalysts showed its effect on the selective reduction of NO with NH_3 at low temperatures.

Li et al. (2013) reported an increase in surface area and narrowing of the band gap of titania codoped with V and Zn. The sample was synthesized by the sol–gel method and evaluated by decomposition of organic dyes in a heterogeneous system under both the UV light and visible light. La-WO_3 codoped TiO_2 was synthesized by Diao and Zhou (2014) via the sol–gel method using butyl titanate, anhydrous ethanol as a solvent, and glacial acetic acid as an inhibitor. Photocatalytic activity was evaluated for degradation of methyl orange. The effect of different operating parameters such as heat treatment temperature, different dopants, pH, dosage of catalyst, and so on, on photooxidation was also investigated.

7.5.2 METAL AND NONMETAL

Metal and nonmetal codoping raises the VB edge significantly and also increases the CB edge. Thus, this change in electronic structure increases the performance of the photocatalyst. The enhanced efficiency of C–Mo, C–W, N–Nb, and N–Ta codoped anatase TiO_2 systems for hydrogen generation from water and degradation of organic pollutants on irradiation was observed by Liu (2012).

Obata et al. (2007) carried out codoping of TiO_2 by Ta and N dopants via a radio-frequency (RF) magnetron sputtering method. Its photoelectrochemical and photocatalytic properties were tested by oleic acid decomposition. Wei et al. (2007) codoped TiO_2 with boron and cerium. This photocatalyst was used to degrade dye acid red B. Increased photocatalytic activity of titania photocatalyst was also observed by codoping it with nitrogen and cerium (Shen et al. 2009). Then it was used for degradation of nitrobenzene under visible light illumination as a probe reaction to evaluate the photoactivity of the codoped photocatalyst.

The synergistic effect of metal and nonmetal dopant not only changes the microstructure or optical band gap, but also prevents the possibility of electron–hole pair recombination. A plate with Ce and F codoped Bi_2WO_6 was synthesized by hydrothermal reaction in a single step (Huang et al. 2014a). Its improved efficiency was evaluated by photodegradation of rhodamine B dye. The increase in the efficiency of codoped Bi_2WO_6 compared with pure Bi_2WO_6 was due to the efficient separation and migration of charge carriers generated on irradiation.

Wang et al. (2013) prepared Eu–B codoped $BiVO_4$ by the sol–gel method. Enhanced photodegradation of methyl orange was reported by codoped ternary oxide compared with $BiVO_4$ and B–$BiVO_4$. The synergistic effects of boron and europium in doped $BiVO_4$ led to more surface oxygen vacancies, high specific surface area, small crystallite size, narrower band gap, and intense light absorbance in the visible region, thus improving the visible light photocatalytic activity of Eu–B codoped $BiVO_4$.

7.5.3 NONMETAL AND NONMETAL

Synergistic effect helps in efficient inhibition of the recombination of photogenerated electrons and holes, increase in visible light absorption ability, surface hydroxyl and specific surface area, as well as the improvement of surface textural properties. Three nonmetals that is, carbon, nitrogen, and sulfur, were used to codope titania through the hydrothermal method. Thiourea was used as a source of C, N, and S and as-prepared sample was tested by degradation of toluene in the gas phase (Dong et al. 2008). TiO_2 photocatalyst was codoped with iodine and boron using the hydrolyzation–precipitation method by Ding et al. (2009).

Xu et al. (2011) used carbon black as the carbon source to synthesize crack-free, high surface roughness, and visible light active C–N codoped TiO_2 films by an organic free sol–gel method. It was also used as a template to increase the roughness of the surface. They found that both calcination temperature and carbon black concentration affect the concentration of carbon and nitrogen dopants in the TiO_2 films. Its photocatalytic activity was examined by taking stearic acid as the model pollutant compound. The maximum performance was observed at 10.0 wt.% carbon, which was just double that of the titania doped with nitrogen.

Zn/ZnO composite was doped with Cu and further modified with carbon through a simple replacement–hydrothermal method. Zn powder and $CuSO_4 \cdot 5H_2O$ were used to prepare the sample. The results showed an increase in crystal growth of ZnO by Cu doping and avoided the situation of phase transfer of metallic Zn to ZnO. This led to an increase in degradation of reactive brilliant blue KN-R dye solution on exposure to sunlight (Ma et al. 2012). XPS data showed deposition of carbon on the surface of composites, which was formed by dissolution of CO_2 in the solution. The enhanced efficiency was observed because of the inhibition of electron–hole pair recombination.

Sulfur and nitrogen codopants were used to dope α-Fe_2O_3, and its efficiency was evaluated by degradation of rhodamine B. Its activity was compared with the bulk material as well as the single nonmetal-doped hematite. The trend of photocatalytic activity of codoped semiconductor was studied by variation of some factors such as particle size, surface area, [110] plane in the sulfur doped material, formation of OH radical, and so on. More than 90% degradation was obtained after 4 hours under natural light. A comparison between the adsorption, Fenton, photo-Fenton, and photocatalytic degradation of rhodamine B was also made by Pradhan et al. (2013).

Diclofenac from water was eliminated by carbon- and nitrogen-doped TiO_2 by Buda and Czech (2013). The synthesized photocatalyst showed reduction of the COD (chemical oxygen demand) value of the wastewater by at least 60%. The process of diclofenac photooxidation followed pseudo-first-order kinetics. In this process, best results were observed during the first 50 minutes of treatment, but after 50 minutes, mineralization of pollutant showed a decline in the rate.

Degradation of methylene blue by N–F codoped TiO_2 was much better under both UV and visible light (Yu et al. 2015). N–F–TiO_2 nanomaterial exhibited different properties than pure TiO_2, that is, smaller crystalline size, broader light absorption spectrum, and lower charge recombination. Jiang et al. (2013) compared the performance of undoped, single doped, codoped, and Sm, N, P-tridoped anatase–TiO_2 nano-photocatalyst (SNPTO) synthesized by some modification in the sol–solvothermal process. The highest degradation of 4-chlorophenol (4-CP) was observed by tridoped photocatalyst with the rate constant at 2.83×10^{-2} per minute, which was 3.98 times more than that with commercial P25 TiO_2, that is, $k_{app} = 7.11 \times 10^{-3}$ per minute (20 mg/L). Nearly 87% degradation of 4-CP in the presence of SNPTO (0.4 g/L) was observed in 2 hours. SNPTO exhibited good photochemical stability also and could be reused five times with less than 1.6% decrease in the efficiency of 4-CP removal.

Various codoped photocatalysts along with their applications are given in Table 7.3.

Some semiconductor (photocatalysts) absorbs in the border area of the UV range (slightly below 400 nm) and therefore these cannot be used efficiently as photocatalysts in the presence of sunlight. The band gap of such materials can be engineered either by metal doping, nonmetal doping, or codoping (metal–nonmetal, metal–metal,

TABLE 7.3

Some Codoped Photocatalysts and Their Applications

Photocatalyst	Codopants	Applications	References
TiO_2	Gd, La	Photodegradation of methyl orange	Chen et al. (2014)
TiO_2	N, F	Decomposition of acetic acid	Dozzi et al. (2013)
TiO_2	Sm, N	Degradation of salicylic acid	Ma et al. (2010)
TiO_2	Nb, N	O_2 sensors	Folli et al. (2015a)
TiO_2	W, N	-	Folli et al. (2015b)
$SrTiO_3$	Rh, Sb	Water splitting	Modak and Ghosh. (2015)
TiO_2	C, N, S	Removal of NO	Wang et al. (2009)
TiO_2	C, N, S	Photocatalytic oxidation of formaldehyde	Zhou and Yu (2008)
Y_2WO_6	(Ln^{3+} = Sm, Eu, Dy) codoped with Gd^{3+}	Luminescence properties	Kaczmarek et al. (2014)

and nonmetal–nonmetal), so that these can be used as effective photocatalysts in solar insolation. The CB will be lowered down by metal doping and the level of VB will be uplifted by nonmetal doping, thus reducing the band gap and making it effective in the visible range.

REFERENCES

Abdollahi, Y., A. H. Abdullah, Z. Zainal, and N. A. Yusof. 2011. Synthesis and characterization of manganese doped ZnO nanoparticles. *Int. J. Basic Appl. Sci.* 11: 44–50.

Ahmad, M., E. Ahmed, Y. Zhang, N. R. Khalid, J. Xu, M. Ullah, and Z. Hong. 2013. Preparation of highly efficient Al-doped ZnO photocatalyst by combustion synthesis. *Curr. Appl. Phys.* 13 (4): 697–704.

Ali, A. M., A. Muhammad, A. Shafeeq, H. M. A. Asghar, S. N. Hussain, and H. Sattar. 2012. Doped metal oxide (ZnO) and photocatalysis: A review. *J. Pak. Inst. Chem. Eng.* 40 (1): 11–19.

An, L. and H. Onishi. 2015. Electron–hole recombination controlled by metal doping sites in NaTaO$_3$ photocatalysts. *ACS Catal.* 5: 3196–3206.

Anpo, M. 2004. Preparation, characterization, and reactivities of highly functional titanium oxide based photocatalysts able to operate under UV-visible light irradiation: Approaches in realizing high efficiency in the use of visible light. *Bull. Chem. Soc. Jpn.* 77 (8):1427–1442.

Apno, M., 2000. Applications of titanium oxide photocatalyst and unique second generation TiO$_2$ photocatalysts able to operate under visible light irradiation for the reduction of environmental toxins on a global scale. *Stud. Surf. Sci. Catal.* 130: 157–166.

Arin, J., P. Dumrongrojthanath, O. Yayapao, A. Phuruangrat, S. Thongtem, and T. Thongtem. 2014. Synthesis, characterization and optical activity of La-doped ZnWO$_4$ nanorods by hydrothermal method. *Superlattices Microst.* 67: 197–206.

Asahi, R., T. Morikawa, T. Ohwaki, K. Aoki, and Y. Taga. 2001. Visible-light photocatalysis in nitrogen-doped titanium oxides. *Science.* 293 (5528): 269–271.

Ashkarran, A. A. 2011. Antibacterial properties of silver-doped TiO_2 nanoparticles under solar simulated light. *J. Theor. Appl. Phys.* 4 (4): 1–8.

Bangkedphol, S., H. E. Keenan, C. M. Davidson, A. Sakultantimetha, W, Sirisaksoontorn, and A. Songsasen. 2010. Enhancement of tributyltin degradation under natural light by N-doped TiO_2 photocatalyst. *J. Hazard Mater.* 184 (1–3): 533–537.

Baruah, S., S. K. Pal, and J. Dutta. 2012. Nanostructured zinc oxide for water treatment. *Nanosci. Nanotechnol. Asia.* 2: 90–102.

Batzill, M., E. H. Morales, and U. Diebold. 2006. Influence of nitrogen doping on the defect formation and surface properties of TiO_2 rutile and anatase. *Phys. Rev. Lett.* 96. doi:org/10.1103/PhysRevLett.96.026103.

Behnajady, M. A., H. Taba, N. Modirshahla, and M. Shokri. 2013. Photocatalytic activity of Cu doped TiO_2 nanoparticles and comparison of two main doping procedures. *Micro Nano Lett.* 8 (7): 345–348.

Bettinelli, M., V. Dallacasa, D. Falcomer, P. Fornasiero, V. Gombac, T. Montini, L. Romanòd, and A. Speghini. 2007. Photocatalytic activity of TiO_2 doped with boron and vanadium. *J. Hazard Mater.* 146: 529–534.

Beydoun, D., H. Tse, R. Amal, G. Low, and S. McEvoy. 2002. Effect of copper (II) on the photocatalytic degradation of sucrose. *J. Mol. Catal. A: Chem.* 177: 265–272.

Buda, W. and B. Czech. 2013. Preparation and characterization of C,N-codoped TiO_2 photocatalyst for the degradation of diclofenac from wastewater. *Water Sci Technol.* 68 (6): 1322–1328. doi:10.2166/wst.2013.369.

Burda, C., Y. Lou, X. Chen, A. C. S. Samia, J. Stout, and J. L. Gole. 2003. Enhanced nitrogen doping in TiO_2 nanoparticles. *Nano Lett.* 3: 1043–1051.

Carvalho, H. W. P., A. P. L. Batista, P. Hammer, and T. C. Ramalho. 2010. Photocatalytic degradation of methylene blue by TiO_2–Cu thin films: Theoretical and experimental study. *J. Hazard Mater.* 184 (1–3): 273–280.

Chang, C.-J., T.-L. Yang, and Y.-C. Weng. 2014. Synthesis and characterization of Cr-doped ZnO nanorod-array photocatalysts with improved activity. *J. Solid State Chem.* 214: 101–107.

Chang, G. S., E. Z. Kurmaev, D. W. Boukhvalov, L. D. Finkelstein, A. Moewes, H. Bieber, S. Colis, and A. Dinia. 2009. Co and Al co-doping for ferromagnetism in ZnO:Codiluted magnetic semiconductors. *J. Phys.: Condens. Matter.* 21 (5): 056002. doi: org/10.1088/0953-8984/21/5/056002.

Channei, D., B. Inceesungvorn, N. Wetchakun, S. Phanichphant, A. Nakaruk, P. Koshy, C. C. Sorrell. 2013. Photocatalytic activity under visible light of Fe-doped CeO_2 nanoparticles synthesized by flame spray pyrolysis. *Ceram. Inter.* 39: 3129–3134.

Chen, R. F., C. X. Zhang, J. Deng, and G. Q. Song. 2009. Preparation and photocatalytic activity of Cu^{2+} doped TiO_2/SiO_2. *Int. J. Min. Met. Mater.* 16: 220–225.

Chen, T., T. Y. Liu, Q. R. Zheng, F. F. Li, D. S. Tian, X. Y. Zhang, and X. Y. Zhang. 2007. First-principles study on the La^{3+} doping $PbWO_4$ crystal for different doping concentrations. *Phys. Lett. A.* 363 (5–6): 477–481.

Chen, X., H. Cai, Q. Tang, Y. Yang, and B. He. 2014. Solar photocatalysts from Gd–La codoped TiO_2 nanoparticles. *J. Mater. Sci.* 49 (9): 3371–3378.

Chen, X. and S. S. Mao. 2007. Titanium dioxide nanomaterials: Synthesis, properties, modification and applications. *Chem. Revs.* 107: 2891–2959.

Cheng, Z. W., L. Feng, J. M. Chen, J. M. Yu, and Y. F. Jiang. 2013. Photocatalytic conversion of gaseous ethylbenzene on lanthanum-doped titanium dioxide nanotubes. *J. Hazard Mater.* 254–255: 354–363.

Cho, S., J.-W. Jang, J. S. Lee, and K.-H. Lee. 2010. Carbon-doped ZnO nanostructures synthesized using vitamin C for visible light photocatalysis. *Cryst. Eng. Comm.* 12: 3929–3935.

Choi, W. Y., A. Termin, and M. R. Hoffmann. 1994. The role of metal ion dopants in quantum-sized TiO_2: Correlation between photoreactivity and charge carrier recombination dynamics. *J. Phys. Chem.* 98: 13669–13679.

Dai, Q., H. Song, X. Bai, G. Pan, S. Lu, T. Wang, X. Ren, and H. Zhao. 2007. Photoluminescence properties of $ZnWO_4:Eu^{3+}$ nanocrystals prepared by a hydrothermal method. *J. Phys. Chem. C.* 111 (21): 7586–7592.

Deng, L., S. Wang, D. Liu, B. Zhu, W. Huang, S. Wu, and S. Zhang. 2009. Synthesis, characterization of Fe-doped TiO_2 nanotubes with high photocatalytic activity. *Catal. Lett.* 129: 513–518.

Diao, X. and D. Zhou. 2014. Preparation and photocatalytic performance study of La-WO_3 Co-doping TiO_2. *Adv. Mater. Res.* 936: 809–813.

Diebold, U. 2003. The surface science of titanium dioxide. *Surf. Sci. Rep.* 48: 53–229.

Ding, B., C. Han, L. Zheng, J. Zheng, R. Wang, and Z. Tang. 2015. Tuning oxygen vacancy photoluminescence in monoclinic Y_2WO_6 by selectively occupying yttrium sites using lanthanum. *Scientific Reports* 5: 9443. doi:10.1038/srep09443.

Ding, J., Y. Yuan, J. Xu, J. Deng, and J. Guo. 2009. TiO_2 nanopowder co-doped with iodine and boron to enhance visible-light photocatalytic activity. *J. Biomed. Nanotechnol.* 5: 521–527.

Dong, F., W. Zhao, and Z. Wu. 2008. Characterization and photocatalytic activities of C, N and S co-doped TiO_2 with 1D nanostructure prepared by the nano-confinement effect. *Nanotechnology.* 19: 365607–365616.

Dozzi, M. V., C. D'Andrea, B. Ohtani, G. Valentini, and E. Selli. 2013. Fluorine-doped TiO_2 materials: Photocatalytic activity vs time-resolved photoluminescence. *J. Phys. Chem. C.* 117 (48): 25586–25595.

Fan, X., X. Chen, S. Zhu, Z. Li, T. Yu, J. Ye, and Z. Zou. 2008. The structural, physical and photocatalytic properties of the mesoporous Cr-doped TiO_2. *J. Mol. Catal. A: Chem.* 284: 155–160.

Feng, Y., W-. X. Ji, B-.J. Huang, X-. L. Chen, F. Li, P. Li, C-.W. Zhang, and P-.J. Wang. 2015. The magnetic and optical properties of 3d transition metal doped SnO_2 nanosheets. *RSC Adv.* 5: 24306–24312.

Folli, A., J.Z. Bloh, A. Lecaplain, R. Walker, and D. E, Macphee. 2015a. Properties and photochemistry of valence-induced-Ti^{3+} enriched (Nb,N)-codopedanatase TiO_2 semiconductors. *Phys. Chem. Chem. Phys.* 17: 4849–4853.

Folli, A., J. Z. Bloh, and D. E. Macphee. 2015b. Band structure and charge carrier dynamics in (W,N)-codoped TiO_2 resolved by electrochemical impedance spectroscopy combined with UV–vis and EPR spectroscopies. *J. Electroanal. Chem.* doi:10.1016/j.jelechem.2015.10.033.

Fu, Y., C. Chang, P. Chen, X. Chu and, L. Zhu. 2013. Enhanced photocatalytic performance of boron doped Bi_2WO_6 nanosheets under simulated solar light irradiation. *J. Hazard Mater.* 254–255: 185–192.

Guo, W., Y. Guo, H. Dong, and X. Zhou. 2015b. Tailoring the electronic structure of β-Ga_2O_3 by non-metal doping from hybrid density functional theory calculations. *Phys. Chem. Chem. Phys.* 17 (8): 5817–5825.

Guo, Y., X. Qiu, H. Dong, and X. Zhou. 2015a. Trends in non-metal doping of the surface: A hybrid density functional study. *Phys. Chem. Chem. Phys.* 17 (33): 21611–21621.

Haibo, O., H. J. Feng, L. Cuivan, C. Liyun, and F. Jie. 2013. Synthesis of carbon doped ZnO with a porous structure and its solar-light photocatalytic properties. *Mater. Lett.* 111: 217–220.

Hameed, A., M. A. Gondal, and Z. H. Yamani. 2004. Effect of transition metal doping on photocatalytic activity of WO_3 for water splitting under laser illumination: Role of 3d-orbitals. *Catal. Comm.* 5: 715–719.

Harikishore, M., M. Sandhyarani, K. Venkateswarlu, T. A. Nellaippan, and N. Rameshbabu. 2014. Effect of Ag doping on antibacterial and photocatalytic activity of nanocrystalline TiO_2. *Proc. Mater. Sci.* 6: 557–566.

Haspulat, B., A. Gulce, and H. Gulce. 2013. Efficient photocatalytic decolorization of some textile dyes using Fe ions doped polyaniline film on ITO coated glass substrate. *J. Hazard. Mater.* 260: 518–526.

Hoffmann, M. R., S. T. Matin, W. Choi, and D. W. Bahnemann. 1995. Environmental applications of semiconductor photocatalysis. *Chem. Revs.* 95: 69–96.

Hsieh, C. T., W. S. Fan, W. Y. Chen, and J. Y. Lin. 2009. Adsorption and visible light derived photocatalytic kinetics of organic dye on Co-doped titania nanotubes prepared by hydrothermal synthesis. *Sep. Purif. Technol.* 67: 312–318.

Hsu, M.-H. and C.-J. Chang. 2014. Ag-doped ZnO nanorods coated metal wire meshes as hierarchical photocatalysts with high visible-light driven photoactivity and photostability. *J. Hazard. Mater.* 278: 444–453.

Hu, S., F. Li, and Z. Fan. 2011b. The influence of preparation method, nitrogen source, and post-treatment on the photocatalytic activity and stability of N-doped TiO_2 nanopowder. *J. Hazard. Mater.* 196: 248–254.

Hu, Y., H. Liu, Q. Rao, X. Kong, W. Sun, and X. Guo. 2011a. Effects of N precursor on the agglomeration and visible light photocatalytic activity of N-doped TiO_2 nanocrystalline powder. *J. Nanosci. Nanotechnol.* 11: 3434–3444.

Huang, H., K. Liu, K. Chen, Y. Zhang, Y. Zhang, and S. Wang. 2014a. Ce and F co-modification on the crystal structure and enhanced photocatalytic activity of Bi_2WO_6 photocatalyst under visible light irradiation. *J. Phys. Chem. C.* 118: 14379–14387.

Huang, M. N., Y. Y. Ma, F. Xiao, and Q. Y. Zhang. 2014b. Bi^{3+} sensitized $Y_2WO_6:Ln^{3+}$ (Ln = Dy, Eu, and Sm) phosphors for solar spectral conversion. Spectrochim. *Acta A.* 120: 55–59.

Ihara, H., M. Miyoshi, Y. Triyama, O. Marsumato, and S. Sugihara. 2003. Visible-light-active titanium oxide photocatalyst realized by an oxygen-deficient structure and by nitrogen doping. *Appl. Catal. B: Environ.* 42: 403–409.

Irie, H., Y. Watanabe, and K. Hashimoto. 2003. Nitrogen concentration dependence on photocatalytic activity of $TiO_{2-x}N_x$ powders. *J. Phys. Chem. B.* 107: 5483–5486.

Janes, R., L. J. Knightley, and C. J. Harding. 2004. Structural and spectroscopic studies of iron (III) doped titania powders prepared by sol-gel synthesis and hydrothermal processing. *Dyes Pigments.* 62: 199–212.

Jiang, H., Q. Wang, S. Zang, J. Li, and Q. Wang. 2013. Enhanced photoactivity of Sm, N, P-tridoped anatase-TiO_2 nano-photocatalyst for 4-chlorophenol degradation under sunlight irradiation. *J. Hazard. Mater.* 261: 44–54.

Jose, M., M. Kumari, R. Karunakaran, and S. Shukla. 2015. Methylene blue adsorption from aqueous solutions using undoped and silver-doped nanotubes of anatase-titania synthesized via modified hydrothermal method. *J. Sol-Gel Sci. Technol.* 75 (3): 541–550.

Julkapli, N. M., S. Bagheri, and S. B. A. Hamid. 2014. Recent advances in heterogeneous photocatalytic decolorization of synthetic dyes. *Sci. World J.* doi:org/10.1155/2014/692307.

Kaczmarek, A. M., K. V. Hecke, and R. V. Deun. 2014. Enhanced luminescence in Ln^{3+}-doped Y_2WO_6 (Sm, Eu, Dy) 3D microstructures through Gd^{3+} codoping. *Inorg. Chem.* 53 (18): 9498–9508.

Kadi, M. W. and R. M. Mohamed. 2013. Enhanced photocatalytic activity of ZrO_2-SiO_2 nanoparticles by platinum doping, *Int. J. Photoenergy.* doi:org/10.1155/2013/812097.

Kandula, S., and P. Jeevanandam. 2015. Sun-light-driven photocatalytic activity by ZnO/Ag heteronanostructures synthesized via a facile thermal decomposition approach. *RSC Adv.* 5: 76150–76159.

Karvinen, S. P. Hirva, and T. A. Pakkanen. 2003. Ab initio quantum chemical studies of cluster models for doped anatase and rutile TiO_2. *J. Mol. Struc. Theochem.* 626: 271–277.

Khairy, M. and W. Zakaria. 2014. Effect of metal-doping of TiO_2 nanoparticles on their photocatalytic activities toward removal of organic dyes. *Egypt. J. Petroleum.* 23: 419–426.

Khataee, A. R. and M. B. Kasiri. 2010. Photocatalytic degradation of organic dyes in the presence of nanostructured titanium dioxide: Influence of the chemical structure of dyes. *J. Mol. Catal. A: Chem.* 328 (1–2): 8–26.

Kirupavasam, E. K. and G. A. G. Raj. 2012. Photocatalytic degradation of amido black-10B catalyzed by carbon doped TiO$_2$ photocatalyst. *Int. J. Green Chem. Bioproc.* 2 (3): 20–25.

Klosek, S. and D. Raftery. 2001. Visible light driven V-doped TiO$_2$ photocatalyst and its photooxidation of ethanol. *J. Phys. Chem. B.* 105 (14): 2815–2819.

Kumaresan, L., A. Prabhu, M. Palanichamy, E. Arumugam, and V. Murugesan. 2011. Synthesis and characterization of Zr^{4+}, La^{3+} and Ce^{3+} doped mesoporous TiO$_2$: Evaluation of their photocatalytic activity. *J. Hazard Mater.* 186 (2–3): 1183–1192.

Kuo, C. Y., C. H. Wu, J. T. Wu, and Y. R. Chen. 2014. Synthesis and characterization of a phosphorus-doped TiO$_2$ immobilized bed for the photodegradation of bisphenol A under UV and sunlight irradiation. *React. Kinet. Mech. Catal.* 114 (2): 753–766.

Kuriakose, S., B. Satpati, and S. Mohapatra. 2014. Enhanced photocatalytic activity of Co doped ZnO nanodisks and nanorods prepared by a facile wet chemical method. *Phys. Chem. Chem. Phys.* 16: 12741–12749.

Labiadh, H., T. B. Chaabane, L. Balan, N. Beckheik, S. Corbel, G. Medjahdi, and R. Schneider. 2014. Preparation of Cu-doped ZnS QDs/TiO$_2$ nanocomposites with high photocatalytic activity. *Appl. Catal. B: Environ.* 144: 29–35.

Lavand, A. B. and Y. S. Malghe. 2015. Synthesis, characterization, and visible light photocatalytic activity of nanosized carbon doped zinc oxide. *Inter. J. Photochem.* doi:org/10.1155/2015/790153.

Lavanya, N., S. Radhakrishnan, N. Sudhan, C. Sekar, S. G. Leonardi, C. Cannilla, and G. Neri. 2014. Fabrication of folic acid sensor based on the Cu doped SnO$_2$ nanoparticles modified glassy carbon electrode. *Nanotechnol.* 25: 295501. doi:10.1088/0957-4484/25/29/295501.

Lee, H. U., S. C. Lee, S. Choi, B. Son, H. Kim, S. M. Lee, H. J. Kim, and J. Lee. 2013. Influence of visible-light irradiation on physicochemical and photocatalytic properties of nitrogen-doped three-dimensional (3D) titanium dioxide. *J. Hazard. Mater.* 258–259: 10–18.

Li, F., L. X. Guan, M. L. Dai, J. J. Feng, and M. M. Yao. 2013. Effects of V and Zn codoping on the microstructures and photocatalytic activities of nanocrystalline TiO$_2$ films. *Ceram. Inter.* 39 (7): 7395–7400.

Li, H., H. K. Yang, B. K. Moon, B. C. Choi, J. H. Jeong, K. Jang, H. S. Lee, and S. S. Yi. 2011. Crystal structure, electronic structure, and optical and photoluminescence properties of Eu(III) ion-doped Lu$_6$Mo(W)O$_{12}$. *Inorg. Chem.* 50 (24): 12522–12530.

Li, N., H. Teng, Li. Zhang, J. Zhou, and M. Liu. 2015. Synthesis of Mo-doped WO$_3$ nanosheets with enhanced visible-light-driven photocatalytic properties. *RSC Adv.* 5: 95394–95400.

Li, W., L. Jie, X. Wang, and Q. Chen. 2012a. Preparation and water-splitting photocatalytic behavior of S-doped WO$_3$. *Appl. Surf. Sci.* 263: 157–162.

Li, Y., D. S. Hwang, N. H. Lee, and S. J. Kim. 2005. Synthesis and characterization of carbon-doped titania as an artificial solar light sensitive photocatalyst. *Chem. Phys. Lett.* 404 (1–3): 25–29.

Li, Y., X. Zhou, W. Chen, L. Li, M. Zen, S. Qin, and S. Sun. 2012b. Photodecolorization of rhodamine B on tungsten-doped TiO$_2$/activated carbon under visible-light irradiation. *J. Hazard. Mater.* 227–228: 25–33.

Liewhiran, C., N. Tamaekong, A. Wisitsoraat, and S. Phanichphant. 2009. H$_2$ sensing response of flame-spray-made Ru/SnO$_2$ thick films fabricated from spin-coated nanoparticles. *Sensors.* 9 (11): 8996–9010.

Lin, J., J. C. Yu, D. Lo, and S. K. Lam. 1999. Photocatalytic activity of rutile Ti$_{1-x}$Sn$_x$O$_2$ solid solutions. *J. Catal.* 183: 368–372.

Lin, Y. G., Y. K. Hsu, Y. C. Chen, L. C. Chen, S. Y. Chen, and K. H. Chen. 2012. Visible-light-driven photocatalytic carbon-doped porous ZnO nanoarchitectures for solar water-splitting. *Nanoscale.* 4: 6515–6519.

Liu, G., P. Niu, C. Sun, S. C. Smith, Z. Chen, G. Q. (Max) Lu, and H. M. Cheng. 2010. Unique electronic structure induced high photoreactivity of sulfur-doped graphitic C_3N_4. *J. Am. Chem. Soc.* 132 (33): 11642–11648.

Liu, J. 2012. Band gap narrowing of TiO_2 by compensated codoping for enhanced photocatalytic activity. *J. Natural Gas Chem.* 21: 302–307.

Liu, L., X. R. Wang, X. Yang, W. Fan, X. Wang, N. Wang, X. Li, and F. Xue. 2014. Preparation, characterization, and biotoxicity of nanosized doped ZnO photocatalyst. *Int. J. Photoenergy.* doi:org/10.1155/2014/475825.

Liu, R., P. Wang, X. Wang, H. Yu, and J. Yu. 2012a. UV- and visible-light photocatalytic activity of simultaneously deposited and doped Ag/Ag(I)-TiO_2 photocatalyst. *J. Phys. Chem. C.* 116 (33): 17721–17728.

Liu, Y. G. Chen, C. Zhou, Y. Hu, D. Fu, J. Liu, and Q. Wang, 2011. Higher visible photocatalytic activities of nitrogen doped In_2TiO_5 sensitized by carbon nitride. *J. Hazard Mater.* 190 (1–3): 75–80.

Lui, Y., X. Chen, J. Li, and C. Burda. 2005. Photocatalytic degradation of azo dyes by nitrogen-doped TiO_2 nanocatalysts. *Chemosphere.* 61: 11–18.

Ma, H., L. Yue, C. Yu, X. Dong, X. Zhang, M. Xue, X. Zhang, and Y. Fu. 2012. Synthesis, characterization and photocatalytic activity of Cu-doped Zn/ZnO photocatalyst with carbon modification. *J. Mater. Chem.* 22: 23780–23788.

Ma, Y., J. Zhang, B. Tian, F. Chen, and L. Wang. 2010. Synthesis and characterization of thermally stable Sm,N co-doped TiO_2 with highly visible light activity. *J. Hazard. Mater.* 182 (1–3): 386–393.

Maeda, M. and T. Yamada. 2007. Photocatalytic activity of metal-doped titanium oxide films prepared by sol-gel process. *J. Phys.: Conf. Ser.* 61: 755–759.

Maheshwary, B. P. Singh, J. Singh, and R. A. Singh. 2014 Luminescence properties of Eu^{3+}-activated $SrWO_4$ nanophosphors-concentration and annealing effect. *RSC Adv.* 4: 32605–32621.

Marschall, R. and L. Wang. 2014. Non-metal doping of transition metal oxides for visible-light photocatalysis. *Catal. Today.* 225: 111–135.

Modak, B. and S. K. Ghosh. 2015. Origin of enhanced visible light driven water splitting by (Rh, Sb)-$SrTiO_3$. *Phys. Chem. Chem. Phys.* 17 (23): 15274–15283.

Mohamed, M. A., W. N. W. Salleh, J. Jaafar, and A. F. Ismail. 2015. Structural characterization of N-doped anatase–rutile mixed phase TiO_2 nanorods assembled microspheres synthesized by simple sol–gel method. *J. Sol-Gel Sci. Technol.* 74 (2): 513–520.

Mohan, R., K. Krishnamoorthy, and S. J. Kim. 2012. Enhanced photocatalytic activity of Cu doped ZnO nanorods. *Solid State Commun.* 152 (5): 375–380.

Munusamy, S., R. S. L. Aparna, and R. G. S. V. Prasad. 2013. Photocatalytic effect of TiO_2 and the effect of dopants on degradation of brilliant green. *Sustain. Chem. Proc.* 1 (4): 1–8.

Nah, Y. C., I. Paramasivam, and P. Schmuki. 2010. Doped TiO_2 and TiO_2 nanotubes: Synthesis and applications. *Chem. Phys. Chem.* 11 (13): 2698–2713.

Ndong, L. B. B., M. P. Ibondou, X. Gu, S. Lu, Z. Qiu, Q. Sui, and S. M. Mbadinga. 2014. Enhanced photocatalytic activity of TiO_2 nanosheets by doping with Cu for chlorinated solvent pollutants degradation. *Ind. Eng. Chem. Res.* 53 (4): 1368–1376.

Neamtu, J. and M. Volmer. 2014. The influence of doping with transition metal ions on the structure and magnetic properties of zinc oxide thin films. *Sci. World J.* doi:org/10.1155/2014/265969.

Nolan, N. T., D. W. Synnott, M. K. Seery, S. J. Hinder, A. V. Wassenhoven, and S. C. Pillai. 2012. Effect of N-doping on the photocatalytic activity of sol–gel TiO_2. *J. Hazard. Mater.* 211–212: 88–94.

Obata, K., H. Irie, and K. Hashimoto. 2007. Enhanced photocatalytic activities of Ta, N co-doped TiO_2 thin films under visible light. *Chem. Phys.* 339: 124–132.

Pal, B. T. Hata, K. Goto, and G. Nogami. 2001. Photocatalytic degradation of o-cresol sensitized by iron-titania binary photocatalysts. *J. Mol. Cat. A: Chem.* 169: 147–155.

Patil, G. E., D. D. Kajale, S. D. Shinde, V. G. Wagh, V. B. Gaikwad, and G. H. Jain. 2013. Synthesis of Cu-doped SnO_2 thin films by spray pyrolysis for gas sensor application. *Adv. Sensing Technol. Smart Sensors, Measur. and Instr.* 1: 299–311.

Paul, A. K., M. Prabu, G. Madras, and S. Natarajan. 2010. Effect of metal ion doping on the photocatalytic activity of aluminophosphates. *J. Chem. Sci.* 122: 771–785.

Pham, T. T., C. Nguyen-Huy, H. J. Lee, L. Phan, T. H. Son, and C. K. Kim. 2015. Cu-doped TiO_2/reduced graphene oxide thin-film photocatalysts: Effect of Cu content upon methylene blue removal in water. *Ceram. Inter.* doi:10.1016/j.ceramint.2015.05.068.

Phuruangrat, A., O. Yayapao, T. Thongtem, and S. Thongtem. 2014. Synthesis and characterization of europium-doped zinc oxide photocatalyst. *J. Nanomater.* doi:org/10.1155/2014/367529.

Pradhan, G. K., N. Sahu, and K. M. Parida. 2013. Fabrication of S, N co-doped α-Fe_2O_3 nanostructures: Effect of doping, OH radical formation, surface area, [110] plane and particle size on the photocatalytic activity. *RSC Adv.* 3: 7912–7920.

Rahimi, R., J. Shokrayian, and M. Rabbani. 2013. Photocatalytic removing of methylene blue by using of Cu doped ZnO, Ag doped ZnO and Cu, Ag codoped ZnO nanostructures. In: *17th International Electronic Conference on Synthetic Organic Chemistry.* doi:10.3390/ecos-17-b018.

Raj, K. J. A., A. V. Ramaswamy, and B. Viswanathan. 2009. Surface area, pore size, and particles size engineering of titania with seeding technique and phosphate modification. *J. Phys. Chem. C.* 113: 13750–13757.

Rajabi, H. R., O. Khani, M. Shamsipur, and V. Vatanpour. 2013. High-performance pure and Fe^{3+}-ion doped ZnS quantum dots as green nanophotocatalysts for the removal of malachite green under UV-light irradiation. *J. Harazd. Mater.* 250–251: 370–378.

Rajkumar, R. and N. Singh. 2015. To study the effect of the concentration of carbon on ultraviolet and visible light photo catalytic activity and characterization of carbon doped TiO_2. *J. Nanomed. Nanotechnol.* 6: 260. doi:10.4172/2157-7439.1000260.

Ran, L., D. Zhao, X. Gao, and L. Yin. 2015. Highly crystalline Ti-doped SnO_2 hollow structured photocatalyst with enhanced photocatalytic activity for degradation of organic dyes. *Cryst. Eng. Comm.* 17: 4225–4237.

Ray, S. S. and J. Podder. 2009. Studies on tin oxide (SnO_2) and Cu doped SnO_2 thin films deposited by spray pyrolysis technique for window materials in solar cells. *Proceedings of the International Conference on Mechanical Engineering (ICME 2009).* 26–28 December. Dhaka, Bangladesh.

Rehman, S., R. Ullah, A. M. Butt, and N. D. Gohar. 2009. Strategies of making TiO_2 and ZnO visible light active. *J. Hazard. Mater.* 170 (2–3): 560–569.

Sakthivel, S. and H. Kisch. 2003. Daylight photocatalysts by carbon-modified titanium dioxide. *Angew. Chem. Int. Ed.* 42: 4908–4911.

Saleh, R. and N. F. Djaja. 2014b. Transition metal-doped ZnO nanoparticles: Synthesis, characterization and photocatalytic activity under UV light. *Spectrochim. Acta A.* 130: 581–590.

Saleh, R. and N. F. Djaja. 2014a. UV light photocatalytic degradation of organic dyes with Fe-doped ZnO nanoparticles. *Superlattices Microst.* 74: 217–233.

Sang, Y., H. Liu, and A. Umar. 2015. Photocatalysis from UV/Vis to near-infrared light: Towards full solar-light spectrum activity. *Chem. Cat. Chem.* 7: 559–573.

Satheesh, R., K. Vignesh, A. Suganthi, and M. Rajarajan. 2014. Visible light responsive photocatalytic applications of transition metal (M = Cu, Ni and Co) doped α-Fe_2O_3 nanoparticles. *J. Environ. Chem. Eng.* 2: 1956–1968.

Seddigi, Z. S., A. Bumajdad, S. P. Ansari, S. A. Ahmed, E. Y. Danish. N. H. Yarkandi, and S. Ahmed. 2014. Preparation and characterization of Pd doped ceria–ZnO nanocomposite catalyst for methyl tert-butyl ether (MTBE) photodegradation. *J. Hazard. Mater.* 264: 71–78.

Sharma, K. G. and N. R. Singh. 2013. Synthesis and luminescence properties of CaMO$_4$:Dy^{3+} (M = W, Mo)nanoparticles prepared *via* an ethylene glycol route. *New J. Chem.* 37: 2784–2791.

Shen, X. Z., Z. C. Liu, S. M. Xie, and J. Guo. 2009. Degradation of nitrobenzene using titania photocatalyst co-doped with nitrogen and cerium under visible light illumination. *J. Hazard. Mater.* 162: 1193–1198.

Singh, B. P., A. K. Parchur, R. S. Ningthouiam, A. A. Ansari, P. Singh, and S. B. Rai. 2014. Enhanced photoluminescence in CaMoO$_4$:Eu^{3+} by Gd^{3+} co-doping. *Dalton Trans.* 43: 4779–4789.

Singkammo, S., A. Wisitsoraat, C. Sriprachuabwong, A. Tuantranont, S. Phanichphant, and C. Liewhiran. 2015. Electrolytically exfoliated graphene-loaded flame-made Ni-doped SnO$_2$ composite film for acetone sensing. *ACS Appl. Mater. Interface.* 7 (5). doi:10.1021/acsami.5b00161.

Siriwong, C., N. Wetchakun, B. Inceesungvorn, D. Channei, T. Samerjai, and S. Phanichphant. 2012. Doped-metal oxide nanoparticles for use as photocatalysts. *Prog. Crystal Growth Charact Mater.* 58: 145–163.

Sivalingam, G., K. Nagaveni, M. S. Hegde, and G. Madras. 2003. Photocatalytic degradation of various dyes by combustion synthesized nanoanatase TiO$_2$. *Appl. Catal. B: Environ.* 45 (1): 23–38.

Solanki, M. S., M. Trivedi, R. Ameta, and S. Benjamin. 2015. Preparation and use of chlorophyll sensitized carbon doped tin (IV) oxide nanoparticles for photocatalytic degradation of azure A. *Int. J. Chem. Sci. Res.* 5: 1–11.

Stankic, S., M. Sterrer, P. Hofmann, J. Bernardi, O. Diwald, and E. Knozinger. 2005. Novel optical surface properties of Ca^{2+}-doped MgO nanocrystals. *Nano Lett.* 5 (10): 1889–1893.

Stoyanova, A. M., H. Y. Hitkova, N. K. Ivanova, A. D. Bachvarova-Nedelcheva, R. S. Iordanova, and M. P. Sredkova. 2013. Photocatalytic and antibacterial activity of Fe-doped TiO$_2$ nanoparticles prepared by non-hydrolytic sol-gel method. *Bulg. Chem. Commun.* 45 (4): 497–504.

Subramanian, M., Y. Akaike, Y. Hayashi, M. Tanemura, H. Ebisu, and D. L. S. Ping. 2012. Effect of defects in ferromagnetic C doped ZnO thin films, *Physica Status Solidi (B).* 249: 1254–1257.

Sun, Y., R. Rajpura, and D. Raftery. 2009. Photoelectrochemical and structural characterization of carbon-doped In$_2$O$_3$ and carbon-doped WO$_3$ films prepared via spray pyrolysis. *Proc. SPIE 7408, Solar Hydrogen Nanotechnol. IV.* doi:10.1117/12.825376.

Suriye, K., P. Praserthdam, and B. Jongsomjit. 2005. Impact of Ti^{+3} present in titania on characterization and catalytic properties of the Co/TiO$_2$ catalyst. *Indust. Eng. Chem. Res.* 44: 6599–6604.

Suwarnkar, M. B., R. S. Dhabbe, A. N. Kadam, and K. M. Garadkar. 2014. Enhanced photocatalytic activity of Ag doped TiO$_2$nanoparticles synthesized by a microwave assisted method. *Ceram. Inter.* 40 (4): 5489–5496.

Takeuchi, M., Y. Shimizu, H. Yamagawa, T. Nakamuro, and M. Anpo. 2011. Preparation of the visible light responsive N^{3-} doped WO$_3$ photocatalyst by a thermal decomposition of ammonium paratungstate. *Appl. Catal. B: Environ.* 110: 1–5.

Tang, X. H., C. H. Wei, J. R. Liang, and B. G. Wang. 2006. Preparation and photocatalytic activity of boron doped CeO$_2$/TiO$_2$ mixed oxides. *Huan Jing Ke Xue.* 27: 1329–1333.

Thirupathi, B. and P. G. Smirniotis. 2011. Co-doping a metal (Cr, Fe, Co, Ni, Cu, Zn, Ce, and Zr) on Mn/TiO$_2$ catalyst and its effect on the selective reduction of NO with NH$_3$ at low-temperatures. *Appl. Catal. B: Environ.* 110: 195–206.

Triantis, T. M., T. Fotiou, T. Kaliudis, A. G. Kontos, P. Falaras, D. D. Dionysiou, M. Pelaez, and A. Hiskia. 2012. Photocatalytic degradation and mineralization of microcystin-LR under UV-A, solar and visible light using nanostructured nitrogen doped TiO_2. *J. Hazard. Mater.* 211–212: 196–202.

Tsuruoka, A., T. Isobe, S. Matsushita, M. Wakamura, and A. Nakajima. 2015. Comparison of photocatalytic activity and surface friction force variation on Ti-doped hydroxyapatite and anatase under UV illumination. *J. Photochem. Photobiol. A: Chem.* 311: 160–165.

Ullah, R. and J. Datta. 2007. Synthesis and optical properties of transition metal doped ZnO nanoparticles. In: *International Conference on Emerging Technologies*, November 2007. Islamabad, Pakistan. 306–311. doi:10.1109/ICET.2007.4516363.

Vignesh, K., R. Hariharan, M. Rajarajan, and A. Suganthi. 2013. Photocatalytic performance of Ag doped SnO_2 nanoparticles modified with curcumin. *Solid State Sci.* 21: 91–99.

Wang, D. H., L. Jia, X. L. Wu, L. Q. Lu, and A. W. Xu. 2012. One-step hydrothermal synthesis of N-doped TiO_2/C nanocomposites with high visible light photocatalytic activity. *Nanoscale.* 4 (2): 576–584.

Wang, D., J. Lu, Z. Zhang, Y. Hu, and Z. Shen. 2011. Upconversion luminescence of Er^{3+}/Yb^{3+} co-doped Sb_2O_3-WO_3-Li_2O antimonate glasses. *New J. Glass Ceram.* 1: 34–39.

Wang, M., Y. Che, C. Niu, M. Dang, and D. Dong. 2013. Effective visible light-active boron and europium co-doped $BiVO_4$ synthesized by sol–gel method for photodegradion of methyl orange. *J. Hazard. Mater.* 262: 447–455.

Wang, Y., W. Duan, B. Liu, X. Chen, F. Yang, and J. Guo. 2014. The effects of doping copper and mesoporous structure on photocatalytic properties of TiO_2. *J. Nanomater.* doi:org/10.1155/2014/178152.

Wang, Y., Y. Huang, W. Ho, L. Zhang, Z. Zou, and S. Lee. 2009. Biomolecule-controlled hydrothermal synthesis of C-N-S-tridoped TiO_2 nanocrystalline photocatalysts for NO removal under simulated solar light irradiation. *J. Hazard. Mater.* 169 (1–3): 77–87.

Wang, W., T. Ding, G. Chen, L. Zhang, Y. Yu, and Q. Yang. 2015. Synthesis of Cu_2SnSe_3–Au heteronanostructures with optoelectronic and photocatalytic properties. *Nanoscale.* 7: 15106–15110.

Wei, C. H., X. H. Tang, J. R. Liang, and S. Y. Tan. 2007. Preparation, characterization and photocatalytic activities of boron- and cerium-codoped TiO_2. *J. Environ. Sci.*, 19: 90–96.

Wong, M. S., H. P. Chou, and T. S. Yang. 2006. Reactivity sputtered N-doped titanium oxide films as visible light photocatalyt. *Thin Solid Films.* 494: 244–249.

Xiao, J., T. Peng, R. Li, Z. Peng, and C. Yan. 2006. Preparation, phase transformation and photocatalytic activities of cerium-doped mesoporous titania nanoparticles. *J. Solid State Chem.* 179: 1161–1170.

Xiao, Q., Z. Si, J. Zhang, C. Xiao, and X. Tan. 2008. Photoinduced hydroxyl radical and photocatalytic activity of samarium-doped TiO_2 nanocrystalline. *J. Hazard. Mater.* 150 (1): 62–67.

Xin, B., P. Wang, D. Ding, J. Liu, Z. Ren, and H. Fu. 2008. Effect of surface species on Cu-TiO_2 photocatalytic activity. *Appl. Surf. Sci.* 254 (9): 2569–2574.

Xu, J. C., M. Lu, X. Y. Guo, and H. L. Li. 2005. Zinc ions surface doped titanium dioxide nanotubes and its photocatalytic activity for degradation of methyl orange in water. *J. Mol. Catal. A: Chem.* 226: 123–127.

Xu, J. J., Y. H. Ao, M. D. Chen, and D. G. Fu. 2009. Low temperature preparation of boron-doped titania by hydrothermal method and its photocatalytic activity. *J. Alloys. Compd.* 484: 73–79.

Xu, Q. C., D. V. Wellia, S. Yan, D. W. Liao, T. M. Lim, and T. T. Y. Tan. 2011. Enhanced photocatalytic activity of C–N-codoped TiO_2 films prepared via an organic-free approach. *J. Hazard. Mater.* 188 (1–3): 172–180.

Yaithongkum, J., K. Kooptarnond, L. Sikong, and D. Kantachote. 2011. Photocatalytic activity against Penicillium Expansum of Ag-doped $TiO_2/SnO_2/SiO_2$. *Adv. Mater. Res.* 214: 212–217.

Yan, H., X. Wang, M. Yao, and X. Yao. 2013. Band structure design of semiconductors for enhanced photocatalytic activity: The case of TiO_2. *Prog. Nat. Sci.: Mater. Int.* 23: 402–440.

Yao, W., C. Huang, N. Muradov, and A. T-Raissi. 2011. A novel $Pd–Cr_2O_3/CdS$ photocatalyst for solar hydrogen production using a regenerable sacrificial donor. *Inter. J. Hydrogen Energy.* 36: 4710–4715.

Yi, Z., G. Chen, Y. Yu, Y. Zhou, and F. He. 2016. Synthesis of carbon doped $WO_3 \cdot 0.33H_2O$ hierarchical photocatalyst with improved photocatalytic activity. *Appl. Surf. Sci.* 362: 182–190.

Yin, C., S. Zhu, Z. Chen, W. Zhang, J. Gu, and D. Zhang. 2013. One step fabrication of C-doped $BiVO_4$ with hierarchical structures for a high-performance photocatalyst under visible light irradiation. *J. Mater. Chem. A.* 1: 8367–8378.

Yu, C., Q. Fan, Y. Xie, J. Chen, Q. Shu, and J. C. Yu. 2012. Sonochemical fabrication of novel square-shaped F doped TiO_2 nanocrystals with enhanced performance in photocatalytic degradation of phenol. *J. Hazard. Mater.* 237–238: 38–45.

Yu, J. C., Jiaguo, Wingkei, Zitao, and Lizhi. 2002. Effects of F^- doping on the photocatalytic activity and microstructures of nanocrystalline TiO_2 powders. *Chem. Mater.* 14 (9): 3808–3816.

Yu, J. G., W. G. Wang, B. Cheng, and B. L. Su. 2009. Enhancement of photocatalytic activity of mesoporous TiO_2 powders by hydrothermal surface fluorination treatment. *J. Phys. Chem. C.* 113: 6743–6750.

Yu, J., Z. Liu, H. Zhang, T. Huang, J. Han, Y. Zhang, and D. Chong. 2015. Synergistic effect of N- and F-codoping on the structure and photocatalytic performance of TiO_2. *J. Environ. Sci.* 28C: 148–156.

Zaleska, A. 2008. Doped-TiO_2: A review. *Recent Patent Engg.* 2: 157–164.

Zhan, C., F. Chen, J. Yang, D. Dai, X. Cao, and M. Zhong. 2014. Visible light responsive sulfated rare earth doped TiO_2@fumed SiO_2 composites with mesoporosity: Enhanced photocatalytic activity for methyl orange degradation. *J. Hazard. Mater.* 267: 88–97.

Zhang, C., Y. Jia, Y. Jing, Y. Yao, J. Ma, and J. Sun. 2013. Effect of non-metal elements (B, C, N, F, P, S) mono-doping as anions on electronic structure of $SrTiO_3$. *Comput. Mater. Sci.* 79: 69–74.

Zhang, D. E., M. Y. Wang, J. J. Ma, G. Q. Han, S. A. Li, H. Zhao, B. Y. Zhao, and Z. W. Tong. 2014b. Enhanced photocatalytic ability from carbon-doped ZnO photocatalyst synthesized without an external carbon precursor. *Funct. Mater. Lett.* 7: 4.1450026. doi:10.1142/S179360471450026X.

Zhang, F., Z. Cheng, L. Kang, L. Cui, W. Liu, X. Xu, G. Hou, and H. Yang. 2015. A novel preparation of Ag-doped TiO_2 nanofibers with enhanced stability of photocatalytic activity. *RSC Adv.* 5: 32088–32091.

Zhang, J. M., D. Gao, and K. W. Xu. 2012. The structural, electronic and magnetic properties of the 3d TM (V, Cr, Mn, Fe, Co, Ni and Cu) doped ZnO nanotubes: A first-principles study. *Sci. China Phys, Mech. Astron.* 55: 428–435.

Zhang, J., Y. Wu, M. Xing, S. A. K. Leghari, and S. Sajjad. 2010. Development of modified N doped TiO_2 photocatalyst with metals, nonmetals and metal oxides. *Energy Environ. Sci.* 3: 715–726.

Zhang, K. J., W. Xu, X. J. Li, S. J. Zheng, G. Xu, and J. H. Wang. 2006. Photocatalytic oxidation activity of titanium dioxide film enhanced by Mn non-uniform doping. *T. Nonferr. Metal Soc.* 16: 1069–1075.

Zhang, L., J. Niu, D. Li, D. Gao, and J. Shi. 2014a. Preparation and photocatalytic activity of Ag modified Ti-doped-Bi_2O_3 photocatalyst. *Adv. Conden. Matter Phys.* 2014. doi:org/10.1155/2014/749354.

Zhang, Q., D. Q. Lima, I. Lee, F. Zaera, M. Chi, and Y. Yin. 2011. A highly active titanium dioxide based visible-light photocatalyst with nonmetal doping and plasmonic metal decoration. *Angew. Chemie Int. Ed.* 50 (31): 7088–7092.

Zhang, X., J. Qin, R. Hao, L. Wang, X. Shen, R. Yu, S. Limpanart, M. Ma, and R. Liu. 2015. Carbon-doped ZnO nanostructures: Facile synthesis and visible light photocatalytic applications. *J. Phys. Chem. C.* 119 (35): 20544–20554.

Zhang, Z., H. Zhang, C. Duan, J. Yuan, X. Wang, D. Xiong et al. 2008. Structure refinement of Lu_2WO_6 and luminescent properties of Eu^{3+}, Pr^{3+} doped Lu_2WO_6. *J. Alloys Compd.* 466 (1–2): 258–263.

Zhao, Z. and Q. Liu. 2008. Mechanism of higher photocatalytic activity of anatase TiO_2 doped with nitrogen under visible-light irradiation from density functional theory calculation. *J. Phys. D: Appl. Phys.* 41: 1–10.

Zhou, M. and J. Yu. 2008. Preparation and enhanced daylight-induced photocatalytic activity of C,N,S-tridoped titanium dioxide powders. *J. Hazard. Mater.* 152 (3): 1229–1236.

Zou, Z., J. Ye, K. Sayama, and H. Arakawa. 2001. Direct splitting of water under visible light irradiation with an oxide semiconductor photocatalyst. *Nature.* 414 (6864): 625–627.

8 Sensitization

8.1 INTRODUCTION

Photocatalysis deals with exposing a semiconductor to sunlight or any other source of light during a reaction, but semiconductor sensitivity to that specific light or radiation is the necessary condition. Therefore, in order to make a semiconductor sensitive toward a major portion of sunlight (visible light), some chemical substance consisting of chromophores such as synthetic dyes, colored semiconductors, or natural pigments could be used. Chromophores are responsible for photosensory processes or may help in generating artificial photoreactive systems as well.

The sensitization of various kinds of metal oxides has been used for different applications such as solar cells, hydrogen production by water splitting, degradation of organic pollutants, and so on.

8.2 PHOTOSENSITIZATION

The photosensitization process includes absorption of light by a molecule, which changes photophysical or photochemical properties of another molecule. It means that a sensitizer absorbs light and transfers energy or electron to semiconductor for a photosensitized chemical reaction to proceed. Thus, photosensitization is a process in which a reaction is initiated by utilization of a substance capable of absorbing light and transferring this absorbed light (energy) to the desired reactants. Photosensitization occurs in the presence of two molecules. These molecules are as follows:

- Photosensitizer (or sensitizer)
- Substrate or acceptor

Photosensitizers (sensitizer) are the molecules absorbing light (energy) of particular wavelength only and these are not consumed in the reaction. They return to their starting (ground) state after the completion of reaction. In this way, they assist in enhancing the efficiency of a semiconductor. On the other hand, the molecule that accepts energy or electrons is referred as the substrate or acceptor.

Photosensitizers allow photochemical reactions of colorless compounds in visible light. They also help in enhancing the triplet excited state population of semiconductors, which is not easily obtained by light absorption because of certain limits of quantum chemical selection rules.

Photosensitization is widely applied in photochemistry, photocatalysis, and photodynamic therapy (Spikes 1989).

A suitable photosensitizer must have the following properties:

- The deep colors of sensitizers are due to the presence of chromophore and auxochrome groups. These groups act as light-harvesting antenna for solar energy conversion.
- It leads to the possibility of reaching spectroscopically hidden, but photo-chemically active, excited state levels by means of spectral sensitization.
- Helps in energy or electron transfer to the semiconductor.
- Finally, it increases the photocatalytic efficiency of a photocatalyst.

Basically, there are the following two types of photosensitization reaction:

1. Type I reaction
2. Type II reaction

8.2.1   TYPE I REACTION

The sensitizer (S) gets excited by absorption of a photon. Then this excited molecule reacts with an acceptor through one electron transfer reaction, which results in formation of a radical or radical ion in both the sensitizer and the acceptor (A). This electron transfer process is governed by the redox potential of both the excited sensitizer as well as its counterpart, the acceptor. This process takes place in either direction:

$$S \xrightarrow{h\nu} S^* \tag{8.1}$$

$$S^* + A \xrightarrow{\text{Type I}} A^{\bullet-} + S^{\bullet+} \tag{8.2}$$

$$A^{\bullet-} + S^{\bullet+} + {}^3O_2 \longrightarrow \text{Oxidized products} \tag{8.3}$$

8.2.2   TYPE II REACTION

The excited sensitizer generates the excited state of oxygen, singlet molecular oxygen ($^1O_2$ or $^1\Delta_g$), and returns to its ground state via transfer of excess energy of the excited state. This excited state oxygen plays an important role in the sensitized oxidation of organic pollutants in the absence of a heterogeneous photocatalyst:

$$S \xrightarrow{h\nu} S^* \tag{8.4}$$

$$S^* + {}^3O_2 \xrightarrow{\text{Type II}} {}^1O_2 + S \tag{8.5}$$

$$\text{Substrate} + {}^1O_2 \longrightarrow \text{Oxidized products} \tag{8.6}$$

8.3   TYPES OF PHOTOSENSITIZERS

Photosensitizers can be mainly classified in the following categories:

1. Synthetic organic dye sensitizers
2. Ru–bipyridyl complex sensitizers
3. Metallophalocynine and metallophorphyrin sensitizers
4. Metal-based organic sensitizers
5. Inorganic sensitizers
6. Quantum dots (QDs) sensitizers
7. Polymer sensitizers
8. Natural sensitizers

This is also referred as push–pull architecture because of the donor–acceptor structure. The organic chromophore groups and photoactive pigments seem to be sufficient for creating the response of the substrate toward natural light.

8.3.1   SYNTHETIC ORGANIC DYE SENSITIZERS

In an organic dye/photocatalyst/visible light system, organic dyes act as both sensitizer and substrate to be degraded. In such a case, once the dye solution is completely decolorized, then no further degradation occurs because the sensitizer itself degrades. This is quite common in treatment of dye pollutants in wastewater effluents. Nonregenerative photosensitized degradation of dye pollutants was carried out by many researchers from time to time. Various dyes such as rhodamine B (Wu et al. 1998), Alizarin red (Liu et al. 1999; Liu et al. 2000a), squarylium cyanine (Wu et al. 1999), sulforhodamine B (Liu et al. 2000b; Liu and Zhao 2000; Chen et al. 2002), and methylene blue (MB) (Zhang et al. 2002) were degraded as organic contaminants in the presence of the visible light (> 420 nm).

Dye sensitization is an efficient and simple method to extend absorption of a photocatalyst toward the visible region of solar spectrum. Watanabe et al. (1977) reported the injection of electrons from the absorbed dye rhodamine B into the conduction band (CB) of the CdS particles. It leads to an efficient photochemical N-deethylation accompanying acetaldehyde formation in the aerated dye solution. Ross et al. (1994) used rose bengal for sensitization of titanium dioxide. It was used for photocatalytic decomposition of terbutylazine. More than 50% degradation was observed under visible light.

Platinized TiO_2 particles were sensitized by xanthene dye (eosin Y [EY]) through silane-coupling reagent to synthesize dye-sensitized photocatalyst. They reported evolution of hydrogen from triethanolamine (TEOA) aqueous solution under visible light. Long-time H_2 evolution was observed using EY fixed Pt–TiO_2. On the other hand, the mixture of EY and Pt–TiO_2 was effective only for >10 hours. The turnover number (TON) of the dye molecule deposited on the surface of the metal oxide (TiO_2) reached >10,000, and the quantum yield of the EY–TiO_2 was found to be approximately 10% at 520 nm (Abe et al. 2000).

The photooxidation was carried out by hybrid compounds as well, which were made up of an organic sensitizer and a polyoxometalate unit (Bonchio et al. 2004). The compound was prepared by two strategies. These are as follows:

1. The covalent functionalization of lacunary decatungstosilicate with organosilylfulleropyrrolidines
2. The charge interaction between cationic sensitizers and the polyoxoanions, yielding electrostatic aggregates

They oxidized phenol within 150 minutes with a COD loss up to 30% (turnover number [TON] up to 50), while L-methionine methyl ester (15 mM) undergoes selective photooxygenation to the corresponding sulfoxide in 90 minutes (TON up to 200). As-prepared photocatalyst was found to be stable and reusable.

Dye molecules adsorbed on TiO_2 act as an antenna to absorb the desired light as titanium dioxide does not absorb light directly from the source (Zhao et al. 2005). They explained in the proposed mechanism that the excited dye molecules transfer their electrons in the CB of the titania while the VB remains unaffected.

The dye sensitization is helpful in oxidative degradation of the dye itself after charge transfer in the absence of any redox couple. The trapped electron reacts with dissolved oxygen and generates a superoxide radical anion ($O_2^{\cdot-}$). It forms a reactive oxygen species such as hydroxyl radical (Chen et al. 2010; Yang et al. 2005). This oxidizing agent is responsible for the degradation of organic pollutants including the dye itself.

Eosin dye sensitized using different noble metal loaded TiO_2 photocatalysts was investigated by Jin et al. (2006) in the presence of TEOA, acetonitrile, and trimethylamine as electron donors. These samples were used for hydrogen production. The highest quantum yield for hydrogen generation was 10.27% under irradiation with a wavelength longer than 420 nm. They also found that activity of samples also increases as the adsorption of dye sensitizer increases.

Later, Jin et al. (2007) also used eosin to sensitize CuO-incorporated TiO_2. Eosin gets strongly adsorbed by multidentate complexation on the catalyst in the presence of CuO. They used it for photocatalytic hydrogen production from water under visible light. The electron excited in the eosin dye was transferred to the CB of TiO_2, which further moves to the CB of CuO. It create excess electrons in the CB of CuO, which resulted in a negative shift in the Fermi level of CuO, thus enhancing quantum yield, that is, ~5.1%. A good stability was observed over the dye-sensitized 1.0 wt.% CuO/TiO_2 photocatalyst.

Li et al. (2007a) made Ti-MCM-41 zeolite sensitive to visible light by sensitizing it with EY. It was used with deposition of metals such as Pt, Ru, and Rh for H_2 production and its quantum efficiency was reported to be around 12%. Here, absorbed light was converted into chemically storable hydrogen through electron transfer from dye molecule to TiO_x clusters in zeolite.

Li et al. (2007b) prepared EY-sensitized multiwalled carbon nanotube (MWCNT)/Pt catalyst using TEOA as the electron donor. It was also used for hydrogen production. They observed that diluted HNO_3-treated sample showed more efficiency in comparison with concentrated HNO_3. It was due to formation of $-COOH$ and $-OH$ on MWCNT, which provided anchoring sites for EY. The electrons produced by light get trapped by MWCNT and avoid the chance of electron–hole pair recombination. Thus, it acts as a charge transfer carrier.

Li et al. (2009a) prepared a three-dimensional (3D) polymeric dye structure using Fe^{3+} ions. EY was used to sensitize titania and Fe^{3+} linkage between TiO_2 and EY and also between different EY molecules resulting in a 3D structure. The multilayer dye-sensitized photocatalyst was found to have high harvesting ability and quantum yield for hydrogen evolution (19%) from aqueous TEOA.

Photosensitizers should have the following characteristics for dye-sensitized photocatalysis (Hagfeldt et al. 2010):

- The dye sensitizer must be sensitive toward a broad range of the solar spectral region, that is, the visible region and also part of the near-infrared (NIR) region.
- The sensitizer should be stable unless self-sensitized degradation is required.
- Unfavorable aggregation of dye in the photocatalyst surface should be avoided.
- The anchoring group, such as –COOH, –SO_3H, H_2PO_3, and so on, should be present on the dye sensitizer to facilitate strong binding on the photocatalyst surface.
- The energy of the excited state of the photosensitizer should be higher than the CB edge of the photocatalyst.

8.3.2 RU–BIPYRIDYL COMPLEX SENSITIZERS

It becomes bit difficult to realize complete degradation of dye solution under visible light illumination in nonregenerative dyes. Therefore, regenerative dyes were used in order to regenerate the sensitizer, which could completely mineralize pollutants, even colorless compounds. Hence, ruthenium(II) complexes and metal porphyrins have been extensively used.

Oxidation of herbicide terbutryne by tris(bipyridyl)ruthenium complex-sensitized titanium dioxide was investigated by Lobedank et al. (1997). The combination of sensitizer tris(4,4'-dicarboxy-2,2'-bipyridyl)ruthenium(II) chloride and titania photocatalyst was found quite helpful in complete degradation of this pollutant under exposure of sunlight.

Os and Ru polypyridyl complexes were used to sensitize titanium dioxide by Sauvé et al. (2000). Sensitized nanoporous titanium dioxide electrode was used in photoelectrochemical cells for solar energy conversion. The polypyridyl complexes have various ground state reduction potentials. The spectral response and current versus potential characteristics of electrodes were modified due to the pressure of dyes.

Fung et al. (2003) observed the excellent redox potential (i.e., about −1.24 V vs. NHE) of $[Ru^{II}(py\text{-}pzH)_3]^{2+}$ complex. Anatase TiO_2 surface was sensitized by Ru^{II} complex. It got attached on the surface via *in situ* silylation. This silyl linkage helps in increasing the stability in a wide range of pH and that too in a common solvent. The sensitized subtrace showed good activity toward degradation of CCl_4 in neutral aqueous medium in the presence of broadband visible irradiation ($\lambda > 450$ nm).

Many researchers have utilized Ru bipyridyl complexes as photosensitizers for solar energy conversion. Bae and Choi (2006) used Ru–bipyridyl complexes as a sensitizer on TiO$_2$. The number and type of anchoring groups, that is, C and P, in Ru–bipyridyl complexes affect their properties and efficiency in different ways. They varied the anchoring group in the following ways:

- Di-, tetra-, and hexacarboxylate (C2, C4, and C6)
- Di-, tetra-, and hexaphosphonate (P2, P4, and P6)

It was observed that P–TiO$_2$ showed much higher performance for hydrogen production than C–TiO$_2$ because of strong P-complex formation. Among P complexes, P2 exhibited maximum hydrogen production. The activity also depends on concentration of sensitizer and electron donors in different ways as well.

Park and Choi (2006) sensitized TiO$_2$ by ruthenium bipyridyl complexes. Hydrogen was produced at TiO$_2$/H$_2$O interface. Here, titanium dioxide was simply coated with perfluorosulfonate polymer (Figure 8.1). Effects of various parameters such as pH, concentration of Ru(bpy)$_3$$^{2+}$, Nafion loading, and the kind of TiO$_2$ were also investigated. The H$_2$ production rate was about 80 μmol/h, which corresponds to an apparent photonic efficiency of 2.6%. The roles of the Nafion layer on TiO$_2$ in the sensitized H$_2$ production were proposed to be twofold that is, to provide binding sites for cationic sensitizers and to enhance the local activity of protons in the surface region.

Probst et al. (2009) sensitized hydrogen evolution reaction using catalyst [Co(dmgH)$_2$] [ReBr(CO)$_3$bipy)], and TEOA as an irreversible reductive quencher and a photosensitizer (Figure 8.2). Its efficiency was observed to be much more than photosensitizer [Ru(bipy)$_3$]$^{2+}$. The quantum yield for hydrogen production was observed to be 26 ± 2% (H produced per absorbed photon) and the rate was found to be 2.5 ± 0.1 × 10^8 per M/s as determined by time-resolved the infrared spectroscopy.

Lee et al. (2010) studied bonding of N719 sensitizer on surface of titanium dioxide in a dye-sensitized solar cell (DSSC). The N719 dye binds on TiO$_2$ photocatalyst through two neighboring carboxylic acid/carboxylate groups via a combination of bidentate-bridging and H-bonding involving a donating group from the N719 (and/ or Ti–OH) units and acceptor from the Ti–OH (and/or N719) groups (Figure 8.3). Raman and IR spectroscopy were used to study the role of surface hydroxyl groups in

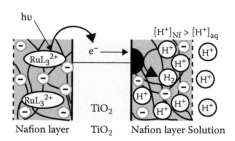

FIGURE 8.1 Perfluorosulfonate polymer–coated TiO$_2$ for hydrogen production. (Adapted from Park, H. and Choi, W., *Langmuir*, 22, 2906–2911, 2006. With permission.)

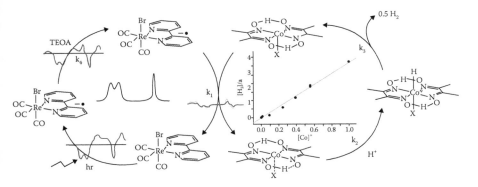

FIGURE 8.2 Rhenium-based system. (Adapted from Probst, B. et al., *Inorg. Chem.*, 48, 1836–1843, 2009. With permission.)

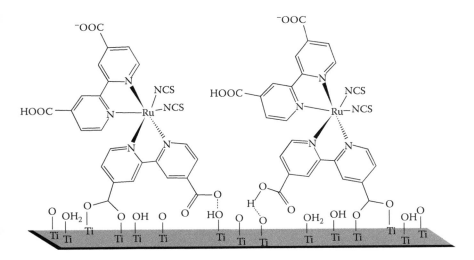

FIGURE 8.3 Adsorption mechanism of N719 on TiO$_2$ film. (Adapted from Lee, K. E. et al., *Langmuir*, 26, 9575–9583, 2010. With permission.)

the anchoring mode. The distribution features of key dye groups (COO–, bipyridine, and C=O) on the anatase surface were investigated by confocal Raman imaging. The Raman imaging distribution of COO$^-_{sym}$ on TiO$_2$ was used to show the covalent bonding while the distribution of C=O mode was applied to observe the electrostatically bonded groups.

Swetha et al. (2015) synthesized heteroleptic complexes of phosphonite coordinated ruthenium(II) sensitizers containing $\hat{C}\hat{N}N$ ligand and/or terpyridine derivatives carboxylate anchor (Figure 8.4). Such complexes extend the adsorption region up to 900 nm and also show metal-to-ligand charge transfer (MLCT) transition. It was tested for hydrogen production over a Pt–TiO$_2$ system. Density functional theory (DFT) results showed that the highest occupied molecular orbitals (HOMO) were distributed over the

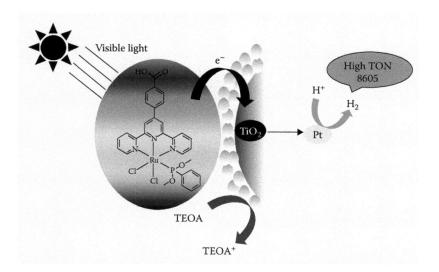

FIGURE 8.4 Efficient hydrogen generation using ruthenium sensitizers. (Adapted from Swetha, T. et al., *ACS Appl. Mater. Interf.*, 35, 19635–19642, 2015. With permission.)

Ru and Cl atom. On the contrary, the lowest unoccupied molecular orbitals (LUMO) were localized on the polypyridil ligand, which were anchored on the TiO_2 surface. A maximum turnover number (TON) of 8605 (for 8 hours) was observed and it was very high compared with the reference sensitizer (N719) with TON 163 under similar conditions.

8.3.3 Metallophthalocyanine and Metalloporphyrin Sensitizers

Phthalocyanines were used by Mele et al. (2007) as sensitizers. They prepared poly-crystalline TiO_2 samples impregnated with double-decker phthalocyanine complexes of the lanthanide metals, such as Ce, Pr, Nd, Sm, Ho, and Gd. The testing involved photodegradation of 4-nitrophenol (4-NP). Enhanced photocatalytic activity of TiO_2 samples impregnated with lanthanide diphthalocyanines in comparison with those impregnated with Cu(II)–porphyrin has been reported in the decomposition of 4-NP.

Nada et al. (2008) studied spectral sensitization of wide band gap semiconductors by dye molecules for photocatalytic H_2 production from water. They used copper phthalocyanine, ruthenium bipyridyl, and EY as photosensitizers. These sensitizers were added to slurry of TiO_2/RuO_2 semiconductor containing methyl viologen (MV^{2+}) as an electron relay. The sensitized photocatalyst showed more efficiency compared with the unmodified one. Among all the sensitizers, copper phthalocyanine exhibited the highest efficiency.

However, these complexes have the following drawbacks:

- High cost
- Need for sophisticated preparation techniques
- Lack of absorption in the visible region

- Potential metal toxicity
- Long-term stability and adaptability in real wastewater

Wang et al. (2010) synthesized a Cu(II) porphyrins composite. They mainly used the following four phorphyrins (sensitizer):

- 5-[3-(3-Phenoxy)-propoxy]phenyl porphyrin
- 5,15-Di-[3-(3-phenoxy)-propoxy]phenyl porphyrin
- 5,10,15-Tri-[3-(3-phenoxy)-propoxy]phenyl porphyrin
- 5,10,15,20-Tetra-[3-(3-phenoxy)-propoxy]phenyl porphyrin

These were used for photodegradation of 4-nitrophenol in aqueous suspensions.

Słota et al. (2011) used a hybrid photocatalyst made up of TiO_2 (anatase) matrix impregnated with lanthanide diphthalocyanine and metalloporphyrin sensitizers effectively. The sensitized TiO_2 degraded 4-nitrophenol in a UV-simulated reaction.

Zhang et al. (2014) investigated the photogenerated electron transfer process using Zn-tri-PcNc-sensitized g-C_3N_4 photocatalyst. It was observed that highly asymmetric sensitizer, that is, zinc phthalocyanine derivative (Zn–tri-PcNc), extended the spectral response region of graphitic carbon nitride (g–C_3N_4) from ∼450 nm to more than 800 nm. The production efficiency and turnover number was observed around 125.2 and 5008 μmol/h, respectively. Dye-sensitized semiconductor exhibited a quantum yield of 1.85% at 700 nm monochromatic light irradiation.

8.3.4 METAL-BASED ORGANIC SENSITIZERS

Wang et al. (2011a) formed platinum(II) terpyridylacetylide charge-transfer complexes, which contain a lone ancillary ligand systematically varied in phenylacetylide π-conjugation length. [Pt($^tBu_3$tpy)([C≡CC$_6$H$_4$]$_n$ H)]ClO$_4$ (n = 1–3) were used as photosensitizers for visible light driven ($\lambda > 420$ nm) hydrogen production in the presence of a cobaloxime catalyst and the sacrificial electron donor TEOA.

A bismuth complex, Bi$_n$(Tu)$_x$Cl$_{3n}$ was found to be a photosensitizer with BiOCl nanosheets. It was used for degradation of RhB under visible light. The superoxide radical was the main active species in the photodegradation process (Ye et al. 2012).

Khnayzer et al. (2013) used [Cu(dsbtmp)$_2$]$^+$ as a photosensitizer (where dsbtmp is 2,9-di(sec-butyl)-3,4,7,8-tetramethyl-1,10-phenanthroline). In this complex, Cu is present in Cu(I) state and forms MLCT complex with 2,9-di(sec-butyl)-3,4,7,8-tetramethyl-1,10-phenanthroline, which shows excellent stability as a visible light-absorbing photosensitizer in hydrogen-evolving homogeneous photocatalysis. The water reduction was observed using catalyst Co(dmgH)$_2$(py)Cl and N,N-dimethyl-p-toluidine sacrificial donor in 1:1 H_2O:CH_3CN. This Cu(I) MLCT complex remained stable for more than 5 days (Figure 8.5).

8.3.5 INORGANIC SENSITIZERS

Both hydrogen peroxide and titanium dioxide do not absorb visible light. But the surface complexation of H_2O_2 with TiO_2 surface exhibits a visible light response. The –OOH group present in H_2O_2 substitutes the basic –OH groups on the surface of

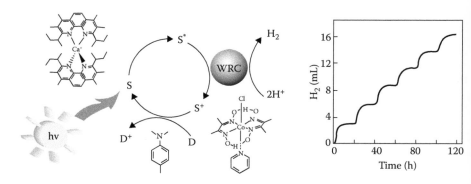

FIGURE 8.5 Cuprous phenanthroline sensitizer. WRC: water reduction catalyst. (Adapted from Khnayzer, R. S. et al., *J. Am. Chem. Soc.*, 135, 14068–14070, 2013. With permission.)

TiO_2. It forms a yellow surface complex of $\equiv$Ti(IV)-OOH. It helps in degradation of salicylic acid (Li et al. 2001).

The electrons were injected to the CB of TiO_2 and generated $\equiv$Ti(IV)-OOH˙ from the excited surface complex ($\equiv$Ti(IV)-OOH)* under visible light exposure. It further results in $\equiv$Ti(IV)-OOH and O_2. The conduction electron generated the ˙OH radicals by reacting with the adsorbed H_2O_2 on the TiO_2 surface. They also reported that the number of H_2O_2 molecules adsorbed on the TiO_2 surface directly depends on the intensity of the visible light.

Metal ion sensitizers such as Cu(II), Cr(III), Ce(III), and Fe(III) were grafted on TiO_2 powder by Yu et al. (2010). Such photocatalysts were prepared simply by the impregnation method without any thermal calcination. These sensitizers were present only on the surface of titania and unlikely to be doped into lattices. They reported a high efficiency of metal ion grafted titanium dioxide. This means electrons present in the VB were not transferred via excited state but transferred directly to Fe(III), which generated Fe(II), whereas the holes created in the VB decompose organic compounds. The adsorbed oxygen could be catalytically reduced to reactive oxygen species (ROS) by photoproduced Fe(II) species.

8.3.6 Quantum Dot Sensitizers

The term QDs is used for particles of the conventional nanosized semiconductor that are subjected to the quantum confinement effect. Even nanoparticles of almost the same optical and redox properties could be placed in this category, although they do not completely define quantum confinement (Fernando et al. 2015).

Li et al. (2009) used cadmium sulfide as a sensitizer and fabricated QDs mesoporous titania. The CdS was obtained by ion exchange technique, where CdO was converted to CdS QDs. The combination of sensitizer and QDs increased the response toward visible light by enhancing the electron transfer generated from light adsorption. As-prepared photocatalyst was applied for oxidation of NO gas in air and elimination of organic compounds in aqueous solution in the presence of visible light.

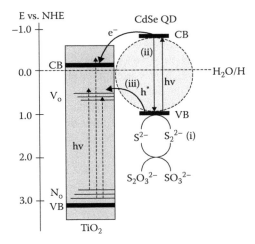

FIGURE 8.6 Synergistic effect of CdSe QDs and N-doped TiO$_2$. (Adapted from Hensel, J. et al., *Nano Lett.*, 10, 478–483, 2010. With permission.)

Hensel et al. (2010) carried out photoelectrochemical hydrogen production by CdSe QDs sensitized as well as N-doped TiO$_2$. The synergistic effect facilitates charge transfer and transport from CdSe to N-TiO$_2$ via an oxygen vacancy state (V$_\circ$) (Figure 8.6).

Wang et al. (2011) studied the effect of PbS QDs sensitizer on the efficiency of TiO$_2$ catalysts in the wavelength range of 420–610 nm, that is, from the violet to the orange–red edge of the electromagnetic spectrum. A comparative study of reduction of CO$_2$ was carried out between the PbS QDs sensitized titania and unsensitized photocatalysts. They reported that prolonged exposure to light leads to deactivation of photocatalyst as evident from X-ray photoelectron spectroscopy (XPS).

Qin and Zhang (2013) coupled the CdS QDs with a Cd$_x$Zn$_{1-x}$S cluster via the two following consecutive methods:

- Coprecipitation method
- Hydrothermal method

They observed photodegradation of rhodamine B in aqueous suspension using as-prepared photocatalyst under visible light illumination. As-prepared sample was found active in visible light because it changed the band gap, which accelerated the photoengraved electron and transporters. Not only this, but it also enhanced incident photon-to-electron conversion efficiency (IPCE).

8.3.7 POLYMER SENSITIZERS

Conducting polymers are widely used in electronics and photonics because of their specific features such as good semiconducting, electronics, and optical properties. Along with these properties, they also possess fair stability, high absorption coefficient, and wide absorption wavelength in the visible region. Therefore, materials such as polythiophene, polyaniline, polyrrole, and their derivatives are considered promising materials for their applications as visible light photosensitizers.

Hara et al. (2001) carried out oxidation of water by colloidal IrO_2. It was sensitized by cationic polymer containing the tris(4,4'-dialkyl-2,2'-bipyridyl)ruthenium group linked by aliphatic spacers, which was used as a photosensitizer. It showed much higher quantum efficiency for O_2 evolution and turnover number with respect to the Ru complex in the polymer. They also studied heterogeneous photocatalysis by adsorption of a mixture of the polymer and colloidal IrO_2 onto SiO_2 in Na_2SiF_6–$NaHCO_3$ buffer solution. This composite exhibited high activity compared with the polymer–colloidal IrO_2 system. Polymer–IrO_2 aggregates retained their activity when immobilized on a support that might be used to organize overall water splitting systems (Figure 8.7).

Polyaniline (PAn) was used as a sensitizer and it showed strong interaction with the catalyst. A comparative study between composite and bare photocatalyst was made taking degradation of MB in aqueous solution as the model reaction. Enhanced photodegradation in the presence of composite was observed due to charge transfer from PAn to TiO_2 and an efficient separation of e^-–h^+ pairs on the interface of PAn and TiO_2 in the excited state. The composite was active in both UV as well as visible light (λ = 190–800 nm), but pure photocatalyst was active only in UV light (λ < 400 nm). Composite of TiO_2 and polyaniline was prepared by Min et al. (2007) through *in situ* chemical oxidation of aniline on the surfaces of the TiO_2 nanoparticles. The same sensitizer was also used for degradation of rhodamine B (Zhang et al. 2008a).

Higher melamine condensation products can be used as a visible light sensitizer. Carbon nitride (CN) conducting polymer was synthesized using different precursors. Urea was used as a source of C and N in CN polymer by Mitoraj and Kisch (2008). They showed 92% removal efficiency of formic acid within 3 hours. Zhou et al. (2011) used dicyandiamide precursor for synthesis of a sensitizer polymer. The CN-sensitized TiO_2 nanotube (TNT) was tested by degradation of acid orange II. Complete degradation of this dye was observed at optimal conditions in the presence of visible light.

Zhang et al. (2009) used poly-fluorine-co-thiophene (PFT) sensitizer to sensitize titanium dioxide. It was found that singlet oxygen was the major reactive oxygen species in anhydrous solution using the NaN_3 quenching method. Song et al. (2007) reported 74.3% degradation of phenol in 10 hours by the TiO_2 PFT sensitizer. The degradation was carried out under GaI_3 lamp irradiation. The degradation was compared with TiO_2/rhodamine B and it was found that TiO_2/PfT showed higher efficiency than TiO_2/rhodamine B. The fluorescent quantum efficiency of TiO_2/PFT and pure PFT were observed as 3.8% and 14.7%, respectively. It showed the electrons transferred from the PFT to the CB.

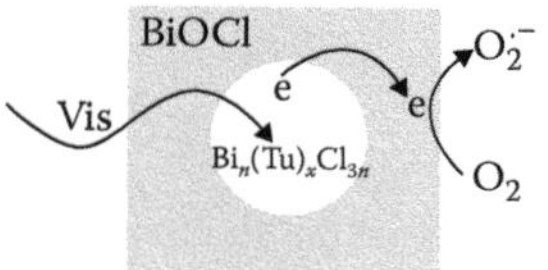

FIGURE 8.7 A novel sensitizer $Bi_n(Tu)_xCl_{3n}$. (Adapted from Ye, L. et al., *J. Chem. Mater.*, 22, 8354–8360. With permission.)

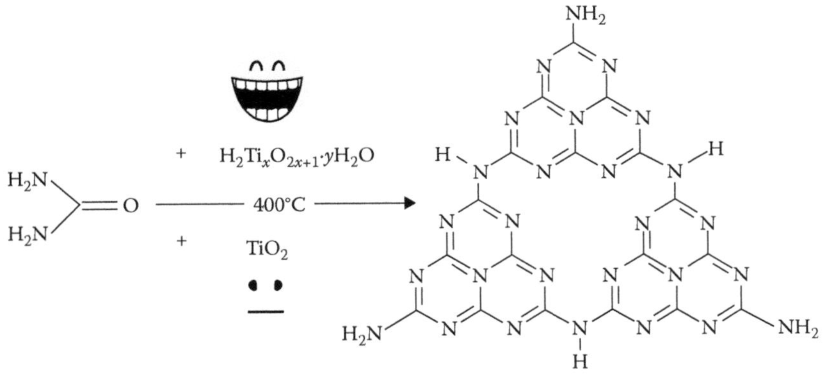

FIGURE 8.8 Production of melon. (Adapted from Gong, D. et al., *ACS Sustainable Chem. Eng.*, 2, 149–157, 2014. With permission.)

Xu et al. (2010) utilized polythiophene for degradation of methyl orange. It showed about 86% degradation of dye solution in 2 hours. Liang and Li (2009) used its derivative along with polythiophene for decomposition of organic pollutant. They observed the following trend for decomposition of 2,3-dichlorophenol:

Poly-3-methylthiophene ≈ Polythiophenecarboxylic acid
> Poly(3-hexylthiophene) (P3HT)

Gong et al. (2014) decreased the band gap of TiO_2 by using melon [poly(tri-s-triazine)] as a visible sensitizer. The composite consists of hydrogen titanate cores bearing shells of melon and the related graphitic CN (g–C_3N_4) as sensitizers. This composite was used for photodegradation of methyl orange dye and photooxidation of ethanol. The melon layer was found to be fully developed after thermal activation at ~400°C and it is photostable under open beam irradiation. However, more severe heat treatment resulted in degradation of melon, which was confirmed by TGA, and loss of visible-responsive photocatalytic activity (Figure 8.8).

8.3.8 NATURAL ORGANIC DYE SENSITIZERS

Xu et al. (2002) isolated hypocrellin B pigment from *Hypocrella bambusae* and sensitized a TiO_2 semiconductor and then used this pigment in order to increase the sensitivity of the semiconductor toward UV light. Hypocrellin B–TiO_2 chelate had a strong redox ability and more generation efficiency of active oxygen species such as $^1O_2$, $O_2^{\bullet-}$, and $^{\bullet}OOH$. Among these active oxygen, singlet oxygen ($^1O_2$) was produced by energy transfer process and superoxide radical anion ($O_2^{\bullet-}$) was produced via an electron transfer step.

The concept of photosensitization by natural pigment was also applied in photoelectrochemical solar cells. Tennakone et al. (1997) prepared flower pigment cyanidine-sensitized nanoporous TiO_2. Bernard et al. (2004) investigated 3.16% photon to current conversion efficiency in photoelectrochemical solar cells with sensitization of colloidal TiO_2 by anthocyanin pigments, delphinidin purple, and cyanidin 3,5-diglucose extracted from *Hibiscus sabdariffa* and *Ribes nigra* plants, respectively. Sensitized solar cells showed high activation energy of about 0.3–0.5 eV.

A lot of work has been done on DSSC, where sensitization was done by natural pigments. Most of the investigations were carried out on natural pigment-sensitized titania-based solar cells. Kumara et al. (2006) coated a solid state solar cell with natural pigment shisonin and further deposited p-CuI on the cell, which exhibited efficiency of about 1.3%. They also used both shisonin and chlorophyll, a cocktail of pigments, for synergistic sensitization study and understanding the role of multiple pigments. The dye cocktail extracted from a single natural resource contributed to light energy harvesting as evident from the photocurrent action spectrum of the cell. Increased performance was reported by broadening the spectral response.

Calogero and Marco (2008) prepared natural pigment (fruit)-sensitized solar cells using sicilian orange juice (*Citrus sinensis*) and the extract of eggplant peels (*Solanum melongena*, L.), where cyanidine-3-glucoside (cyanine) and delphinidin3-[4-(*p*-coumaroyl)-L-rhamnosyl(1-6)-glucopyranoside]-5 glucopyranoside (nasunin) were the main pigments. It was observed that red orange juice shows more efficiency in solar cells than eggplant peel extract.

Betalain is a water-soluble pigment that can be found in roots, fruits, and flowers. Its functional group –COOH is responsible for providing better sensitization of the TiO_2 nanostructure. Betalain and betaxanthin pigments were isolated from red beet roots. These water-soluble pigments are easily oxidized. They show high sensitivity toward visible light because of their excellent molar absorptivity of about 65,000/M/cm at 535 nm. Therefore, betalain pigment was used to sensitize TiO_2 film in the presence of methoxypropionitrile containing I^-/I_3^- redox mediator. The incident IPCEs were reported to be 14% and 8% for betaxanthin- and betanin-based solar cells, respectively (Zhang et al. 2008b).

Ito et al. (2010) prepared an extract of Monascus yellow as a food pigment from Monascus fermentations, which is commonly known as red yeast rice, and used it for increasing the efficiency of titania-based solar cells. The power conversion efficiency of such a cell was observed at more than 2% at optimum condition. Calogero et al. (2010) fabricated dye-based solar cells with some natural sensitizer such as bougainvillea flowers, red turnip, and the purple wild Sicilian prickly pear fruit juice extracts. In these pigments, the red purple betacyanins and yellow orange indicaxanthin were the core cocktail components. The performance was obtained to be approximately 2% under AM 1.5 irradiation with the red turnip extract and approximately 1.3% with the purple extract of the wild Sicilian prickly pear fruit.

Benjamin et al. (2011) studied the increase in the photocatalytic efficiency of zinc oxide semiconductor by coating it with extracts of natural pigments such as chlorophyll and anthocyanin from China rose. Coated photocatalyst was tested by photobleaching of rose bengal dye. They also observed the effect of variation of different parameters such as pH, concentration of dye, amount of semiconductor, and light intensity on the rate of photobleaching. A tentative mechanism for the reaction was also proposed involving ˙OH radicals.

The increasing order of the rate of degradation of rose bengal with different sensitizers was as follows:

China rose-coated ZnO > Chlorophyll-coated ZnO > Pure ZnO

Anthocyanin is a natural molecular dye that is used for photocatalytic sensitization of TiO_2 particles. It was applied for photodegradation of organic contaminated industrial effluent. Anthocyanin dye is a substitute of costly and harmful sensitizers consisting of metals such as CdS or ruthenium compounds. The natural dyes are more promising because after complete mineralization of anthocyanin, no traces of organic pollutants are left in an aqueous system. Therefore, the catalyst can be reused with a fresh dye (Zyoud et al. 2011).

Porphyrins and related tertrapyrrole pigments are well-known sensitizers in natural and artificial photosynthesis. These compounds are mainly responsible for harnessing far-red and NIR light in the solar spectral region. Therefore, natural pigments extracted from fruits and vegetables, such as chlorophyll and anthocyanins, have been extensively investigated as sensitizers in DSSCs. The effect of natural pigment on performance of DSSCs under various factors was studied by Narayan (2012). Numerous kinds of natural dyes or natural pigments such as anthocyanin, carotenoid, chlorophyll, flavonoid, carotene, betalain, tannin, and so on, have been used as sensitizers.

A natural DSSC achieved efficiency of around 7%, which is near to that obtained by use of synthetic dyes in DSSCs. Okoli et al. (2012) explained that although a TiO_2 electrode has a high surface area, even then it decreases the injection efficiency of carriers, and as a result, leads to lower photovoltaic performance. On the contrary, *H. sabdariffa*-extracted anthocyanin dye-stained TiO_2 electrode had excellent optical absorbance within the wavelength range of 283–516 nm. Hence, the anthocyanin-dyed cell showed 20 times higher photoconversion efficiency in comparison with the unsensitized one. Mostly, it has been observed that unstained TiO_2 showed no absorption beyond the UV region, that is, 200–350 nm. Though such natural dyes show lower light to electricity conversion efficiency, the natural dyes (found locally) have considerably low cost than the ruthenium complexes.

Hernández-Martínez et al. (2012) extracted five natural dyes from *Festuca ovina*, *H. sabdariffa*, *Tagetes erecta*, *Bougainvillea spectabilis*, and *Punica granatum* peel and used these as sensitizers in solar cells. As the sources of these pigments are inexpensive with no nutritional value, their use is no problem in energy production. The best efficiency was reported with *P. granatum* with power conversion efficiency of 1.86%. Ozuomba et al. (2013) also reported an increase in cell performance in the near visible region using anthocyanin dye. Such results were observed for both stained and unstained cells with conversion efficiency as 0.58% and 0.03%, respectively.

Generally, silicon cells have a longer lifetime and cells sensitized by synthetic dyes based on ruthenium compounds also show a relatively longer lifetime. Short lifetime is a major disadvantage with a natural sensitizer. Solvent evaporation and high dye photoreactivity are the major factors on which lifetime of DSSCs depends and this also leads to fast degradation of dye. A number of investigations have been carried out using flowering plants of the order *Caryophyllales*. Nitrogen-containing betalain pigments were used as a coloring agent or sensitizer. It has desired light-absorbing and antioxidant properties. It is commonly present in nature and also associated with different copigments that change their light-absorption properties. The functional groups (–COOH) present in pigment easily bind with the surface of TiO_2. Therefore, Hernández-Martínez et al. (2013) studied efficiencies of solar energy conversion by betalains (0.90%) compared with *Beta vulgaris* extract (BVE) with tetraethylorthosilicate (TEOS) (0.68%). They proposed that TEOS can be used as a photodegradation inhibiting agent to avoid UV degradation with a concomitant increment in the lifetime of the cell. Thus, BVE/ TEOS increased stability, UV resistance, and the lifetime of solar cells.

Li et al. (2013) fabricated DSSC and studied the photoelectrochemical optimal conditions for red cabbage extract as a natural dye. It was concluded that cell performance increases as purification, immersion time, and extract content had a great impact on the total load volume and specific activity.

Kushwaha et al. (2013) extracted a natural pigment from the leaves of *Tectona grandis*, *Tamarindus indica*, *Eucalyptus globulus*, and the flower of *Callistemon citrinus*. They used sensitized TiO_2 in the DSSC. Such a cell showed an IPCE from 12% to 37%. Among the four dyes studied, the extract obtained from teak showed the best photosensitization effects in terms of cell output.

Natural dyes are environmentally and economically superior to ruthenium-based dyes because they are nontoxic and low cost. Hemmatzadeh and Mohammadi (2013) fabricated a DSSC that was sensitized by extracted dye from *Pastinaca sativa* and *B. vulgaris*. They observed that these natural extracts show less increase in efficiency. Thus, they studied different factors that may affect the efficiency of sensitized cells. These are the following:

- Utilization of different extraction approaches
- The acidity of the extraction solvent
- Different extraction solvents affecting the optical absorption spectra of dye

On studying all these parameters, it was concluded that better performance and optical characteristics could be obtained by choosing a proper mixture of ethanol and water as extracting solvent and also maintaining the acidity of dye solution.

Hug et al. (2014) prepared a DSSC with natural sensitizers such as carotenoids, polyphenols, and chlorophylls. They named such cells biophotovoltaic cells. These compounds provided excellent photon sensitivity because they increase charge carrier efficiency, enhance sustainability, reduce production cost, and provide easy waste management.

Isah et al. (2014) extracted red *Bougainvillea glabra* flower to obtain a natural dye using water as a dye-extracting solvent. This sensitizer was evaluated at three different pH values of 1.23, 3.0, and 5.7, and maximum power values of 0.50, 1.64, and 0.94 mW/cm^2 were observed with this dye sensitizer.

Ananth et al. (2014) prepared pure TiO_2 by the sol–gel method. They also sensitized it by *Lawsonia inermis* seed extract by doing some modification in the sol–gel method used for preparation of pure titania. Structural, optical, spectral, and morphological studies were carried out for both photocatalysts, and a comparative study was conducted. It was observed that pre–dye-treated semiconductor exhibited 47% more efficiency in comparison with the traditional DSSC.

Yan et al. (2014) used anthocyanin as a template for synthesis of highly crystalline mesoporous titania. Natural food coloring agent leads to red shift of band gap absorption and increases its performance under visible light irradiation. It was examined in degradation of dyes and phenols.

Maabong et al. (2015) observed 0.036%, 0.023%, 0.005%, and 0.0008% power conversion efficiency of TiO_2 solar cell sensitized by lemon (*Citrus limon*) dye cell, red and orange bougainvillea (*B. glabra*), and morula (*Sclerocarya birrea*), respectively. Despite the low efficiencies, natural dyes showed a great photoelectrochemical conversion because they provide a low cost and environment-friendly alternative to commercial synthetic dyes. These natural dyes exhibited absorption under 400–750 nm (visible light).

Strawberry and turmeric powder dyes were used by Mohammed et al. (2015) in a titanium dioxide-based cell. They reported that a lamp emitting all wavelengths in the visible spectrum does not give consistent data because of substantial heating of the cell. However, tests carried out in natural sunlight provide a steady voltage at much higher level.

Recently, Solanki et al. (2015a) prepared pure SnO_2, carbon-doped SnO_2, and carbon-doped chlorophyll-sensitized SnO_2 by the precipitation method. They used glucose as a carbon precursor, and natural pigment (chlorophyll) extracted from garden plant Madhumamalti (*Combretum indicum*) leaves was used to sensitize carbon-doped SnO_2 photocatalyst. The performance of these photocatalysts was evaluated under visible light. Efficiency of prepared samples was studied by degradation of toluidine blue and azure A in a synthetic waste-water system (Solanki et al. 2015b). The optimum conditions were obtained by varying parameters such as pH, concentration of dye, amount of photocatalyst, and light intensity. A comparative study was conducted between these prepared samples.

Photocatalysts find wide applications in artifical photosynthesis and solar fuel production. The photocatalyst must be active in the far or NIR region for such applications. Thus, longer wavelength spectral sensitization of a photocatalyst by various types of sensitizers (chromphore group) is of prime interest. Natural dyes seem to be a viable substitute of high-cost and rare organic sensitizers. These dyes have some advantages over organic dyes such as low cost, easy availability, abundance in nature and also no environmental threat. It is expected that different components of plants such as flower petals, leaves, and bark could be used as sensitizers in years to come.

REFERENCES

Abe, R., K. Hara, K. Sayama, K. Domen, and H. Arakawa. 2000. Steady hydrogen evolution from water on eosin Y-fixed TiO_2 photocatalyst using a silane-coupling reagent under visible light irradiation. *J. Photochem. Photobiol. A: Chem.* 137 (1): 63–69.

Ananth, S., P. Vivek, T. Arumanayagam, and P. Murugakoothan. 2014. Natural dye extract of *Lawsonia inermis* seed as photo sensitizer for titanium dioxide based dye sensitized solar cells. *Spectrochim. Acta A.* 128C: 420–426.

Bae, E. and W. Choi. 2006. Effect of the anchoring group (carboxylate vs phosphonate) in Ru-complex-sensitized TiO$_2$ on hydrogen production under visible light. *J. Phys. Chem. B.* 110 (30): 14792–14799.

Benjamin, S., D. Vaya, P. B. Punjabi, and S. C. Ameta. 2011. Enhancing photocatalytic activity of zinc oxide by coating with some natural pigments. *Arab. J. Chem.* 4: 205–209.

Bernard, J. S., O. A. Julius, and M. Mwabora. 2004. Anthocyanin-sensitized nanoporous TiO$_2$ photoelectrochemical solar cells prepared by a sol–gel process, *Progr. Coll. Polym. Sci.* 125: 34–37.

Bonchio, M., M. Carraro, G. Scorrano, and A. Bagno. 2004. Photooxidation in water by new hybrid molecular photocatalysts integrating an organic sensitizer with a polyoxometalate core. *Adv. Synth. Catal.* 346 (6): 648–654.

Calogero, G. and G. D. Marco. 2008. Red Sicilian orange and purple eggplant fruits as natural sensitizers for dye-sensitized solar cells. *Sol. Energy Mater. Solar Cells.* 92: 1341–1346.

Calogero, G., G. D. Marco, S. Cazzanti, S. Caramori, R. Argazzi, A. D. Carlo, and C. A. Bignozzi. 2010. Efficient dye-sensitized solar cells using red turnip and purple wild sicilian prickly pear fruits, *Int. J. Mol. Sci.* 11: 254–267.

Chen, C. C., W. H. Ma, and J. C. Zhao. 2010. Semiconductor-mediated photodegradation of pollutants under visible light irradiation. *Chem. Soc. Rev.* 39: 4206–4219.

Chen, C. C., W. Zhao, J. Y. Li, and J. C. Zhao. 2002. Formation and identification of intermediates visible-light photodegradation sulforhodamine-B dye in aqueous TiO$_2$ dispersion. *Environ. Sci. Technol.* 36: 3604–3611.

Fernando, S., K. A., S. Sahu, Y. Liu, W. K. Lewis, E. A. Guliants, A. Jafariyan, P. Wang, C. E. Bunker, and Y.-P. Sun. 2015. Carbon quantum dots and applications in photocatalytic energy conversion. *ACS Appl. Mater. Interf.* 7 (16): 8363–8376.

Fung, A. K. M., B. K. W Chiu, and M. H. W. Lam. 2003. Surface modification of TiO$_2$ by a ruthenium(II) polypyridyl complex via silyl-linkage for the sensitized photocatalytic degradation of carbon tetrachloride by visible irradiation. *Water Res.* 37 (8): 1939–1947.

Gong, D., J. G. Highfield, S. Z. E. Ng, Y. Tang, W. C. J. Ho, Q. Tay, and Z. Chen. 2014. Poly tri-s-triazines as visible light sensitizers in titania-based composite photocatalysts: Promotion of melon development from urea over acid titanates. *ACS Sustain. Chem. Eng.* 2 (2): 149–157.

Hagfeldt, A., G. Boschloo, L. Sun, L. Kloo, and H. Pettersson. 2010. Dye-sensitized solar cells. *Chem. Rev.* 110: 6595–6663.

Hara, M., J. T. Lean, and T. E. Mallouk. 2001. Photocatalytic oxidation of water by silica-supported tris(4,4′-dialkyl-2,2′-bipyridyl)ruthenium polymeric sensitizers and colloidal iridium oxide. *Chem. Mater.* 13 (12): 4668–4675.

Hemmatzadeh, R. and A. Mohammadi. 2013. Improving optical absorptivity of natural dyes for fabrication of efficient dye-sensitized solar cells. *J. Theor. Appl. Phys.* 7. 10.1186/2251-7235-7-57.

Hensel, J., G Wang, Y Li, and J. Z. Zhang. 2010. Synergistic effect of CdSe quantum dot sensitization and nitrogen doping of TiO$_2$ nanostructures for photoelectrochemical solar hydrogen generation. *Nano Lett.* 10 (2): 478–483.

Hernández-Martínez, A. R., A. Estevez, S. Vargas, and R. Rodriguez. 2013. Stabilized conversion efficiency and dye-sensitized solar cells from Beta vulgaris pigment. *Int. J. Mol. Sci.* 14: 4081–4093.

Hernández-Martínez, A. R., M. Estevez, S. Vargas, F. Quintanilla, and R. Rodríguez. 2012. Natural pigment-based dye-sensitized solar cells. *J. Appl. Res. Technol.* 10: 38–47.

Hug, H., M. Bader, P. Mair, and T. Glatzel. 2014. Biophotovoltaics: Natural pigments in dye-sensitized solar cells. *Appl. Energy.* 115: 216–225.

Isah, K. U., U. Ahmadu, A. Idris, M. I. Kimpa, U. E. Uno, M. M. Ndamitso and N. Alu. 2014. Betalain pigments as natural photosensitizers for dye-sensitized solar cells: The effect of dye pH on the photoelectric parameters, *Mater. Renew. Sustain. Energy.* 4. 10.1007/s40243-014-0039-0.

Ito, S., T. Saitou, H. Imahori, H. Uehara and N. Hasegawa. 2010. Fabrication of dye-sensitized solar cells using natural dye for food pigment: Monascus yellow. *Energy Environ. Sci.* 3: 905–909.

Jin, Z., X. Zhang, G. Lu, and S. Li. 2006. Improved quantum yield for photocatalytic hydrogen generation under visible light irradiation over eosin sensitized TiO_2—Investigation of different noble metal loading. *J. Mol. Catal. A: Chem.* 259 (1–2): 275–280.

Jin, Z., X. Zhang, Y. Li, S. Li, and G. Lu. 2007. 5.1% Apparent quantum efficiency for stable hydrogen generation over eosin-sensitized CuO/TiO_2 photocatalyst under visible light irradiation. *Catal. Commun.* 8 (8): 1267–1273.

Khnayzer, R. S., C. E. McCusker, B. S. Olaiya, and F. N. Castellano. 2013. Robust cuprous phenanthroline sensitizer for solar hydrogen photocatalysis. *J. Am. Chem. Soc.* 135 (38): 14068–14070.

Kumara, G. R. A., S. Kaneko, M. Okuya, B. Onwona-Agyeman, A. Konno, and K. Tennakone. 2006. Shiso leaf pigments for dye-sensitized solid-state solar cell. *Solar Energy Mater. Solar Cells.* 90: 1220–1226.

Kushwaha, R., P. Srivastava, and L. Bahadur. 2013. Natural pigments from plants used as sensitizers for TiO_2 based dye-sensitized solar cells. *J. Energy.* org/10.1155/2013/654953.

Lee, K. E., M. A. Gomez, S. Elouatik, and G. P. Demopoulos. 2010. Further understanding of the adsorption mechanism of N719 sensitizer on anatase TiO_2 films for DSSC applications using vibrational spectroscopy and confocal Raman imaging. *Langmuir.* 26 (12): 9575–9583.

Li, G.-S., D.-Q. Zhang, and J. C. Yu. 2009. A new visible-light photocatalyst: CdS quantum dots embedded mesoporous TiO_2. *Environ. Sci. Technol.* 43 (18): 7079–7085.

Li, Q., L. Chen, and G. Lu. 2007b. Visible-light-induced photocatalytic hydrogen generation on dye-sensitized multiwalled carbon nanotube/Pt catalyst. *J. Phys. Chem. C.* 111 (30): 11494–11499.

Li, Q., Z. Jin, Z. Peng, Y. Li, S. Li, and G. Lu. 2007a. High-efficient photocatalytic hydrogen evolution on eosin Y-sensitized Ti–MCM41 zeolite under visible-light irradiation. *J. Phys. Chem. C.* 111 (23): 8237–8241.

Li, X. Z., C. C. Chen, and J. C. Zhao. 2001. Mechanism of photodecompositon of H_2O_2 on TiO_2 surface under visible light irradiation. *Langmuir.* 17: 4118–4122.

Li, Y., M. Guo, S. Peng, G. Lu, and S. Li. 2009a. Formation of multilayer-eosin Y-sensitized TiO_2 via Fe^{3+} coupling for efficient visible-light photocatalytic hydrogen evolution. *Int. J. Hydrogen Energy.* 34 (14): 5629–5636.

Li, Y., S.-H. Ku, S.-M. Chen, M. A. Ali, and F. M. A. AlHemaid. 2013. Photoelectrochemistry for red cabbage extract as natural dye to develop a dye-sensitized solar cells, *Int. J. Electrochem. Sci.* 8: 1237–1245.

Liang, H. and X. Li. 2009. Visible induced photocatalytic reactivity of polymer sensitized titania nanotube films. *Appl. Catal. B.* 86: 8–17.

Liu, G. M., X. Z. Li, J. C. Zhao, H. Hidaka, and N. Serpone. 2000b. Photooxidation pathway of sulforodamine B. Dependence on the asorption mode TiO_2 exposed to visible light radiation. *Environ. Sci. Technol.* 34: 3982–3990.

Liu, G. M., X. Z. Li, J. C. Zhao, S. Horikoshi, and H. Hidaka. 2000a. Photooxidation mechanism of dye alizarin red in TiO_2 dispersions under visible light illumination: An experimental and theoretical examination. *J. Mol. Catal. A: Chem.* 153: 221–229.

Liu, G. M., T. X. Wu, J. C. Zhao, H. Hidaka, and N. Serpone. 1999. Photoassisted degradation of dye pollutants. 8. Irreversible degradation of alizarine red under visible light radiation in air equilibrated aqueous TiO_2 dispersions. *Environ. Sci. Technol.* 33: 2081–2087.

Liu, G. M., and J. C. Zhao. 2000. Photocatalytic degradation of dye sulforhodamine B: A comparative study of photocatalysis with photosensitizations. *New J. Chem.* 24: 411–417.

Lobedank, J., E. Bellmann, and J. Bendig. 1997. Sensitized photocatalytic oxidation of herbicides using natural sunlight. *J. Photochem. Photobiol. A: Chem.* 108 (1): 89–93.

Maabong, K., C. M. Muiva, P. Monowe, T. S. Sathiaraj, H. Hopkins, L. Nguyen et al. 2015. Natural pigments as photosensitizers for dye-sensitized solar cells with TiO_2 thin films. *Int. J. Renew. Energy Res.* 5 (2): 501–506.

Mele, G., E. Garcìa-Lòpez, L. Palmisano, G. Dyrda, and R. Słota. 2007. Photocatalytic degradation of 4-nitrophenol in aqueous suspension by using polycrystalline TiO_2 impregnated with lanthanide double-decker phthalocyanine complexes. *J. Phys. Chem. C.* 111 (17): 6581–6588.

Min, S., F. Wang, and Y. Han. 2007. An investigation on synthesis and photocatalytic activity of polyaniline sensitized nanocrystalline TiO_2 composites. *J. Mater. Sci.* 42 (24): 9966–9972.

Mitoraj, D. and H. Kisch. 2008. The nature of nitrogen-modified dioxide photocatalysts active in visible light. *Angew. Chem. Int. Ed.*, 47: 9975–9978.

Mohammed, A. A., A. S. S. Ahmad, and W. A. Azeez. 2015. Fabrication of dye sensitized solar cell based on titanium dioxide (TiO_2). *Adv. Mater. Phys. Chem.* 5: 316–367.

Nada, A. A., H. A. Hamed, M. H. Barakat, N. R. Mohamed, T. N. Veziroglu. 2008. Enhancement of photocatalytic hydrogen production rate using photosensitized TiO_2/RuO_2-MV^{2+}. *Int. J. Hydrogen Energy.* 33 (13): 3264–3269.

Narayan, M. R. 2012. Review: Dye sensitized solar cells based on natural photosensitizers. *Renew. Sustain. Energy Rev.* 16: 208–215.

Okoli, L. U., J. O. Ozuomba, A. J. Ekpunobi, and P. I. Ekwo. 2012. Anthocyanin-dyed TiO_2 electrode and its performance on dye-sensitized solar cell. *Res. J. Recent Sci.* 1: 22–27.

Ozuomba, J. O., L. U. Okoli, and A. J. Ekpunobi. 2013. The performance and stability of anthocyanin local dye as a photosensitizer for DSSCs. *Adv. Appl. Sci. Res.* 4: 60–69.

Park, H. and W. Choi. 2006. Visible-light-sensitized production of hydrogen using perfluorosulfonate polymer-coated TiO_2 nanoparticles: An alternative approach to sensitizer anchoring. *Langmuir.* 22 (6): 2906–2911.

Probst, B., C. Kolano, P. Hamm, and R. Alberto. 2009. An efficient homogeneous intermolecular rhenium-based photocatalytic system for the production of H_2. *Inorg. Chem.* 48 (5): 1836–1843.

Qin, Z.-Q. and F.-J. Zhang. 2013. Surface decorated $Cd_xZn_{1-x}S$ cluster with CdS quantum dot as sensitizer for highly photocatalytic efficiency. *Appl. Surf. Sci.* 285: 912–917.

Ross, H., J. Bendig, and S. Hecht. 1994. Sensitized photocatalytical oxidation of terbutylazine. *Sol. Energy Mater. Solar Cells.* 33 (4): 475–481.

Sauvé, G., M. E. Cass, G. Coia, S. J. Doig, I. Lauermann, K. E. Pomykal, and N. S. Lewis. 2000. Dye sensitization of nanocrystalline titanium dioxide with osmium and ruthenium polypyridyl complexes. *J. Phys. Chem. B.* 104 (29): 6821–6836.

Słota, R., G. Dyrda, K. Szczegot, G. Mele, and I. Pio. 2011. Photocatalytic activity of nano and microcrystalline TiO_2 hybrid systems involving phthalocyanine or porphyrin sensitizers. *Photochem. Photobiol. Sci.* 10: 361–366.

Solanki, M. S., R. Ameta, and S. Benjamin. 2015a. Sensitization of carbon doped tin (IV) oxide nanoparticles by chlorophyll and its application in photocatalytic degradation of toluidine blue. *Int. J. Adv. Chem. Sci. Appl.* 3 (3): 24–30.

Solanki, M. S., M. Trivedi, R. Ameta, and S. Benjamin. 2015b. Preparation and use of chlorophyll sensitized carbon doped tin(IV) oxide nanoparticles for photocatalytic degradation of azure A. *Int. J. Chem. Res.* 5 (4): 1–11.

Song, L., R. Qiu, Y. Mo, D. Zhang, H. Wei, and Y. Xiong. 2007. Photodegradation of phenol in a polymer-modified TiO$_2$ semiconductor particulate system under the irradiation of visible light. *Catal. Commun.* 8: 429–433.

Spikes, J. D. 1989. Photosensitization. In: *The Science of Photobiology*, C. S. Kendric (Ed.), New York and London: Plenum Press, pp. 79–110.

Swetha, T., I. Mondal, K. Bhanuprakash, U. Pal, and S. P. Singh. 2015. First study on phosphonite-coordinated ruthenium sensitizers for efficient photocatalytic hydrogen evolution. *ACS Appl. Mater. Interf.* 7 (35): 19635–19642.

Tennakone, K., A. R. Kumarasinghe, G. R. R. A. Kumara, K. G. U. Wijayantha, and P. M. Sirimanne. 1997. Nanoporous TiO$_2$ photoanode sensitized with the flower pigment cyanidine, *J. Photochem. Photobiol. A: Chem.* 108: 193–195.

Wang, C., R. L. Thompson, P. Ohodnicki, J. Baltrus, and C. Matranga. 2011. Size-dependent photocatalytic reduction of CO$_2$ with PbS quantum dot sensitized TiO$_2$ heterostructured photocatalysts. *J. Mater. Chem.* 21: 13452–13457.

Wang, F., S. Min, Y. Han, and L. Feng. 2010. Visible light induced photocatalytic degradation of methylene blue with polyaniline sensitized TiO$_2$ composite photoctalysts. *Superlatttices Microstruct.* 48: 170–180.

Wang, X., S. Goeb, Z. Ji, N. A. Pogulaichenko, and F. N. Castellano. 2011a. Homogeneous photocatalytic hydrogen production using π-conjugated platinum(II) arylacetylide sensitizers. *Inorg. Chem.* 50 (3): 705–707.

Watanabe, T., T. Takizawa, and K. Honda. 1977. Photocatalysis through excitation of adsorbates. 1. Highly efficient N-deethylation of rhodamine B adsorbed to cadmium sulfide. *J. Phys. Chem.* 81: 1845–1851.

Wu, T. X., G. M. Zhao, J. C. Zhao, H. Hidaka, and N. Serpone. 1998. Photoassisted degradation of dye pollutants. V. Self-photosensitized oxidative transformation of rhodamine under visible light radiation in aqueous TiO$_2$ dispersions. *J. Phys. Chem. B.* 102: 5845–5851.

Wu, T. X., G. M. Zhao, J. C. Zhao, H. Hidaka, and N. Serpone. 1999. TiO$_2$-Photoassisted degradation of dye pollutants. 9. Photooxidation of a squarylium cyanine dye in aqueous dispersion under visible light irradiation. *Environ. Sci. Technol.* 33: 1379–1387.

Xu, S. H., S. Y. Li, Y. X. Wei, L. Zhang, and F. Xu. 2010. Improving the photocatalytic performance of conducting polymer polythiophene sensitized TiO$_2$ nanoparticles under sunlight irradiation. *React. Kinet. Mech. Catal.* 101: 237–249.

Xu, S., J. Shen, S. Chen, M. Zhang, and T. Shen. 2002. Active oxygen species (^{1}O$_2$, O$_2$$^{\bullet-}$) generation in the system of TiO$_2$ colloid sensitized by hypocrellin B. *J. Photochem. Photobiol. B: Biol.* 67: 64–70.

Yan, Z., W. Gong, Y. Chen, D. Duan, J. Li, W. Wang, and J. Wang. 2014. Visible-light degradation of dyes and phenols over mesoporous titania prepared by using anthocyanin from red radish as template. *Int. J. Photoenergy.* org/10.1155/2014/968298.

Yang, J., C. C. Chen, H. W. Ji, W. H. Ma, and J. C. Zhao. 2005. Mechanism of TiO$_2$ assisted photocatalytic degradation of dyes under visible irradiation of dyes under visible irradiation: Photoelectrocatalytic study by TiO$_2$ film electrodes. *J. Phys. Chem. B.* 109: 21900–21907.

Ye, L., C. Gong, J. Liu, L. Tian, T. Peng, K. Deng and L. Zan. 2012. Bi$_n$(Tu)$_x$Cl$_{3n}$: A novel sensitizer and its enhancement of BiOCl nanosheets' photocatalytic activity. *J. Mater. Chem.* 22: 8354–8360.

Yu, H., H. Irie, Y. Shimodaira, Y. Hosogi, Y. Kuroda, M. Miyauchi, and K. Hashimoto. 2010. An efficient visible light sensitive Fe(III) grafted TiO$_2$ photocatalyst. *J. Phys. Chem. C.* 114: 16481–16487.

Zhang, D., S. M. Lanier, J. A. Downing, J. L. Avent, J. Lum, and J. L. McHale. 2008a. Betalain pigments for dye-sensitized solar cell. *J. Photochem. Photobiol. A: Chem.* 195: 72–80.

Zhang, D., R. Qiu, L. Song, B. Eric, Y. Mo, and X. Huang. 2009. Role of oxygen active species in the photocatalytic degradation of phenol using polymer sensitized TiO_2 under visible light irradiation. *J. Hazard. Mater.* 163: 843–847.

Zhang, H., R. Zong, J. Zhao, and Y. Zhu. 2008a. Dramatic visible photocatalytic degradation performance due to synergistic effect of TiO_2 with PANI. *Environ. Sci. Technol.* 42: 3803–3807.

Zhang, T. Y., T. Oyama, S. Horikoshi, H. Hidaka, J. C. Zhao, and N. Serpone. 2002. Photocatalyzed N-demethylation and degradation of methylene blue in titania dispersion exposed to concentrated sunlight. *Sol. Energy Mater. Sol. Cells.* 73: 287–303.

Zhang, X., L. Yu, C. Zhuang, T. Peng, R. Li, and X. Li. 2014. Highly asymmetric phthalocyanine as a sensitizer of graphitic carbon nitride for extremely efficient photocatalytic H_2 production under near-infrared light. *ACS Catal.* 4 (1): 162–170.

Zhao, J. C., C. C. Chen, and W. H. Ma. 2005. Photocatalytic degradation of organic pollutants under visible light irradiation. *Top. Catal.* 35: 269–278.

Zhou, X., F. Peng, H. Wang, H. Yu, and Y. Fan. 2011. Carbon nitride polymer sensitized TiO_2 nanotube arrays with enhanced visible light photoelectrochemical and photocatalystic performance. *Chem. Commun.* 47: 10323–10325.

Zyoud, A., N. Zaatar, I. Saadeddin, M. H. Helal, G. Campet, M. Hakim et al. 2011. Alternative natural dyes in water purification: Anthocyanin as TiO_2-sensitizer in methyl orange photo-degradation. *Solid State Sci.* 13: 1268–1275.

9 Composites

9.1 INTRODUCTION

A composite material or composition material or simply composite is a material made from two or more constituent materials that possess different physical or chemical properties, which on combination produce a material with characteristics different from its individual components. The individual components remain as separate and distinct within the finished (final) structure. The new material is preferred for many reasons, such as the materials are stronger, lighter, or of low cost, and so on compared to the traditional materials.

9.1.1 NATURAL COMPOSITES

Natural composites exist in both the animal as well as plant kingdoms. A most common example of a natural composite is wood that is made from long cellulose fibers (a polymer) held together by a much weaker substance called lignin. Cellulose is also found in cotton, but it does not contain lignin and therefore it is relatively much weaker than wood. Lignin and cellulose are individually two weak substances but when they are present together, they are quite strong. In the human body, the bone is also a composite made up of hard but brittle material known as hydroxyapatite that is basically calcium phosphate and collagen, which is protein, a soft and flexible material. Collagen is also found in hair and fingernails. Collagen does not have the hardness that is required in a skeleton, but combined with hydroxyapatite it develops properties like hardness that is essential for body support.

9.1.2 TRADITIONAL COMPOSITES

People have been making composites from last thousands of years. One of the earliest examples of a composite is mud bricks that are used as a building material. Mud has good compressive strength, but lacks the desired tensile strength. On the other hand, straw is strong enough to stretch out, but it gets crumpled up easily. If straw and mud is combined in the form of a composite, it is possible to make bricks that are resistant to properties such as squeezing and tearing. As such, these bricks can be used for making excellent building blocks. Another old example of a composite is concrete, which is a mixture of small stones or gravel, cement, and sand and has good compressive strength. It is known that addition of some metal rods or wires to this concrete increases its tensile strength, and this composite is named reinforced concrete.

9.1.3 RECENT COMPOSITES

Fiberglass can be considered the first modern composite material, and is widely used today for preparing boat hulls, sports equipment, building panels, car bodies, and so on. The basic matrix is plastic and the reinforcement is glass that has been made into fine threads and often woven into a sort of cloth. Glass itself is very strong, but it is brittle in nature and breaks on bending sharply. The plastic matrix holds these glass fibers together and also protects them from any damage by sharing out the forces acting on them. Carbon nanotubes (CNTs) have also been used successfully to make new composites.

9.2 COMPOSITE PHOTOCATALYSTS

TiO_2 is the most common photocatalyst with wide applications in environmental purification, energy conversion and storage, self-cleaning, and so on due to its low cost, nontoxicity, high oxidizing power, chemical stability, and environment-friendly characteristics (Cao et al. 2000). It is a good choice for making composites with materials for different purposes.

Composite photocatalysts are defined as materials comprised of an active photocatalyst with another material that is made either from some inert compounds or other photoactive compounds. Out of several types of composites reported, one of the types consists of composites made of an inert skeleton onto which titanium dioxide is coated. Choi et al. (2006) reported on a composite membranes made of a porous alumina skeleton with mesoporous TiO_2 grown onto it by the sol–gel method using a mixture of titanium tetraisopropoxide, acetic acid, and a surfactant. Another type of composite photocatalyst consists of two (or more) domains, where both are exposed to the outer environment and non-TiO_2 domains serve as a part of the photocatalytic process. Examples of this type are as follows:

- Partial coverage of activated carbon with titanium dioxide (by atomic layer deposition) (Dey et al. 2011)
- By physical contact (Avraham-Shinman and Paz 2006)
- Structures made from titanium dioxide and inert oxides such as zeolites (Liu et al. 1998)

9.3 TITANIUM DIOXIDE–BASED COMPOSITES

Titanium dioxide has been effectively used in the fields of photocatalytic application for environmental purification, recovery of metal ions at lower concentrations, decomposition of toxic substances, utilization of solar energy, and so on. TiO_2 materials have to be fabricated with high specific surface area and highly porous structures for effective contact with reacting substances to acquire high photocatalysis in decomposing various harmful substances in gas or liquid. Various TiO_2 nanomaterials with large surface areas have been prepared in different forms and used as powder (Znaidi et al. 2001), nanotubules (Martin 1994), nanofibers (Terashima et al. 2001) thin films comprised of nanocrystals by many methods such as the sol–gel process, pulsed laser deposition, electrodeposition (Ishikawa and Matsumoto 2001), and so on. Nanostructured TiO_2 materials (nanotubules and nanofibers) fabricated through template synthesis method have attracted much interest because of their large surface areas and various potential applications.

9.3.1 METAL OXIDE–BASED COMPOSITES

Highly porous three-dimensional composite nanostructures (30–120 nm) of TiO_2–$4\%SiO_2$–$1\%TeO_2$ (TST)Al_2O_3/TiO_2 were fabricated by anodization of a superimposed Al/Ti layer sputter deposited on glass and the sol–gel process (Chu et al. 2003). The porous composite nanostructures exhibited an increase in photocatalytic performance in decomposing acetaldehyde gas under UV illumination, which may be due to the combination of their large surface areas (7750–14,770 m^2), high porosities (34.2%–45.6%), and transparency. Particularly, the composite nanostructure with ~120 nm pores calcined at 500°C showed the highest photocatalytic activity, which is 6–10 times higher than commercial P25 TiO_2.

The TST/Al_2O_3/TiO_2 composite nanostructures on glass synthesized upon phosphoric-anodized films (PF) and oxalic-anodized films (CF) are shown in Figure 9.1. The composite nanostructures comprised numerous open pores that are distributed uniformly on the surface irrespective of the large roughness, even though some of the pores in the CF specimens are covered by the calcined TST layer. A porous alumina with coral-like structure was formed during anodization because of the induction of aluminum crystal grains that lead to high porosities or large surface areas. It was observed that the TST remain adhered to the alumina walls and form a hollow structure. The TST layers (~20 nm for the PF and ~10 nm for the CF specimens) were continuous within the channels and exhibited a rough inner surface, thus leading to larger surface areas than the other conventional TiO_2 films. A joined alumina–titania structure was formed on the Ti/glass substrate as the titanium layer was partly oxidized into a dense titania layer in anodization. The anodic titania may work as a binding layer and provide a good adherence between the porous nanostructures and glass substrates.

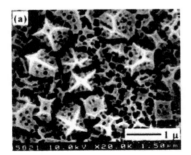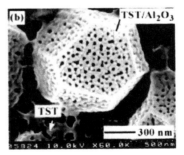

FIGURE 9.1 Field emission scanning electron microscopy (FESEM) images of surface morphologies. (a) PF specimens anodized in 10% H_3PO_4 and pore widened for 40 minutes and (b) CF specimens anodized in 3% $(COOH)_2$ and pore widened for 20 minutes. (Adapted from Chu, S. et al., *J. Phys. Chem. B.*, 107, 6586–6589, 2003. With permission.)

The shell TiO_2 nanocrystals can be derived via sol–gel technology by heat treatment at 450°C and the core $CoFe_2O_4$ nanoparticles (NPs) were synthesized via the coprecipitation method (Fu et al. 2005). As-synthesized composite particles can be utilized as a magnetic photocatalyst. It can be fluidized and recovered by an applied magnetic field. Both separation and mixing efficiencies for recyclable fluids are enhanced.

Boron- and cerium-codoped TiO_2 photocatalysts were synthesized by Wei et al. (2007) using the modified sol–gel reaction process. The photocatalytic activities were evaluated by monitoring the degradation of dye acid red B (ARB). It was observed that the prepared photocatalysts were mixed oxides mainly consisting of titania, ceria, and boron oxide. The structure of TiO_2 could be transformed from amorphous to anatase and then to rutile by increasing the calcination temperature. The transformation was accompanied by the growth of particle size without any change in the phase structure of CeO_2. $B_{1.6}Ce_{1.0}$–TiO_2 showed that a few boron atoms were incorporated into titania and ceria lattice, whereas others existed as B_2O_3. There are two oxidation states of cerium ions, that is, Ce^{3+} and Ce^{4+}, and the atomic ratio of Ce^{3+}/Ce^{4+} was 1.86.

The UV–Vis adsorption band undergoes a shift toward the visible range (≤526 nm), when titania was doped with boron and cerium. When the atomic ratio of Ce/Ti was elevated to 1.0, the absorbance edge wavelength increased to 481 nm. But as the doping level of (Ce/Ti = 2.0) was increased, the absorbance edge wavelength decreases. The degradation of ARB dye indicated that the photocatalytic activities of boron- and cerium-codoped TiO_2 were much higher than that of P25. The photocatalytic activity was found to increase as the boron doping was increased, whereas it was decreased when the Ce/Ti atomic ratio was greater than 0.5. The optimum atomic ratio of B/Ti and Ce/Ti was 1.6 and 0.5, respectively.

Another catalyst, TiO_2/V_2O_5, was prepared by Liu et al. (2007) via the sol–gel method doped with rare earth ions. Here, titanium tetrapropoxide and vanadium pentoxide were used as initial components of the composite catalyst and rare earth ions were used as dopant. Degradation of methyl orange (MO) in aqueous solution was carried out under UV irradiation to evaluate the photoactivity of the prepared photocatalyst. It was concluded that the catalyst was composed of anatase and rutile, and rare earth metal ions were highly dispersed in the composite catalyst.

Mn–TiO_2/SiO_2 (silica gel loaded with manganese-doped TiO_2) photocatalysts have been prepared by Xu et al. (2008) via the sol–gel method. Photocatalytic activities were found to be enhanced in photocatalytic degradation of MO over Mn–TiO_2/SiO_2. It was also observed that a Ti–O–Si or Ti–O–Mn bond was formed on the surface of the photocatalyst. Mn is doped as a mixture of Mn^{2+} and Mn^{3+} on the surface of 1.0 mol% Mn–TiO_2/SiO_2, where Mn^{3+} appears to trap electrons and prohibit the electron–hole recombination. The electrons trapped in the Mn^{3+} site are subsequently transferred to the adsorbed O_2, which results in the reduction of electron–hole pair combination.

Titania nanocomposite codoped with metallic silver and vanadium oxide can be prepared by a one-step sol–gel–solvothermal method in the presence of a triblock copolymer surfactant (P123) (Yang et al. 2010). It results in an Ag/V–TiO_2 three-component junction system that has an anatase/rutile (weight ratio of 73.8:26.2) mixed phase structure with narrower band gap (2.25 eV), and metallic Ag particles of extremely small sizes (~12 nm) well distributed on the surface of the composite. The Ag/V–TiO_2 nanocomposite can be used as a visible and UV light driven photocatalyst to degrade dyes like rhodamine B and coomassie brilliant blue G-250 in aqueous solutions. The highest visible and UV-light photocatalytic activity was observed at 1.8% Ag and 4.9% V doping, respectively.

Antibacterial surface coating is the best way to suppress microbial infections. TiO_2-based coating is quite commonly used because photocatalysis inactivation has proved to be a capable mechanism for bacteria and fungi peroxidation (Lu et al. 2011). Pure $BaTiO_3$, TiO_2, or their simple mixture do not show any significant photocatalytic effect under visible light, but when these oxides are synthesized as nanoparticulate composite TiO_2–$BaTiO_3$ film under controlled conditions, antibacterial photocatalytic activity on *Staphylococcus aureus* was observed for the resulting crystals under visible light.

The glass crystallization method (GCM) was used for the preparation of TiO_2–$BaTiO_3$. GCM allows preparation of mixtures of particles much more homogeneous than is possible with ceramic preparation methods since both the phases are formed simultaneously. When the synthesis was carried out at high temperature, the oxygen can evaporate, creating oxygen vacancies with +2 charges in the composite structure. The band gap is also lowered by 0.2 eV, leading to trapping of one or two electrons. The TiO_2–$BaTiO_3$ powder was physically characterized and utilized for film preparation. The film was further physically characterized and investigated for its antibacterial activity (Stanca et al. 2012).

9.3.2 Carbon Nanotube–Based Composites

The mixture of titania and CNT also provides a larger surface area, where pollutants (organic or inorganic reactants) can be adsorbed, which is a key process in the photocatalytic destruction of pollutants. Thus, CNT–TiO_2 mixtures and composites exhibit appreciable photocatalytic activities beyond the anatase/rutile composites.

There are two mechanisms proposed to explain the enhancement of photocatalytic properties of CNT–TiO_2 composites. The first one is by Hoffmann et al. (1995). According to this mechanism, a high-energy photon excites an electron from the valence band (VB) to the conduction band (CB) of anatase TiO_2. As a result, photogenerated electrons are formed in the space charge regions, which are transferred to the CNTs, and holes remain on the TiO_2 to take part in redox reactions.

Wang et al. (2005a) proposed the second mechanism, where the CNTs act as sensitizers and transfer electrons to the TiO_2. A photogenerated electron was injected in the CB of TiO_2 allowing for the formation of superoxide radicals (O_2^-) by adsorbed molecular oxygen. The positively charged nanotubes removed an electron from the VB of the TiO_2, leaving a hole there. Thus, TiO_2 is then positively charged, which reacts with adsorbed water to form hydroxyl radicals.

A mixture of two types of semiconductor particles, and semiconductor particles with metal particles have been tried, but carbon particles with anatase showed photocatalytic enhancements in most cases. This concept can then be further extended to clear carbon structures with modified electronic properties. Carbon nanotube-anatase titanium dioxide (CNT–TiO_2) composite systems have been considered for many applications including their latent use to deal with environmental problems. Fullerenes and CNTs with metallic or semiconducting carbons are good candidates for a deeper approach into the semiconductor junction of titania. In addition, CNTs have excellent mechanical properties and a large specific surface area (>150 m²/g) (Lee et al. 2005).

However, the CNT–TiO₂ nanocomposite system is more complex than expected. The two distinct contributions from the CNT–TiO₂ composite were reported by Pyrgiotakis et al. (2005). One is that the C–O–Ti bond extends the light absorption to longer wavelengths, that is, similar to carbon-doped titania, thus potentially leading to improvement in the photocatalytic activity of titania. The second is the electronic configuration of the CNTs. Arc-discharge grown and chemical vapor deposition (CVD)-grown CNTs were coated with TiO₂ via the sol–gel processes. Both nanocomposites seem to be structurally similar, but the photocatalytic dye degradation rate for the arc-discharge CNTs was found to be 10 times higher than with the CVD-grown CNT nanocomposite.

The difference in their activity was therefore considered to be due to the electronic nature of the CNTs. The electronic-band structure of the CNT is an important factor than the chemical bond between the CNT and TiO₂ particularly in reference to the photocatalysis. It was observed that during photocatalytic degradation, oxidation of CNT is expected up to some extent. The oxidized portions of the CNTs may initially show opportunities for defect states permitting better photogeneration of electron–hole pairs; however, complete degradation of the CNTs in long term is likely to reduce the photocatalytic ability of the composite system.

Since p–n heterojunction photocatalysts have higher photocatalytic activities than single-phase catalysts, a novel Cu₂O octadecahedron/TiO₂ quantum dot (Cu₂O–O/ TiO₂–QD) p–n heterojunctions composite was designed and synthesized by Xu et al. (2016). Cu₂O octadecahedra (Cu₂O–O) with exposed {110} facets and {100} facets were synthesized and then highly dispersed TiO₂ QDs (TiO₂–QDs) were loaded on Cu₂O–O by the precipitation of TiO₂–QDs sol in the presence of absolute ethanol. These exhibited high stability in the MO degradation process. This high visible light photocatalytic activity was attributed to more utilization of light, effective separation of photoexcited electron–hole pairs, and instant scavenging of holes in the unique heterojunction structure.

Deng et al. (2016) successfully synthesized Ag/Cu₂O microstructures with diverse morphologies by a facile one-step solvothermal method. The morphologies transformed from microcubes for pure Cu₂O to microspheres with rough surfaces for Ag/Cu₂O. As-prepared sample was used in the photocatalytic degradation of MO in aqueous solution under visible light irradiation. The photocatalytic efficiencies of the dye were found to increase up to a maximum, but it decreases with increased amount of AgNO₃. This indicates that the photocatalytic activities were significantly influenced by the amount of AgNO₃ during the preparation process.

9.4 GRAPHENE-BASED COMPOSITES

Graphene (GR) has attracted much attention among the more recently discovered carbonaceous materials (Novoselov et al. 2004) with a unique sp² hybrid carbon network. Carbonaceous nanomaterials have unique structures and properties that can add attractive features to some photocatalysts (Eder 2010). Generally, enhancement

in photocatalytic activity is ascribed to the suppressed recombination of photogenerated electron–hole pairs, extended excitation wavelength, and increased surface-
adsorbed reactant. Leary and Westwood (2011) have reported recent progress in the
development of TiO_2/nanocarbon photocatalysts covering activated carbon, fullerenes, CNTs, GR, and other novel carbonaceous nanomaterials. It shows wide applications in fields such as nanoelectronics, sensors, catalysts, and energy conversion
(Johnson et al. 2010). GR-based architectures are highly desirable in the field of photocatalysis as they have promising energy and environmental applications. Single-
layer GR sheets not only provide high-quality two-dimensional (2D) photocatalyst
support, but also a 2D circuit board, with an attractive potential to connect their
perfect electrical and redox properties (Figure 9.2).

Graphene oxide (GO) suspended in ethanol undergoes reduction as it accepts
electrons from the UV-irradiated TiO_2 suspensions. This reduction is accompanied
by changes in the absorption of the GO, as evident from shifts in the color of the
suspension from brown to black. A direct interaction between TiO_2 particles and GR
sheets hinders the collapse of exfoliated sheets of GR. Solid films cast on a borosilicate glass gap separated by gold-sputtered terminations show an order of magnitude
decrease in lateral resistance following reduction with the TiO_2 photocatalyst. The
photocatalytic methodology not only provides an on-demand UV-assisted reduction
technique, but it also opens up new avenues to obtain photoactive GR-semiconductor
composites (Williams et al. 2008).

Charge separation was observed upon the UV irradiation of a deaerated suspension of TiO_2 colloids. The holes are scavenged to produce ethoxy radicals in the
presence of ethanol, thus leaving the electrons to accumulate within TiO_2 particles.
The accumulated electrons serve to interact with the GO sheets in order to reduce
certain functional groups.

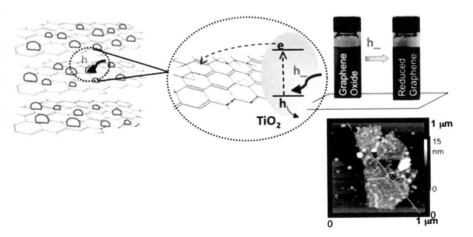

FIGURE 9.2 TiO_2–GR nanocomposite and its response under UV excitation (left and center).
Color of TiO_2 NPs changes with GO before and after UV irradiation for 2 hours (upper right);
atomic force microscopy (AFM) image of single GR sheet with bound TiO_2 NPs (bottom
right). (Adapted from Williams, G. et al., *ACS Nano*, 2, 1487–1491, 2008. With permission.)

$$TiO_2(e) + h\nu \longrightarrow TiO_2(h + e) \xrightarrow{\text{C}_2\text{H}_5\text{OH}} TiO_2(e) + {}^{\cdot}C_2H_4OH + H^+ \quad (9.1)$$

$$TiO_2(e) + Graphene\,oxide\,(GO) \longrightarrow TiO_2 + Reduced\,graphene\,oxide\,(rGO) \quad (9.2)$$

A dilute suspension of TiO_2–GR was placed in the form of a drop onto a heated mica substrate. It was evident that the individual GR sheets are coupled to TiO_2 NPs. NPs such as TiO_2 can interact with the GO sheets through physisorption, electrostatic binding, or through charge transfer interactions. Gold NPs are deposited on GR sheets through physisorption (Muszynski et al. 2008). TiO_2–GR composite samples before and after the UV irradiation exhibited similar particle density, which indicated that the binding of the TiO_2 particles to GR is not affected by the UV-assisted reduction. The initial binding of the TiO_2 particle through the carboxylate linkage remains unaffected and thus is likely to keep them intact in the composite.

A chemically bonded TiO_2 (P25)–GR nanocomposite photocatalyst with GO and P25 has been obtained using a facile one-step hydrothermal method (Zhang et al. 2010). Both reduction of GO and loading of P25 can be achieved during the hydrothermal reaction. The prepared P25–GR photocatalyst possessed great adsorptivity of dyes, extended light absorption range, and efficient charge separation properties, which was rarely reported in other TiO_2–carbon photocatalysts. A significant enhancement in the photodegradation of methylene blue (MB) was observed with P25–GR compared to the bare P25 and P25–CNTs with the same carbon content (Figure 9.3).

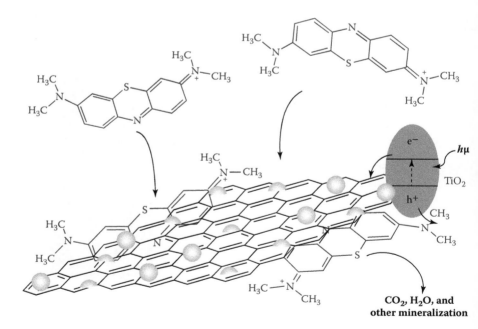

FIGURE 9.3 Structure of P25–GR and photodegradation of methylene blue (MB) over P25–GR. (Adapted from Zhang, H. et al., *ACS Nano*, 4, 380–386, 2010. With permission.)

P25 NPs are dispersed on the GR support, and the carbon platform plays the following important roles during the photodegradation of MB:

- *Increases the adsorptivity of catalyst:* MB molecules could transfer from the solution to the catalyst surface and be adsorbed with offset face-to-face orientation via π–π conjugation between MB and aromatic regions of the GR. As a result, the adsorptivity of dyes increases as compared to P25 alone.
- *Light absorption is extended:* The chemical bonds of Ti–O–C and good transparency of GR result in a red shift in the photoresponding range and facilitate a more efficient utilization of light for the catalyst.
- *Charge recombination is suppressed:* GR acts as an acceptor of the photogenerated electrons by P25 and ensures fast charge transportation because of its high conductivity. Hence, an effective charge separation can be achieved.

$BaTiO_3$ is a large band gap semiconductor and therefore absorbs the UV light only in the solar spectrum. A series of $BaTiO_3$–GR nanocomposites with different weight addition ratios of GO have been synthesized by Wang et al. (2015) by means of a facile one-pot hydrothermal approach. Reduced GO (rGO) and the intimate interfacial contact between $BaTiO_3$ NPs and the GR nanosheets could be obtained. The photocatalytic activity of the as-prepared $BaTiO_3$–GR composites for the degradation of MB was observed under visible light irradiation. A higher photocatalytic activity of $BaTiO_3$–GR composite was obtained compared to that of pure $BaTiO_3$. A new photocatalytic mechanism was suggested, where the role of GR in the $BaTiO_3$–GR nanocomposites was to act as an organic dye-like photosensitizer for a large band gap $BaTiO_3$. The photosensitization process of $BaTiO_3$ by GR transforms the wide band gap $BaTiO_3$ semiconductor into a visible light photoactive material. Such reactions can be very useful in finding suitable applications of $BaTiO_3$–GR nanocomposites in solar energy conversion.

GR has proved to be a promising candidate to construct effective GR-based composite photocatalysts with enhanced catalytic activities for solar energy conversion. Various GR-based composite photocatalysts have been developed and applied in a variety of fields during the past few years. The main focus of researchers was in the applications of GR-based composite photocatalysts for selective organic transformations such as reduction of CO_2 to renewable fuels and nitroaromatic compounds to amino compounds, oxidation of alcohols to aldehydes and acids, oxidation of tertiary amines, epoxidation of alkenes hydroxylation of phenol, and so on as compared with the traditional applications of GR-based nanocomposites for the nonselective degradation of pollutants, photodeactivation of bacteria, and water splitting to H_2 and O_2. GO as a cocatalyst in GO–organic species photocatalysts and GO itself as a photocatalyst have also been used for selective reduction of CO_2 (Yanga and Xu 2013).

GR–CNTs–TiO_2 composites have been synthesized (Wang et al. 2014) by using a one-pot solvothermal method. The photocatalytic activity of prepared composite photocatalysts was tested by degradation of dye MB and photoreduction of Cr(VI) in aqueous solution under UV light irradiation. The results showed that the composites exhibited an enhanced photocatalytic performance compared to a binary one (i.e., GR–TiO_2) as well as pristine TiO_2. The effect of CNT contents on the photocatalytic activity was also studied. It was observed that the photocatalytic performance of the GR–CNTs–TiO_2 hybrids was dependent on the proportion of CNTs in the composite. The mass ratio of CNTs:TiO_2 (5%) was proved to be optimal. The rate constants for MB degradation and Cr(VI) reduction were 2.2 and 1.9 times on the GR–TiO_2 composite, respectively. The enhanced activity can be attributed to the addition of CNTs, which can serve as charge transmitting paths,thus decreasing the recombination rate of photoinduced electron–hole pairs.

GR-like BN/AgBr hybrid materials were synthesized using the facile water bath method (Chen et al. 2014). The photocatalytic activity of the GR-like BN/AgBr hybrid materials was evaluated using MO as a target organic pollutant. The results indicated that the photocatalytic activity of AgBr could be significantly improved by coupling with a proper amount of GR-like BN. The optical loading amount of GR-like BN was observed to be 1 wt.%. MO molecules can be decomposed completely within 15 minutes under visible light irradiation, and the photocatalyst can be reused five times without losing any activity. No diffraction peaks of metallic Ag were present in the XRD pattern of the cycling sample, which indicated that introducing a small amount of GR-like BN can effectively suppress the reduction of silver ions. It was observed that a small amount of GR-like BN was quite beneficial for the separation of photogenerated electrons and holes, and as a consequence, the photoactivity of GR-like BN/AgBr was enhanced.

rGO–CdS nanorod composites were successfully prepared by a one-step microwave-hydrothermal method in an ethanolamine–water solution (Yu et al. 2014). These composites exhibited a high activity for the photocatalytic reduction of CO_2 to CH_4, even in the absence of a noble metal Pt cocatalyst. A high CH_4 production rate of 2.51 μmol/h/g was obtained with an rGO–CdS nanorod composite at an rGO content of 0.5 wt.%. This rate was 10 times more than with pure CdS nanorods and was also better than that observed for an optimized Pt–CdS nanorod composite. This high photocatalytic activity was due to the deposition of CdS nanorods onto the rGO sheets, which act as an electron acceptor and transporter and thus efficiently separated the photogenerated charge carriers. It was suggested that an inexpensive carbon material can be utilized as a substitute for noble metals like Pt in the photocatalytic reduction of CO_2.

Zhai et al. (2015) fabricated homogeneous GO-wrapped Bi_2WO_6 microspheres (GO/Bi_2WO_6). Optical properties of GO/Bi_2WO_6 composites were considered responsible for enhanced photocatalytic activity, which was also attributed to the synergetic effect between GO and Bi_2WO_6, causing fast generation and separation of photogenerated charge carriers.

Graphitic carbon nitride nanosheet (g-C_3N_4 or GCN NS) hybridized nitrogen-doped titanium dioxide (N–TiO_2 or NT) nanofibers (GCN/NT NFs) have been synthesized in situ by Han et al. (2015) via a simple electrospinning process combined with a modified heat-etching method and used for degradation of rhodamine B in aqueous solution. It was found that the GCN/NT NFs have a mesoporous structure, composed of g–C_3N_4 NSs, and N-doped TiO_2 crystallites. A high photocatalytic H_2 production rate of 8.931.3 μmol/h/g in aqueous methanol solution under simulated solar light was observed with GCN/NT NFs. This can be ascribed to the combined effects of g–C_3N_4 NSs and N-doped TiO_2 with enhanced light absorption intensity and improved electron transport ability. Such quasi-one-dimensional hybrid materials have potential in the field of solar energy conversion.

9.5 ZINC OXIDE–BASED COMPOSITES

Novel cobalt oxide–doped $ZnFe_2O_4$–Fe_2O_3–ZnO mixed oxides were synthesized with a citric acid complex method by Bai (2009). Mainly $ZnFe_2O_4$, α-Fe_2O_3, amorphous ZnO, and Fe_2O_3 in the 6 mol% cobalt oxide–doped products were there in the sample calcined at 500°C. It was observed that 5–10 mol% cobalt oxide doping could significantly improve the formation of $ZnFe_2O_4$ with varied phase composition of the mixed oxides. Cobalt oxide doping was found to improve the photocatalytic activity of the mixed oxides remarkably for degradation of phenol.

ZnO–rGO composites can be effectively synthesized via UV-assisted photocatalytic reduction of GO by ZnO NPs in ethanol. Their morphology and structure were characterized by different techniques. The photocatalytic performance of the composite was observed for reduction of Cr(VI). The rGO nanosheets were decorated densely by ZnO NPs in the composites, which displays a good combination between rGO and ZnO. ZnO–rGO composites exhibit an enhanced photocatalytic performance in reduction of Cr(VI) with a maximum removal rate of 96% under UV light irradiation as compared with pure ZnO (67%), which was attributed to the increased light absorption intensity and range as well as the reduction of electron–hole pair recombination in ZnO with the introduction of rGO (Liu et al. 2012).

The rGO-hierarchical ZnO hollow sphere composites can also be prepared via simple ultrasonic treatment of the solution, which contains GO, $Zn(CH_3COO)_2$, dimethyl sulfoxide (DMSO), and H_2O (Luo et al. 2012). There is an effective reduction of GO to rGO and uniform dispersion of ZnO hollow spheres consisting of NPs on the surface of rGO sheets during the ultrasonic process (Figure 9.4). The optimum synergetic effect of rGO–ZnO composites was found at an rGO mass ratio of 3.56%. The photocurrent and photodegradation efficiency of rGO–ZnO composites for MB

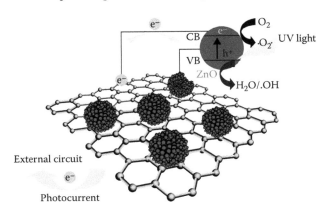

FIGURE 9.4 ZnO–rGO hollow sphere composites. (Adapted from Luo, Q. P. et al., *J. Phys. Chem. C*, 116, 8111–8117, 2012. With permission.)

was improved by five times and 67%, respectively, compared with those of pure ZnO hollow spheres. The enhancement of photocurrent and photocatalytic activity can be attributed to the suppression of charge carriers recombination resulting from the interaction between ZnO and rGO.

Synthesis of $ZnFe_2O_4/ZnO$ nanocomposites immobilized on GR was reported by Sun et al. (2013). It has enhanced photocatalytic activity under solar light irradiation. The molar ratio of $ZnFe_2O_4$ to ZnO and the content of GR could be controlled by adjusting the amount of zinc salts and GO dispersions. The best photocatalytic activity was observed under solar light irradiation, when the molar ratio and the weight ratio of GR to $ZnFe_2O_4/ZnO$ was kept at 0.1 and 0.04, respectively.

$CuInSe_2$–ZnO nanocomposites were prepared by Bagheri et al. (2014) via a solvothermal process. It was observed that crystallite size, BET surface area, and optical absorption of the samples varied with the addition of $CuInSe_2$ to ZnO. The optical band gap values of these nanocomposites were calculated to be about 3.37–2.1 eV, which means that there is a red shift from that of pure ZnO. These red shifts support the incorporation of $CuInSe_2$ in the zinc oxide lattice. The photocatalytic activity of prepared samples was observed in the degradation of Congo red. The effect of the different parameters such as pH, Congo red concentration, $CuInSe_2$ content, and irradiation sources of UV and visible light has been studied. The activity of this sample was improved confirming that the addition of $CuInSe_2$ was effective in enhancing the photocatalytic activity of ZnO.

An enhanced photocatalytic activity and antiphotocorrosion property of semiconductor ZnO was reported by Han et al. (2014) by coupling it with versatile carbon materials such as C_{60}, CNT, GR, and other carbon materials. The primary roles of carbon materials were in boosting the photoactivity and photostability of ZnO. Three main kinds of mechanisms with regard to antiphotocorrosion of ZnO by coupling with carbon have been discussed.

An ultrasonic-assisted method was used by Sara et al. (2015) for preparation of Ag/AgCl sensitized ZnO nanostructures by a one-pot procedure in water without any postpreparation treatments. ZnO and AgCl have wurtzite hexagonal structure in the nanocomposites and cubic crystalline phases, respectively, in the nanocomposites. It was found that their surface morphologies remarkably change with increasing mole fraction of silver chloride. Photocatalytic activity of the nanocomposites was evaluated using aqueous solution of methylene blue under visible light irradiation. Enhancement in activity of the nanocomposite was attributed to its ability to absorb the visible light and separation of electron–hole pairs. It was also reported that the photocatalyst has good activity even after five successive cycles.

9.6 METALLATE COMPOSITES

The Cr-doped $Ba_2In_2O_5/In_2O_3$ (C–BIO) oxide semiconductors were synthesized by Wang et al. (2005b) via a solid-state reaction method. This novel composite photocatalyst system has enhanced activity for water splitting. The C–BIO powder samples were duly characterized with different techniques. The photocatalytic activities of Pt- or NiO-loaded C–BIO and individual precursor materials were evaluated by H_2 evolution.

Aqueous CH_3OH solution and pure water splitting were selected as models under visible light and by UV light irradiation, respectively. It was found that the composite C–BIO showed a higher H_2 evolution rate as compared to individual components. They also discussed overall band structure, charge carrier excitation, separation, transportation, and the redox reactions for H_2 and O_2 evolution in the C–BIO system.

$Co_3O_4/BiVO_4$ composite photocatalyst has been synthesized by Long et al. (2006) with a p–n heterojunction semiconductor structure via impregnation method. It was observed that Co is present as p-type Co_3O_4 and it disperses on the surface of n-type $BiVO_4$ to constitute a heterojunction composite. The photocatalyst exhibited enhanced photocatalytic activity for phenol degradation under visible light irradiation. The highest efficiency was observed with the sample calcined at 300°C with 0.8 wt.% of cobalt content.

A silver mirror reaction (SMR) was used to prepare nano silver particles loaded on micrometer-sized TiO_2, nanosized TiO_2, and $BiVO_4$, with an old and well-known method (Shan et al., 2008). The photocatalytic activities of these photocatalysts were evaluated by the degradation of MO and hydrogen production under UV and visible light irradiation, respectively. It was found that the photocatalytic activities of Ag/TiO_2 and Ag/$BiVO_4$ composites prepared by the SMR were remarkably higher than those prepared by the photoinduced deposition method.

Various compositions of nanosized $(NiMoO_4)_x$-doped $Bi_2Ti_4O_{11}$ ($x = 0.01, 0.05, 0.1$) composites have been prepared by a chemical solution decomposition method using triethanolamine as a complexing agent (Ghorai et al. 2008). In this sample, Ni(II) was one of the reactive species on the catalyst surface and the Mo(VI) ion helped to compensate the charge of the lattice. The photocatalysts based on these compositions have been tested for photobleaching of MO solution under Hg lamp. The average particle size of nickel molybdate doped bismuth titanate was around 30 ± 2 nm. It was observed that nickel molybdate-doped bismuth titanate $(NiMoO_4)_x$ $(Bi_2Ti_4O_{11})_{1-x}$ ($NM_x BT_{1-x}$; $x = 0.01$) composite was found to be more photoactive as compared to all other compositions.

$V_2O_5/BiVO_4$ composite photocatalysts were prepared by the one-step solution combustion synthesis method (Jiang et al. 2009). The physical properties and photophysical properties of $V_2O_5/BiVO_4$ composite photocatalysts were observed. The composite photocatalysts exhibit the enhanced photocatalytic properties for degradation of MB. The mechanism of improved photocatalytic activity has been suggested (Figure 9.5).

A new photocatalyst Er^{3+}:$YAlO_3$/ZnO composite was prepared by the ultrasonic dispersion and liquid boiling method by Wang et al. (2009), which could effectively utilize visible light. ARB dye was used as a model compound for photocatalytic degradation under solar light irradiation to evaluate the photocatalytic activity of the Er^{3+}:$YAlO_3$/ZnO composite. The effects of Er^{3+}:$YAlO_3$ content, heat-treated temperature, and time of photocatalytic activity of Er^{3+}:$YAlO_3$/ZnO composite were observed through the degradation of ARB dye under the solar light. It was found that the photocatalytic activity of Er^{3+}:$YAlO_3$/ZnO composite was much higher than that of pure ZnO powder. This Er^{3+}:$YAlO_3$/ZnO composite may provide a new way to take advantage of ZnO in sewage treatment aspects using solar energy.

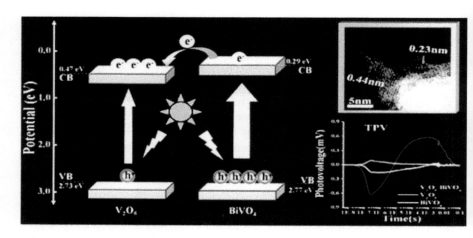

FIGURE 9.5 Enhanced photocatalytic activity of $V_2O_5/BiVO_4$. (Adapted from Jiang, H. et al., *J. Alloys Compd.*, 479, 821–827, 2009. With permission.)

Cheng et al. (2010) prepared nanostructured AgI/BiOI composites by a facile, one-step, and low temperature chemical bath method using $Bi(NO_3)_3$, $AgNO_3$, and KI. The phase structures, morphologies, and optical properties of the samples were characterized. The photoluminescence (PL) intensity of AgI was greatly decreased on combining it with BiOI, which indicates a corresponding decreased recombination of the carriers. The photocatalytic properties of the as-prepared products were observed with the degradation of MO and phenol at room temperature under visible light irradiation. The AgI/BiOI composites showed much higher photocatalytic performances as compared to BiOI as well as AgI.

It was also found that the AgI amount in the AgI/BiOI composites played an important role in deciding photocatalytic properties, and the optimized ratio was obtained at 20%. This dramatic enhancement in visible light photocatalytic performance of the AgI/BiOI composites could be attributed to the effective electron–hole separations at the interfaces of the two semiconductors, which facilitates the transfer of the photoinduced carriers. The photoinduced holes (h_{VB}^+) were considered to be the dominant active species in the photodegradation process. The photocatalytic performances of the AgI/BiOI composites were maintained and the sample can be recycled. AgI was found stable in the composites under visible irradiation. AgI/BiOI composites could be used as stable and efficient visible-light-induced photocatalysts.

Chrysanthemum-analogous Bi_2O_3–Bi_2WO_6 composite microspheres, assembled by nanosheets, were synthesized through a one-step hydrothermal route with the aid of surfactant templates (Ge et al. 2011). The photocatalytic activity of the Bi_2O_3–Bi_2WO_6 composite microspheres was evaluated using rhodamine B as a model contaminant. It was observed that over 99% of rhodamine B was degraded within 10 minutes under the exposure of sunlight. The Bi_2O_3–Bi_2WO_6 composite microspheres show enhanced photocatalytic performances as compared to separate Bi_2O_3, Bi_2WO_6, and the conventional P25.

Nashim et al. (2013) prepared a series of $Gd_2Ti_2O_7/In_2O_3$ heterojunction-based composite photocatalysts using solid-state reaction methods. Their activity was also evaluated by monitoring the photocatalytic H_2 production under visible light ($\lambda \geq 400$ nm). It was found that the modification of $Gd_2Ti_2O_7$ with In_2O_3 resulted in a remarkable enhancement of the activity for H_2 production with respect to the individual components under visible light. It was found that $Gd_2Ti_2O_7$ and In_2O_3 have a well-matched band potential to exhibit synergistic effects. The enhanced activity may be due to light absorption properties and the reduction in the recombination of charge carriers at the composite interface. The catalyst $Gd_2Ti_2O_7/In_2O_3$ with 60 at% indium (60 IGTO) showed the highest activity in comparison to the other composite materials prepared.

A single compound may have one desired property, but it may also be associated with some undesired quality. Another component has some other property, which may compensate for the demerit of the first component. If these two different components form a composite, then it is likely to develop into a more efficient photocatalyst. It is well known that CdS has a disadvantage in that it is photocorroded, but the use of a carbon composite of CdS can overcome this problem.

REFERENCES

Avraham-Shinman, A. and Y. Paz. 2006. Photocatalysis by composite particles containing inert domains. *Isr. J. Chem.* 46 (1): 33–43.

Bagheri, M., A. R. Mahjoub, and B. Mehri. 2014. Enhanced photocatalytic degradation of Congo red by solvothermally synthesized $CuInSe_2$-ZnO nanocomposites. *RSC Adv.* 4 (42): 21757–21764.

Bai, J. 2009. Synthesis and photocatalytic activity of cobalt oxide doped $ZnFe_2O_4$–Fe_2O_3–ZnO mixed oxides. *Mater. Lett.* 63 (17): 1485–1488.

Cao, L., Z. Gao, S. L. Suib, T. N. Obee, S. O. Hay, and J. D. Freihaut. 2000. Photocatalytic oxidation of toluene on nanoscale TiO_2 catalysts: Studies of deactivation and regeneration. *J. Catal.* 196 (2): 253–261.

Chen, J., J. Zhu, Z. Da, H. Xu, J. Yan, H. Ji, H. Shu, and H. Li. 2014. Improving the photocatalytic activity and stability of graphene-like BN/AgBr composites. *Appl. Surf. Sci.* 313: 1–9.

Cheng, H., B. Huang, Y. Dai, X. Qin, and X. Zhang. 2010. One-step synthesis of the nanostructured AgI/BiOI composites with highly enhanced visible-light photocatalytic performances. *Langmuir.* 26 (9): 6618–6624.

Choi, H., E. Stathatos, and D. D. Sionysiou. 2006. Sol gel preparation of mesoporous photocatalytic TiO_2 film and TiO_2/Al_2O_3 composite membranes for environmental applications. *Appl. Catal. Environ.* 63 (1–2): 60–67.

Chu, S. Z., S. Inoue, K. Wada, D. Li, H. Haneda, and S. Awatsu. 2003. Highly porous $(TiO_2$-SiO_2-$TeO_2)/Al_2O_3/TiO_2$ composite nanostructures on glass with enhanced photocatalysis fabricated by anodization and sol-gel process. *J. Phys. Chem. B.* 107 (27): 6586–6589.

Deng, X., C. Wang, E. Zhou, J. Huang, M. Shao, X. Wei, X. Liu, M. Ding, and X. Xu. 2016. One-step solvothermal method to prepare Ag/Cu_2O composite with enhanced photocatalytic properties. *Nanoscale Res. Lett.* 11: 29. 10.1186/s11671-016-1246-7.

Dey, N. K., M. J. Kim, K. D. Kim, H. O. Seo, D. Kim, Y. D. Kim et al. 2011. Adsorption and photocatalytic degradation of methylene blue over TiO_2 films on carbon fibre prepared by atomic layer deposition. *J. Mol. Catal. A: Chem.* 337 (1–2): 33–38.

Eder, D. 2010. Carbon nanotube-inorganic hybrids. *Chem. Rev.* 110 (3): 1348–1385.

Fu, W., H. Yang, M. Li, M. Li, N. Yang, and G. Zou. 2005. Anatase TiO_2 nanolayer coating on cobalt ferrite nanoparticles for magnetic photocatalyst. *Mater. Lett.* 59 (27): 3530–3534.

Ge, M., Y. Li, L. Liu, Z. Zhou, and W. Chen. 2011. Bi_2O_3–Bi_2WO_6 Composite microspheres: Hydrothermal synthesis and photocatalytic performances. *J. Phys. Chem. C.* 115 (13): 5220–5225.

Ghorai, T. K., D. Dhak, S. Dalai, and P. Pramanik, 2008. Preparation and photocatalytic activity of nano-sized nickel molybdate ($NiMoO_4$) doped bismuth titanate ($Bi_2Ti_4O_{11}$) (NMBT) composite. *J. Alloys. Compd.* 463 (1–2): 390–397.

Han, C., Y. Wang, Y. Lei, B. Wang, N. Wu, Q. Shi, and Q. Li. 2015. *In situ* synthesis of graphitic-C_3N_4 nanosheet hybridized N-doped TiO_2 nanofibers for efficient photocatalytic H_2 production and degradation. *Nano Res.* 8 (4): 1199–1209.

Han, C., M.-Q. Yang, B. Weng, and Y.-J. Xu. 2014. Improving the photocatalytic activity and anti-photocorrosion of semiconductor ZnO by coupling with versatile carbon. *Phys. Chem. Chem. Phys.* 16: 16891–16903.

Hoffmann, M. R., S. T. Martin, W. Choi, and D. W. Bahnemann. 1995. Environmental applications of semiconductor photocatalysis. *Chem. Rev.* 95 (1): 69–96.

Ishikawa, Y. and Y. Matsumoto. 2001. Electrodeposition of TiO_2 photocatalyst into nano-pores of hard alumite. *Electrochim. Acta.* 46 (18): 2819–2824.

Jiang, H., M. Nagai, and K. Kobayashi. 2009. Macroporous V_2O_5–$BiVO_4$ composites: Effect of heterojunction on the behavior of photogenerated charges. *J. Phys. Chem. C.* 479 (1–2): 821–827.

Johnson, J., A. Behnam, S. J. Pearton, and A. Ural. 2010. Hydrogen sensing using Pd-functionalized multi-layer graphene nanoribbon networks. *Adv. Mater.* 22 (43): 4877–4880.

Leary, R. and A. Westwood. 2011. Carbonaceous nanomaterials for the enhancement of TiO_2 photocatalysis. *Carbon.* 49 (3): 741–772.

Lee, S. H., S. Pumprueg, B. Moudgil, and W. Sigmund. 2005. Inactivation of bacterial endospores by photocatalytic nanocomposites. *Colloid. Surf. B.* 40 (2): 93–98.

Liu, B. J., T. Torimoto, and H. Yoneyama. 1998. Photocatalytic reduction of carbon dioxide in the presence of nitrate using TiO_2 nanocrystal photocatalyst embedded in SiO_2 matrices. *J. Photochem. Photobiol. A.* 115 (3): 227–230.

Liu, J., R. Yang, and S. Li. 2007. Synthesis and photocatalytic activity of TiO_2/V_2O_5 composite catalyst doped with rare earth ions. *J. Rare Earth.* 25 (2): 173–178.

Liu, X., L. Pan, Q. Zhao, T. Lv, G. Zhu, T. Chen, T. Lu, Z. Sun, and C. Sun. 2012. UV-assisted photocatalytic synthesis of ZnO–reduced graphene oxide composites with enhanced photocatalytic activity in reduction of Cr(VI). *Chem. Eng. J.* 183: 238–243.

Long, M., W. Cai, J. Cai, B. Zhou, X. Chai, and Y. Wu. 2006. Efficient photocatalytic degradation of phenol over Co_3O_4/$BiVO_4$ composite under visible light irradiation. *J. Phys. Chem. B.* 110 (41): 20211–20216.

Lu, S. Y., D. Wu, Q. L. Wang, J. Yan, A. G. Buekens, and K. F. Cen. 2011. Photocatalytic decomposition on nano-TiO_2: Destruction of chloroaromatic compounds. *Chemosphere.* 82 (9): 1215–1224.

Luo, Q. P., X. Y. Yu, B. X. Lei, H. Y. Chen, D. B. Kuang, and C. Y. Su. 2012. Reduced graphene oxide-hierarchical ZnO hollow sphere composites with enhanced photocurrent and photocatalytic activity. *J. Phys. Chem. C.* 116 (14): 8111–8117.

Martin, C. R. 1994. Nanomaterials: A membrane-based synthetic approach. *Science.* 266 (5193): 1961–1966.

Muszynski, R., B. Seger, and P. Kamat. 2008. Decorating graphene sheets with gold nanoparticles. *J. Phys. Chem. C.* 112 (14): 5263–5266.

Nashim, A., S. Martha, and K. M. Parida. 2013. $Gd_2Ti_2O_7/In_2O_3$: Efficient visible-light-driven heterojunction-based composite photocatalysts for hydrogen production. *ChemCatChem.* 5 (8): 2352–2359.

Novoselov, K. S., A. K. Geim, S. V. Morozov, D. Jiang, Y. Zhang, S. V. Dubonos et al. 2004. Electric field effect in atomically thin carbon films. *Science.* 306 (5696): 666–669.

Pyrgiotakis, G., S. H. Lee, and W. M. Sigmund. 2005. Advanced photocatalysis with anatase nano-coated multi-walled carbon nanotubes, presented at MRS Spring Meeting, San Francisco, CA.

Sara, N.-A., H.-Y. Aziz, and P. Mahsa. 2015. One-pot ultrasonic-assisted method for preparation of Ag/AgCl sensitized ZnO nanostructures as visible-light-driven photocatalysts. *Solid State Sci.* 40: 111–120.

Shan, Z., J. Wu, F. Xu, F. Q. Uang, and H. Ding. 2008. Highly effective silver/semiconductor photocatalytic composites prepared by a silver mirror reaction. *J. Phys. Chem. C.* 112 (39): 15423–15428.

Stanca, S. E., R. Muller, M. Urban, A. Csaki, F. Froehlich, C. Krafft, J. Popp, and W. Fritzsche. 2012. Photocatalyst activation by intrinsic stimulation in TiO_2–$BaTiO_3$. *Catal. Sci. Technol.* 2 (7): 1472–1479.

Sun, L., R. Shao, L. Tang, and Z. Chen. 2013. Synthesis of $ZnFe_2O_4/ZnO$ nanocomposites immobilized on graphene with enhanced photocatalytic activity under solar light irradiation. *J. Alloys Compd.* 564: 55–62.

Terashima, M., N. Inoue, S. Kashiwabara, and R. Fujumoto. 2001. Photocatalytic TiO_2 thin-films deposited by a pulsed laser deposition technique. *Appl. Surf. Sci.* 167: 535–538.

Wang, C., M. Cao, P. Wang, Y. Ao, J. Hou, and J. Qian. 2014. Preparation of graphene–carbon nanotube–TiO_2 composites with enhanced photocatalytic activity for the removal of dye and Cr (VI). *Appl. Catal. A: Chem.* 473: 83–89.

Wang, D., Z. Zou, and J. Ye. 2005b. Photocatalytic water splitting with the Cr-doped $Ba_2In_2O_5/In_2O_3$ composite oxide semiconductors. *Chem. Mater.* 17 (12): 3255–3261.

Wang J., Y. Xie, Z. Zhang, J. Li, X. Chen, L. Zhang, R. Xu, and X. Zhang. 2009. Photocatalytic degradation of organic dyes with Er^{3+}:$YAlO_3/ZnO$ composite under solar light. *Sol. Energy Mater. Solar Cells.* 93 (3): 355–361.

Wang, R. X., Q. Zhu, W. S., Wang, C. M. Fan, and A. W. Xu. 2015. $BaTiO_3$–graphene nanocomposites: Synthesis and visible light photocatalytic activity. *New J. Chem.* 39 (6): 4407–4413.

Wang, W. D., P. Serp, P. Kalck, and J. L. Faria. 2005a. Visible light photodegradation of phenol on MWNT-TiO_2 composite catalysts prepared by a modified sol–gel method. *J. Mol. Catal. A: Chem.* 235(1–2): 194–199.

Wei, C. H., X. H. Tang, J. R. Liang, and S. Y. Tan. 2007. Preparation, characterization and photocatalytic activities of boron- and cerium-codoped TiO_2. *J. Environ. Sci.* 19 (1): 90–96.

Williams, G., B. Seger, and P. V. Kamat. 2008. TiO_2-graphene nanocomposites. UV-assisted photocatalytic reduction of graphene oxide. *ACS Nano.* 2 (7): 1487–1491.

Xu, X., Z. Gao, Z. Cui, Y. Liang, Z. Li, S. Zhu, X. Yang, and J. Ma. 2016. Synthesis of Cu_2O octadecahedron/TiO_2 quantum dot heterojunctions with high visible light photocatalytic activity and high stability. *ACS Appl. Mater. Interfaces.* 8 (1): 91–101.

Xu, Y., B. Lei, L. Guo, W. Zhou, and Y. Liu, 2008. Preparation, characterization and photocatalytic activity of manganese doped TiO_2 immobilized on silica gel. *J. Hazard. Mater.* 160 (1): 78–82.

Yang, X., F. Ma, K. Li, Y. Guo, J. Hu, W. Li, M. Huo, and Y. Guo. 2010. Mixed phase titania nanocomposite codoped with metallic silver and vanadium oxide: New efficient photocatalyst for dye degradation. *J. Hazard Mater.* 175 (1–3): 429–438.

Yanga, M. Q. and Y. J. Xu. 2013. Selective photoredox using graphene-based composite photocatalysts. *Phys.Chem. Chem. Phys.* 15 (44): 19102–19118.

Yu, J., J. Jin, B. Cheng, and M. Jaroniec. 2014. A noble metal-free reduced graphene oxide–CdS nanorod composite for the enhanced visible-light photocatalytic reduction of CO_2 to solar fuel. *J. Mater. Chem. A.* 2 (10): 3407–3416.

Zhai, J., H. Yu, H. Li, L. Sun, K. Zhang, and H. Yang. 2015. Visible-light photocatalytic activity of graphene oxide-wrapped Bi_2WO_6 hierarchical microspheres. *Appl. Surface Sci.* 344: 101–106.

Zhang, H., X. Lv, Y. Li, Y. Wang, and J. Li. 2010. P25-graphene composite as a high performance photocatalyst. *ACS Nano.* 4 (1): 380–386.

Znaidi, L., R. Seraphimova, J. F. Bocquet, C. Colbeau-Justin, and C. Pommier. 2001. A semicontinuous process for the synthesis of nanosize TiO_2 powders and their use as photocatalysts. *Mat. Res. Bull.* 36 (5–6): 811–825.

10 Immobilization

10.1 INTRODUCTION

Photocatalytic oxidation has been used to remove organic pollutants from drinking water and this field has attracted considerable interest (Malato et al. 2009). TiO_2 is commonly used as a photocatalyst for purification of water because it is stable, harmless, inexpensive, and can be potentially activated by solar insolation. TiO_2 in suspension is quite effective in utilizing sunlight, because suspended TiO_2 particles have a high specific surface area in the range from 50 to 300 m^2/g (Balasubramanian et al. 2004; Gumy et al. 2006) and therefore, mass transfer limitations are avoided and high photocatalytic activity is obtained (Mehrotra et al. 2003). However, a small transport limitation appeared on high catalyst loading. One of the major disadvantages is that it is difficult to separate small TiO_2 particles from water after treatment (Feitz et al. 2000; McCullagh et al. 2011; Thiruvenkatachari et al. 2008). The TiO_2 particles can be immobilized on a suitable support to overcome this problem. A number of attempts have been made to immobilize TiO_2 photocatalyst over different supports and at the same time increase the surface/volume ratio, resulting in an enhancement in photocatalytic oxidation efficiency.

Different types of materials have been used as a support to immobilize TiO_2, such as stainless steel (Chen and Dionysiou 2006; Yanagida et al. 2005), glass (Parra et al. 2004), cellulose fibers (Goetz et al. 2009), and so on. Stainless steel is an excellent support as it maintains structural integrity under high temperature (calcination), whereas other supports may soften and are deformed. It is also not susceptible to attack by different chemicals during the coating process. Glass, carbon-based materials, polymers, inorganic supports, and so on are among other good choices for immobilizing TiO_2. An overview of the field of immobilizing TiO_2 on some supports has been given by Pozzo et al. (1997) and Shan et al. (2010).

10.2 GLASS

Phosphate-containing intermediates are more stable as compared to chlorine-containing compounds and therefore a longer illumination is required for their complete mineralization. Lu et al. (1993) studied various factors affecting the photocatalytic degradation of organic pollutant such as dichlorovos, which is an organophosphorus insecticide. They utilized TiO_2 immobilized on glass for this purpose. The photoreactor coated with TiO_2 was illuminated with a 20 W black–light UV fluorescent tube. The aqueous solution containing dichlorvos was continuously pumped through this photoreactor. Various factors including, for example, initial dichlorvos concentration, dissolved oxygen, electrolytes, flow rate, and temperature affected the oxidation rate of dichlorvos. The initial quantum yield for the destruction of dichlorvos was 2.67%. Dichlorvos degradation rate was increased on increasing the flow rate and initial dichlorvos concentration. The photocatalytic oxidation of dichlorvos followed the Langmuir–Hinshelwood-type behavior. It was also found that reaction byproducts show an inhibiting effect on rate of degradation.

Fernández et al. (1995) synthesized TiO_2 supported on several rigid substrates such as glass and quartz using a dip coating procedure, and the deposition on stainless steel by an electrophoretic deposition process. As-synthesized samples were then tested against the photocatalytic degradation of malic acid. The sample supported on quartz showed the highest catalytic activity. The order of photocatalytic activity followed the decreasing order:

$$TiO_2/Quartz > TiO_2/Steel \approx TiO_2/Glass$$

It may be correlated with the presence of cationic impurities (Si^{4+}, Na^+, Cr^{3+}, Fe^{3+}) in the layer as a result of the necessary thermal treatments, which was necessary to improve the cohesion of the titania layer and its adhesion onto the support.

Nogueira and Jardim (1996) studied a photocatalytic reactor on a bench scale with solar light as the source of radiation using immobilized TiO_2 (Degussa P25) on a glass plate. Dichloroacetic acid (DCA) was selected as a model compound for this study. A linear dependence of degradation was observed with solar light intensity. Experiments with a single pass of solution as well as recirculation showed no mass transfer limitations in this case. Quantitative amounts of chloride ions were produced as a result of mineralization of DCA. An initial concentration of 5 mmol/L of DCA was reduced to 2 mmol/L after 2-minute irradiation. An exponential decay of degradation was observed with an increase of the molar flow rate, and a saturation was achieved around 1.5-mmol DCA per minute.

Ma et al. (2001) synthesized TiO_2 thin films coated on glass, indium tin oxide (ITO) glass, and p-type monocrystalline silicon. They investigated the photocatalytic degradation of rhodamine B (RhB) in an aqueous medium. It was observed that the microstructures of TiO_2 films were greatly affected by the substrate materials. The rutile form of TiO_2 was easily formed on the surface of ITO glass, as TiO_2 grows as closely packed particles in the form of elongated strips with an average size of 20 nm. Surface photovoltage spectra indicated charge transfer between the film and silicon substrate. This may be the main reason for higher photocatalytic reactivity of TiO_2 films grown on ITO glass and silicon substrates than film on glass substrate. Moreover, the different surface properties also affect the activity.

Photocatalyst TiO_2 was supported on a glass fiber by Ao et al. (2003) for indoor air purification under different humidity levels and residence time. TiO_2 was prepared by the sol–gel process. The conversion of the synthetic photocatalyst and P25 was adversely affected by increasing humidity and decreasing residence time. As-prepared photocatalyst was found to be highly active in photodegradation of indoor air pollutants such as benzene, toluene, o-xylene, and ethyl benzene even at very low concentrations that is, part per billion (ppb) levels. The feasibility of photocatalytic technology was also investigated by coinjecting other indoor air pollutants such as CO and NO_2 at ppb levels along with NO. It was observed that the conversion of CO was not at all promoted by the photodegradation of NO and no competitive effect was observed between these gases; however, the presence of NO promoted the conversion of NO_2. On the other hand, the conversion of NO is decreased due to the competition between NO and NO_2 for adsorption site present on catalyst.

Ryu et al. (2003) synthesized TiO_2 by the modified sol–gel method, immobilized into onto a glass tube, and used it for alachlor photodegradation. The thickness of immobilized TiO_2 film was 174 nm and the average diameter of TiO_2 particles was about 10–15 nm by a 5-time dip-coating. A typical anatase type was obtained, when TiO_2 film was calcined at 300°C for 1 hour. The stability of TiO_2 film was 4% better when prepared by a modified sol–gel method than the typical sol–gel method. The removal rate of alachlor with both Fe^{3+} and UV radiation in the absence and presence of TiO_2 were 0.28 mg/L/h and 0.32 g/L/h, respectively, which was higher by 14%. TOC concentration during the alachlor degradation decreased with both TiO_2 and UV radiation in the absence of added Fe^{3+} from 100% to 81% and 51% to 44% within a time period of 4–10 hours, while TOC concentration with Fe^{3+} and UV radiation in the absence of TiO_2 decreased from 100% to 70% in 10 hours.

Doll and Frimmel (2004) investigated two main routes for the preparation of photocatalytically active TiO_2 films on glass substrates: (1) use of TiO_2 powder and (2) *in situ* generation of the catalyst via hydrolysis of titanium tetraisopropoxide (TTIP) or $TiCl_4$. The activities of the catalyst films were estimated by measuring the degradation of DCA, clofibric acid, and terbuthylazine as model organic compounds. The decrease in concentration of DCA and the increase in concentration of chloride ions indicated the photocatalytic degradation of DCA and adsorption onto the TiO_2 films. These immobilization techniques were easy to handle without need of any expensive equipment. All TiO_2 coatings showed good photocatalytic activities and efficient mechanical stabilities for a long term. The best immobilization reproducibility was shown by the spray coating technique and by the *in situ* method with the dipping sol–gel process.

Na et al. (2005) reported the photocatalytic oxidation of RhB using a newly developed immobilized photocatalyst TiO_2. TiO_2 was immobilized by support consisting of a perlite and silicone sealant. It was used in a fluidized-bed reactor. When this photocatalyst was employed in a batch process, it completely decolorized RhB in less than 60 minutes. The optimum dosage of photocatalyst required was 33.8 g/L. The initial rate of RhB decolorization on the immobilized TiO_2 was higher than that of the suspended TiO_2, which did not follow pseudo-first-order kinetics due to the adsorption onto the surface of the immobilized TiO_2.

Mozia et al. (2007) investigated the photocatalytic oxidation of azo dye acid red 18 in water using immobilized catalyst (Degussa 25) in the quartz labyrinth flow reactor. It was observed that the reaction rate was affected by the circulation flow rate, particularly when the flow rate was low (i.e., 4.3 dm^3/h). The rate constants calculated for dye concentrations for 10 and 30 mg/dm^3 were 0.228 and 0.176 per hour, respectively. Both decolorization and mineralization were effective in this system. After a total fading of the dye solution, about 98% of TOC disappeared. An important advantage of this is that no separation of the catalyst is necessary and as a consequence, the size of installation could be minimized and running cost was also found to decrease.

TiO_2 supported on glass beads was prepared by Daneshvar et al. (2010) and its photocatalytic activity was determined by photooxidation of the commercial textile dye (direct red 23) in aqueous solution illuminated by a UV-C lamp (30 W). Both UV light and TiO_2 were required for the effective destruction of the dye. The effect of pH on the rate of decolorization efficiency was observed in the pH range 2–12, where acidic pH range was found to be favorable. It was found that addition of a proper amount of hydrogen peroxide improved the decolorization, while the excess of hydrogen peroxide retarded it. Photocatalytic decolorization of the dye was lower in the $UV/TiO_2/H_2O_2$ process than that in the UV/TiO_2 process. The efficiency of this method was determined by measuring the changes in the absorption spectra of the dye solution during photodegradation in the real wastewater sample. It was indicated that the decolorization efficiency was more than 80% for 3-hour irradiation time.

Wang et al. (2011) synthesized expanded perlite (EP) modified TiO_2 with different loading times by the sol–gel method and used it for photocatalytic mineralization of RhB in polluted water. They observed that photocatalyst modified three times with TiO_2 had the highest catalytic activity. Degradation ratios of RhB by EP-nanoTiO_2 (modified three times) were 98.0, 75.6, and 63.2 for 10, 20, and 30 mg/L, respectively, under 6-hour irradiation. The activity of photocatalyst showed only a little change after five-time recycling of the photocatalyst, and the degradation rate of RhB decreased by less than 8%. Photocatalysis showed a first-order kinetic model within the initial concentration range of RhB between 10 and 30 mg/L, and the EP-nanoTiO_2 photocatalyst showed a higher activity in photodegradation of RhB in aqueous solution and was found to be stable as well.

Zhang et al. (2011) synthesized TiO_2 film immobilized on the surface of glass as well as stainless steel tubes by the sol–gel method and used it for photodegradation of phenol. It was surprising to learn that stainless steel as support showed higher photocatalytic activity than glass even though its surface is more even. According to the X-ray photoelectron spectroscopy (XPS) results, appearance of the element Na decreases photocatalytic activity. Phenol removal rates were 36.58% and 98.36% for initial concentrations of 118.90 and 1.83 mg/L, respectively, for photodegradation with TiO_2 film immobilized on a stainless steel tube. It was concluded that TiO_2 film immobilized on the inner wall of a stainless steel tube is a better choice to treat contaminants with a low concentration.

Khataee et al. (2012) discussed the combined heterogeneous and homogeneous photodegradation of a dye using immobilized TiO_2 nanophotocatalyst and modified graphite electrode with carbon nanotubes (CNTs). The efficiency of the photoelectro-Fenton (PEF) process along with the photocatalytic process for decolorization of acid yellow 36 (AY36) was studied using TiO_2 nanoparticles (Degussa P25) immobilized on glass plates. The efficiency of bare graphite, activated carbon immobilized onto graphite surface (AC/graphite), and CNTs immobilized onto graphite surface (CNT/graphite) was investigated for H_2O_2 electrogeneration and dye removal. The CNT/graphite electrode showed the best efficiency for H_2O_2 production in the presence of air and decolorization of AY36 solution. A comparison of electro-Fenton (EF), UV/TiO_2, PEF, photolysis, and PEF/TiO_2 processes for dye removal showed that PEF/TiO_2 is the most efficient; the order of their efficiency is as follows:

$$PEF/TiO_2 > PEF > EF > UV/TiO_2 > Photolysis$$

Phenol degradation was carried out by Mozia et al. (2012) in a photocatalytic pilot plant reactor with a UV/vis mercury lamp. The photocatalyst used was TiO_2 P25, which was immobilized on two different supports: (1) a steel mesh and (2) a fiberglass cloth. The total volume of treated water was equal to $1.35\,m^3$. The performance of a commercially available photospheres-40 was examined in addition to an experiment without a photocatalyst. Photospheres-40 was not found very useful for this application because of fragility due to mechanical destruction and because of loss of floating abilities due to vigorous mixing and pumping. The highest effectiveness of phenol decomposition and mineralization was obtained with TiO_2 supported on the fiberglass cloth. Phenol and total organic carbon concentrations decreased by about 80% and 50%, respectively, after 15 hours.

UV/TiO_2/H_2O_2 degradation of basic yellow 28 dye (BY28) in aqueous solutions was reported by Cherif et al. (2014) with immobilized P25 TiO_2 powder on a glass plate by a heat attachment method. The effects of various parameters such as flow rate, initial dye concentration, solution pH, and initial H_2O_2 concentration were observed to know the interactions between them. It was found that solvent type and thickness of the coating were very effective in deciding the photoactivity of immobilized TiO_2. They highlighted a relevant interaction between the initial concentrations of dye and H_2O_2. Effective decolorization (96%) of BY28 was obtained at optimal conditions.

TiO_2 was used by Shavisi et al. (2014) as a photocatalyst immobilized on perlite granules as a supporter. As-prepared catalysts were characterized by scanning electron microscopy (SEM) and Fourier transform infrared spectroscopy (FTIR) analysis that showed TiO_2/perlite catalyst has mesoporous structures and uniform coating of TiO_2 on the support. Photocatalytic removal of ammonia from synthetic wastewater under UV irradiation was observed. The optimum efficiency of photocatalytic degradation of ammonia was obtained at pH 11 for UV irradiation with intensity 125 W. About 68% degradation of ammonia in wastewater was achieved within a 180-minute irradiation at optimized reaction conditions.

Khalilian et al. (2015) carried out the photodegradation of organic compound by a fixed bed photoreactor. This photoreactor consisted of a cylindrical glass tube that was filled by S and N codoped TiO_2 coated glass beads. The photoactive layer of TiO_2 was deposited on glass beads using the sol–gel dip-coating technique. The S and N codoped TiO_2 film was characterized by various techniques after thermal treatment at 500°C for 1 hour. The XRD data exhibited an anatase structure of TiO_2 with particle size about 35 nm. The photocatalytic activity was determined by degradation of methyl orange (MO) (7 mg/L at pH 2). Visible light photocatalytic activity of TiO_2 was increased using codoped TiO_2 sol. Degradation of MO solution (95%) was achieved in sunlight irradiation after only 2 hours. The use of immobilized S and N codoped TiO_2 on glass beads was found to be a promising technology for water treatment.

He et al. (2016) investigated the effect of solar irradiation on photocatalytic degradation of pharmaceutical active compounds (PhACs) in wastewater using immobilized TiO_2 on quartz sand. They also reported the potential of this technique as a posttreatment process for wastewater effluent. A mixture of PhACs spiked in the wastewater effluent and in deionized water (as a control) was treated with simulated solar irradiation for 96 hours with immobilized TiO_2 (photocatalysis) and without it (photolysis). Photocatalysis effectively removed poorly biodegradable PhACs in wastewater effluent, for example 100% propanol, 100% diclofenac, and 76% $\pm$ 3% carbamazepine. Photodegradation of all PhACs followed pseudo-first-order kinetics, and the kinetic constant of photocatalysis was much higher than that of photolysis in the absence of a catalyst. Dissolved organic matter also inhibited photodegradation, possibly because this reforms the oxidation intermediates of PhACs into parent compound. Toxicity of PhACs was found to decrease and biodegradability of the wastewater effluent increased slightly after photocatalysis.

10.3   INORGANIC-BASED SUPPORTS

Pozzo et al. (2000) studied the photocatalytic activity of TiO_2 (Degussa P25) immobilized onto quartz sand using a dry/wet physical deposition method in a fluidized-bed reactor to mineralize dilute oxalic acid at room temperature. A slurry of the powdery P25 was used to compare the activity. Reactor apparent captured power (P_C) and the apparent quantum efficiency (η_{VR}) were taken as the main performance criteria for both the free as well as the immobilized titania. The following variables were investigated:

- Degree of coverage on the support by catalyst
- Aggregation state of the catalyst particles
- Texture of the supporting surface

The apparent captured power increased moderately by increasing amounts of P25 deposited onto the support. However, the η_{VR} of the slurried P25 could not be matched even for better conditions, when the amount of photocatalyst inside the photoreactor was taken eight times higher. The η_{VR} of the fluidized bed was 41% than that of the slurry under these conditions, but the specific quantum efficiency was only 6% of that achieved with the free catalyst. High degree of roughness of support and use of deionized water were essential conditions to prevent the coalescence of titania during immobilization.

Electrodeposition of titania into the pores of hard alumite ($Al/Al_2O_3/TiO_2$) by alternative electrolysis in $(NH_4)_2[TiO(C_2O_4)_2]$ solution was made by Ishikawa and Matsumoto (2001). Alumite was prepared by anodic oxidation of aluminum in sulfuric acid. A cathodic current due to the reduction of H^+ and/or H_2O was observed at about -10 V, when the electrolysis was carried out in an $(NH_4)_2[TiO(C_2O_4)_2]$ solution under AC bias for the Al/Al_2O_3. This caused the deposition of TiO_2 in the pores of the alumite due to an increase in pH. As-prepared TiO_2 in the pores ($Al/Al_2O_3/TiO_2$) showed higher photocatalytic activity for the decomposition of acetaldehyde than that of TiO_2 deposited directly on an aluminum surface (Al/TiO_2).

Subba Rao et al. (2003) developed an easy method to immobilize TiO_2 for photocatalytic transformations of organic pollutants in aqueous solution. They impregnated pumice stone pellets with commercially available TiO_2. Although pumice stone is basically a soft material, this disadvantage can be eliminated by fixing pellets on a hard surface (cement or polycarbonate). Degradation of 3-nitrobenzenesulfonic acid (3-NBSA), Acid Orange 7 (AO7 a dye), and real wastewater collected after biological treatment was observed.

Lee et al. (2004) carried out photodegradation of bisphenol-A (BPA) using TiO_2 particles immobilized with titanium sol-solution synthesized by the sol–gel method. It was observed that the apparent rate constant of the first order increased with increasing

TiO_2 coating time from 1 to 3, but it decreased over a coating time of 4. Rate of degradation increased with increasing UV light intensity. The rate constant was also increased on shifting pH value. Four byproducts were detected during the photodegradation of BPA and these are 1,1-ethenylidenebis-benzene, 4-isopropylphenol, 4-*tert*-butylphenol, and phenol.

Lee et al. (2012) obtained iron oxide nanoparticles immobilized on sand (INS) by a simple impregnation process and employed it in the removal of several toxic heavy metal ions such as Cu(II), Cd(II), and Pb(II) from aqueous solutions under both experimental conditions, static and dynamic. The equilibrium state data obtained by a concentration dependence study followed the Langmuir and Freundlich adsorption model. The breakthrough data were obtained by column studies that were then utilized to model it with the Thomas equation to estimate the loading capacity of Cu(II), Cd(II), or Pb(II) under the specific column conditions. It was concluded that INS is one of the promising and effective solid materials that could be used in several wastewater treatment strategies, particularly in the treatment of wastewater contaminated with these heavy metal toxic ions.

Zhang et al. (2015) discussed the immobilization of predispersed TiO_2 colloidal particles over the external surface of the clay mineral montmorillonite (MMT) *via* a self-assembly method utilizing the cationic surfactant cetyltrimethylammonium bromide (CTAB). They synthesized a series of TiO_2–CTAB–MMT composites (TCM) with various CTAB doses. The role of CTAB was also examined in the process of synthesis and it was concluded that a uniform and continuous TiO_2 film was deposited on the external surface of MMT in the composite synthesized with 0.1 wt.% of CTAB. The TCM nanocomposites showed much higher values for specific surface area, average pore size, and pore volume than the raw MMT clay. 2,4-dichlorophenol (2,4-DCP) was photocatalytically degraded efficiently by this TCM material in aqueous solution and the degradation efficiency reached as high as 94.7%.

TiO_2 nanopowder was immobilized on concrete as a substrate for heterogeneous photocatalytic degradation (Delnavaz et al. 2015). TiO_2 immobilization on the concrete surface was carried out by three different procedures: the (1) slurry method (SM), (2) cement mixed method (CMM), and (3) different concrete sealer formulations. Irradiation of TiO_2 was done by UV-A and UV-C lamps and phenolic wastewater was selected as a model system. The efficiency of the process was determined in various operating conditions such as influent phenol concentration, pH, TiO_2 concentration, immobilization method, and UV lamp intensity. This photocatalytic process removed more than 80% phenol concentration (25–500 mg/L) in 4 hours of irradiation time.

Tian et al. (2016) synthesized a novel "Dumbbell-like" magnetic Fe_3O_4/halloysite nanohybrid (Fe_3O_4/halloysite nanohybrid at C) with oxygen-containing organic group grafted on the surface of natural halloysite nanotubes (HNTs). Homogeneous Fe_3O_4 nanospheres were selectively aggregated at the tips of modified HNTs. Cr(VI) ion adsorption experiments were carried out using Fe_3O_4/HNTs. It was observed that Fe_3O_4/halloysite nanohybrid exhibited higher adsorption ability with a maximum adsorption capacity of 132 mg/L at 303 K, which is about 100 times higher than that of unmodified HNTs. More importantly, with the reduction of Fe_3O_4 and the electron donor effect of oxygen containing organic groups, Cr(III) was adsorbed onto the surface of halloysite nanohybrid. Reduction from Cr(VI) to Cr(III) is beneficial as it is less toxic than Cr(VI). Appreciable magnetization was noticed due to the aggregation of magnetite nanoparticles, which facilitates the separation of adsorbent from aqueous solutions after Cr pollution treatment.

10.4   CARBON-BASED SUPPORTS/NATURAL PRODUCTS

Anatase TiO_2 supported on porous solids was synthesized by Ding et al. (2001) by chemical vapor deposition (CVD). Anatase TiO_2 was coated onto AC, γ-alumina (Al_2O_3), and silica gel (SiO_2). It has been observed that introduction of water vapor during CVD or adsorbed water before CVD was crucial to obtain anatase TiO_2 on the surface of the supports. The main parameters to obtain more uniform and repeatable TiO_2 coating were evaporation temperature of precursor, deposition temperature in the reactor, flow rate of carrier gas, and the length of coating time, whereas high inflow precursor concentration, high CVD reactor temperature, and long coating time cause blocking. The mechanism of the CVD process includes pyrolysis and hydrolysis and it was interesting to note that one of them was dominant in the CVD process under different synthesis routes. Silica gel, with higher surface hydroxyl groups and macropore surface area, was the most efficient support in terms of both anatase TiO_2 coating and photocatalytic reaction in all three types of supports used.

TiO_2 and Fe–C–TiO_2 photocatalysts have been immobilized on the cotton material by Tryba (2008) and used in a flow photocatalytic reactor for the decomposition of phenol. Cotton has been used as a support for photocatalysts, because it can be easily removed and replaced in a reactor, facilitating the performance of the photocatalytic process. An Fe–C–TiO_2 photocatalyst has been prepared by modifying TiO_2 (anatase) fine particles with FeC_2O_4 through heating in Ar at 500°C. These immobilized photocatalysts could efficiently decompose phenol. Fe–C–TiO_2 showed higher photocatalytic activity than TiO_2 above. Around 15–18 mg and 15–16 mg of phenol were decomposed on 5-hour UV irradiation in the presence of Fe–C–TiO_2 and TiO_2, respectively. The phenol decomposition and the mineralization were accelerated after the addition of H_2O_2, especially with immobilized Fe–C–TiO_2. About 26–28 mg and 21–24 mg of phenol was decomposed on Fe–C–TiO_2 and TiO_2, respectively, on UV irradiation in the presence of H_2O_2 exposure for 5 hours.

Albarelli et al. (2009) introduced a novel support Ca–alginate for TiO_2 immobilization to be used in an industrial process. Methylene blue was chosen as the model dye to evaluate this novel immobilization system. The results showed that TiO_2 immobilized in Ca–alginate bead retained its photoactivity during all of the experiments and the TiO_2-gel beads presented good stability in water for maintaining shape after several uses. When a proportion of 10% (v/v) of these beads was used, the configuration system demonstrated an improved mass transfer and consequently enhanced degradation efficiency. Experiments were also performed using "recycled" beads. The results showed an increase in the degradation efficiency, when the beads were reused, with a self-destructive effect. These studies showed great promise regarding the recyclable reagents with a reduction in waste at no greater cost or reduction in efficiency. Therefore, the potential of TiO_2-gel beads as a simple and environment-friendly catalyst for continuous use was suggested.

The immobilization of TiO_2 on activated carbon fiber (TiO_2/ACF) was studied by Yao et al. (2010) by the sol–gel-adsorption method followed by calcination at temperatures varying from 300°C to 600°C in an argon atmosphere. The effects of calcination temperature, photocatalyst dosage, initial solution pH, and radiation time on the degradation of organic pollutants were studied. It was observed that organic pollutants could be removed rapidly from water by the TiO_2/ACF photocatalyst. The temperature of calcination also affected the efficiency of these nanoparticles for removal of pollutants. This immobilized catalyst can be reused continuously. In this case, TiO_2 is tightly bound to ACF so that it can be easily handled and recovered from water. It can therefore be potentially applied for the treatment of water contaminated by organic pollutants such as phenol and MO.

TiO_2/Luffa composites have been successfully prepared by El-Roz et al. (2013) *via* the sol–gel method from the hydrolysis of a precursor of TiO_2. Fibers were successfully used as a biotemplate to self-support hierarchical TiO_2 macrostructures. Photocatalytic activities have been investigated in the photodegradation of methanol, a model molecule for volatile organic compounds (VOCs) in air. TiO_2/Luffa composites exhibited good stability and photocatalytic activity under UV light irradiation, and it has been suggested as giving rise to a new generation of green photocatalysts that are easy to shape and manufacture.

Gadiyar et al. (2013) reported the UV photocatalytic decomposition of acid yellow 17 dye by TiO_2 immobilized on chitosan and cellulose acetate. Nanosized TiO_2 was prepared by the sol–gel technique. Batch studies with free catalyst and immobilized catalysts were conducted and it was observed that immobilization does not offer diffusion limitations. Kinetics of degradation of this dye followed first-order kinetics. The effect of various factors such as catalyst loading, initial dye concentration, and liquid flow rate on the degradation of acid yellow 17 was studied in a fluidized-bed flow reactor operated in batch recycle mode with these immobilized nanoparticles.

A photocatalytic reactor with thin film of TiO_2 on cellulose fibers was used by Costa and Alves (2013) for the posttreatment of olive mill wastewater after anaerobic digestion. The effect of initial chemical oxygen demand (COD), pH, treatment time, recirculation flow, and possible interactions on phenols, color, and COD removal was observed. Removal efficiencies of $90.8 \pm 2.7\%$, $79.3 \pm 1.9\%$, and $50.3 \pm 6.3\%$ were obtained for total phenols (TPh), color, and COD, respectively. It was observed that TPh and color removal were almost complete after 24 hours of treatment, while the removal of COD was partial. Increase in the treatment time was not feasible economically and therefore, a recirculation to the anaerobic reactor should be considered. TPh removal efficiency was found to be dependent on the initial COD concentration, while efficiency of COD removal was directly linked with the treatment time.

Dávila-Jiménez et al. (2015) proposed photodegradation of the anthraquinonic dye, acid green 25, by TiO_2 immobilized on carbonized avocado kernels. They synthesized C–TiO_2 composite from a waste material. Composites were obtained from carbonized avocado kernels, and sols of TiO_2 were prepared from NH_4OH and $TiCl_4$ with glycerol as a binder and heat treatment of the composites at 500°C. XRD studies confirmed that TiO_2 was present in anatase phase while SEM images showed isolated TiO_2 agglomerates attached on the carbon surface with a Ti:C ratio of 1:3.3.

The photocatalytic activities of as-prepared composites were tested in decomposition of acid green 25 and it was found that this sample of TiO_2 and the composite eliminated almost 100% of acid green 25 under UV irradiation while only the composite demonstrated degradation efficiency under solar light.

Antonio-Cisneros et al. (2015) reported the immobilization of TiO_2 on carbon prepared from residues of the plant Manihot, commercial TiO_2, and glycerol to obtain a moderate loading of the anatase phase by preserving the carbonaceous external surface and micropores of the composite. Two preparation methods were used. These were mixing dry precursors and immobilization using a glycerol slurry and the samples were compared for their activity. Manihot residues and glycerol can be used to prepare an anatase containing material with a basic surface and a significant SBET value. It was observed that this TiO_2/carbon eliminated nearly 100% of the indigo carmine dye under UV irradiation using the optimal conditions.

Chen et al. (2015) designed a facile immobilization method of $LnVO_4$ (Ln = Ce, Nd, Gd) via a combined alcohol-thermal and carbon nanofibers (CNFs) template route. The physicochemical properties of these samples were characterized in detail and their photocatalytic activities were observed under UV light for photocatalytic degradation of methylene blue. The results showed that $LnVO_4$ effectively degraded the dye photcatalytically in the following order:

$$GdVO_4 > CeVO_4 > NdVO_4$$

It was suggested that such unique immobilized $LnVO_4$ material may possess many potential applications in photocatalysis.

Pant et al. (2015) proposed a simple and efficient approach to immobilize TiO_2 NFs onto reduced graphene oxide (rGO) sheets. They prepared TiO_2 NF intercalated rGO sheets produced by two-step procedure with the use of electrospinning process to fabricate TiO_2 precursor comprising polymeric fibers on the surface of GO sheets. It was followed by simultaneous TiO_2 NF formation and GO reduction by calcination. TiO_2 precursor containing polymer NF deposited on GO sheets resulted in the formation of TiO_2 NF doped rGO sheets on calcination. Photocatalytic activity was remarkably increased by TiO_2/rGO composite compared to pristine TiO_2 NFs. Thus, TiO_2 NF intercalated rGO sheets can be considered for a promising method for catalytic and other applications.

10.5 POLYMERIC SUPPORTS

As the separation of TiO_2 particles from treated wastewater creates problems in practical application of this process. Tennakone et al. (1995) observed the degradation of organic contaminants in water using TiO_2 on polythene films. It was reported that TiO_2 can be readily supported on polythene films without inhibiting the photocatalytic activity. They studied photocatalytic degradation of phenol by TiO_2 supported on polythene films.

Iguchi et al. (2003) prepared a novel paper-based material containing TiO_2 photocatalyst. They used a paper-making technique with the internal addition of inorganic fibers (as a support for TiO_2 particles). Photodegradation efficiency of acetaldehyde vapor, an indoor pollutant, under UV irradiation was observed. The durability of hese TiO_2-containing papers was also investigated. Ceramic fiber suspension and polydiallyldimethylammonium chloride (PDADMAC) were used as cationic flocculant. It was followed by the addition of TiO_2 suspension and anionic polyacrylamide. Then TiO_2-hand sheets were prepared by the paper-making method. It was observed that the tensile strength of TiO_2-containing paper without a ceramic carrier decreased by about one-third after 240-hour UV irradiation (2 mW/cm^2), but on the contrary, TiO_2 sheets with ceramic fibers remained reasonably stable.

Pan et al. (2006) immobilized TiO_2 onto polymer fibers such as composite nonwoven fiber textile using silica sol and tested it for the degradation of gaseous 2-propanol in a batch photoreactor for 3 months. It was observed that composite nonwoven fiber textile had high TiO_2 immobilization ability and the photocatalytic efficiency of TiO_2 was dependent on the amount of silica sol added. The photoactivity of 0.5 wt.% SiO_2 nonwoven fiber textile becomes higher after 3 months of usage than that of TiO_2 textile, which had relatively poor photocatalytic efficiency due to the reduction of the retained amount of TiO_2.

Matsuzawa et al. (2008) reported a new method for immobilizing TiO_2 nanoparticles on polymeric substrates to facilitate photocatalytic purification of contaminated air and water. These particles were immobilized by dipping a polymeric substrate, treated with polyvinyl chloride–polyvinyl acetate (PVC–PVA) copolymer and/or SiO_2, into a TiO_2/water suspension. This method is based on an electrostatic attraction between positively charged TiO_2 and the negatively charged treated surface of the polymeric substrate. This method avoids the enwrapping of TiO_2 particles in binding components, which was a drawback of conventional methods. It helped in bonding of the TiO_2 nanoparticles on the substrate surface at a high density. The TiO_2-immobilized nonwoven polyester (PES) prepared by this method exhibited high photocatalytic activity for decomposing the air contaminants, particularly toluene. This method is also applicable to polypropylene (PP) nonwoven, polyethylene (PE) nets as well as to PE and PP films. However, TiO_2 bonding was inhibited on PP by treatment with PVC–PVA copolymer, but better TiO_2 immobilization was observed on SiO_2-treated PP.

Priya et al. (2009) reported layer-by-layer (LbL) fabrication of poly(styrene sulfonate)/TiO_2 multilayer thin films for environmental applications. They fabricated multilayer ultrathin composite films composed of nanosized TiO_2 particles (P25, Degussa) and polyelectrolytes (PELs), such as poly(allyl amine hydrochloride) (PAH) and poly(styrene sulfonate sodium salt) (PSS). These were used for the photodegradation of RhB under UV irradiation. PELs and TiO_2 were supported on glass substrates at pH 2.5. The efficiency of both the catalysts immobilized by this technique and prepared by drop casting and spin coating methods was compared in terms of film stability and photodegradation of RhB. It was observed that the degradation efficiency increased with increase in number of catalyst slides (total surface area) and bilayers. A total of 100 mL of 10 mg/L dye solution could be degraded completely in 4 hours using maximum loaded TiO_2, with five catalyst slides having 20 bilayers of PEL/TiO_2 on each. It was encouraging to observe that the same slides could be reused with almost the same efficiency without any loss for several cycles.

The LbL method was also used by Nakajima et al. (2009) for preparation of transparent TiO_2 nanosheet thin films with PDADMAC as a counter polymer. Photocatalytic activity in the higher wavelength region was examined by grafting Cu on the film.

It was observed that photocatalytic decomposition activity of TiO_2 for gaseous 2-propanol under UV illumination was enhanced by Cu grafting. It may be attributed to the fact that the interfacial charge transfer mechanism is also effective for nanosheet films.

Lei et al. (2012) discussed the immobilization of TiO_2 nanoparticles in polymeric substrates by chemical bonding for multicycle photodegradation of organic pollutants. Generally, nano-TiO_2 photocatalyst is immobilized onto some matrix either through physical absorption, hydrogen bonding, or chemical bonding, and used in wastewater treatment. TiO_2 nanoparticles were immobilized in polyvinyl alcohol (PVA) matrix via solution-casting combined with a heat treatment method. It was observed that the Ti–O–C chemical bond was formed via dehydration reaction between TiO_2 and PVA during the heat treatment process, and TiO_2 nanoparticles were chemically immobilized in the PVA matrix. They also observed the efficiency of these nanoparticles by photodegradation of MO and it was concluded that the film with 10 wt.% TiO_2 treated at 140°C for 2 hours exhibited a remarkable UV photocatalytic activity. It is approximately quite close to the activity of the TiO_2 slurry system due to fixation by Ti–O–C chemical bonds; also the TiO_2 photocatalyst loses its activity only slightly even after 25 cycles. PVA matrix provides better opportunities to fully contact with TiO_2 due to its good swelling ability, and thus enhances the efficiency of the photocatalyst. It is a simple and low-cost method to prepare polymer/TiO_2 hybrid materials with high photocatalytic activity for multicycle use, which is significant for its practical applications.

Yousef et al. (2012) encapsuled CdO/ZnO NPs in polyurethane NFs by a simple, effective, and low-cost electrospinning method for effective immobilization of the photocatalysts. Nanostructural photocatalysts effectively eliminated different organic pollutants; however, they create secondary pollution as these are very difficult to separate from treated water especially in large-scale processes. ZnO has been doped with CdO to improve its photocatalytic activity and these CdO/ZnO nanoparticles were successful in eliminating organic pollutants under the visible region. Moreover, it can also be easily separated after use. These nanoparticles were used in photocatalytic degradation of methylene blue and reactive black 5 dyes.

Zhang and Yang (2012) immobilized titania nanoparticles on polyamide 6 (PA6) fiber in the form of thin layers using titanium sulfate and urea at a low-temperture hydrothermal condition. Properties of these nanoparticles were also examined before and after degradation of methylene blue dye under UV irradiation such as optical and mechanical properties and water absorption. The anatase form of nanocrystalline TiO_2 was synthesized, when PA6 fabric was treated in titanium sulfate and aqueous solution of urea, which simultaneously adhered onto the fiber surface. The average crystal size of TiO_2 nanoparticles was found to be 13.2 nm. The thermal behavior of PA6 fiber distinctly changed and the onset decomposition temperature also decreased. Thus, the protection of treated fabric against UV radiation was improved as compared to the untreated fabric. The water absorbency was also found to be enhanced slightly. The TiO_2-coated fabric could degrade methylene blue dye successfully under UV irradiation.

PET fabric was first modified with silane coupling agent KH-560, and then it was loaded with a layer of nanoscaled TiO_2 particles using tetrabutyl titanate as a precursor by a low-temperature hydrothermal method, followed by dyeing with disperse blue 56 (Zhang et al. 2013). It was observed that TiO_2-coated fabric without modification with KH-560 was less effective than TiO_2-coated fabric modified with KH-560. The pure anatase TiO_2 nanoparticle was grafted onto the fiber surface. The absorbing capability for UV radiation was also enhanced. The UV protection ability as well as photodegradation of MO under UV illumination were enhanced.

Immobilization of photocatalytic powder is very important to obtain industrially relevant purification processes. Self-supporting TiO_2 foams were manufactured by a polyacrylamide gel process by Tytgat et al. (2014). These gels were calcined at different temperatures to study the effect of the calcination temperature on foam characteristics, that is, rigidity, crystallinity, and porosity, and its influence on photocatalytic activity. An optimal degradation was achieved for foams calcined between 700°C and 800°C. However, calcination at higher temperatures resulted in a sharp decrease in activity, and it was explained by stability of the material due to formation of Na_2SO_4 phases and a larger rutile fraction.

Cantarella et al. (2016) reported immobilization of TiO_2 nanomaterials in poly(methyl methacrylate) (PMMA) composites and their use for photocatalytic removal of dyes, phenols, and bacteria from water. Nanomaterials represent a solution to solve many of the current problems involving water quality, but there are also certain limitations for their efficient application. These are primarily concerned with dispersion in water, their recovery after water treatment, and the ultimate impact on human health and ecosystems. The incorporation of several nanomaterials into polymeric composites may prove to be a valid solution to these problems. Active TiO_2 nanostructures were embedded in PMMA to avoid recovery of the nanoparticles after water treatment. They combined TiO_2 nanoparticles with single-walled CNT, as acceptor of electrons, and observed a significantly higher photocatalytic efficiency under UV irradiation compared to the systems with TiO_2 only. Photoactive materials were also synthesized with *meso*-tetraphenylporphyrin-4,4',4'',4'''-tetracarboxylic acid (TCPP) as a dye sensitizer and these were found effective even under visible light.

The problem of suspension of titania particles in aqueous solution can be solved by immobilizing these particles on some supports such as glass, stainless steel, polymers, inorganic supports, as well as carbon-based supports including natural products. Many more systems can be tried in the future for this problem with some eco-friendly supports so that this system can be applied on a large scale in treating industrial effluents.

REFERENCES

Albarelli, J. Q., D. T. Santos, S. Murphy, and M. Oelgemöller. 2009. Use of Ca–alginate as a novel support for TiO_2 immobilization in methylene blue decolorisation. *Water Sci. Technol.* 60 (4): 1081–1087.

Antonio-Cisneros, C. M., M. M. Dávila-Jiménez, M. P. Elizalde-González, and E. García-Díaz. 2015. TiO_2 immobilized on Manihot carbon: Optimal preparation and evaluation of its activity in the decomposition of indigo carmine. *Int. J. Mol. Sci.* 16: 1590–1612.

Ao, C. H., S. C. Lee, and J. C. Yu. 2003. Photocatalyst TiO_2 supported on glass fiber for indoor air purification: Effect of NO on the photodegradation of CO and NO_2. *J. Photochem. Photobiol. A: Chem.* 156 (1–3): 171–177.

Balasubramanian, G., D. D. Dionysiou, M. T. Suidan, I. Baudin, and J. M. Laîné, 2004. Evaluating the activities of immobilized TiO$_2$ powder films for the photocatalytic degradation of organic contaminants in water. *Appl. Catal. B: Environ.* 47: 73–84.

Cantarella, M., R. Sanz, M. A. Buccheri, F. Ruffino, G. Rappazzo, S. Scalese, G. Impellizzeri, L. Romano, and V. Privitera. 2016. Immobilization of nanomaterials in PMMA composites for photocatalytic removal of dyes, phenols and bacteria from water. *J. Photochem. Photobiol. A: Chem.* 321: 1–11.

Chen, P., Q. Wu, L. Zhang, and W. Yao. 2015. Facile immobilization of LnVO$_4$ (Ln = Ce, Nd, Gd) on silica fiber via a combined alcohol-thermal and carbon nanofibers template route. *Catal. Commun.* 66: 6–9.

Chen, Y. and D. D. Dionysiou. 2006. TiO$_2$ photocatalytic films on stainless steel: The role of degussa P-25 in modified sol–gel methods. *Appl. Catal. B: Environ.* 62: 255–264.

Cherif, L. Y., I. Yahiaoui, F. Aissani-Benissad, K. Madi, N. Benmehdi, F. Fourcade, and A. Amrane. 2014. Heat attachment method for the immobilization of TiO$_2$ on glass plates: Application to photodegradation of basic yellow dye and optimization of operating parameters, using response surface methodology. *Ind. Eng. Chem. Res.* 53 (10): 3813–3819.

Costa, J. C. and M. M. Alves. 2013. Post treatment of olive mill wastewater by immobilized TiO$_2$ photocatalysis. *Photochem. Photobiol.* 89: 545–551.

Daneshvar, N., D. Salari, A. Niaei, M. H. Rasoulifard, and A. R. Khataee. 2010. Immobilization of TiO$_2$ nanopowder on glass beads for the photocatalytic decolorization of an azo dye C.I. direct red 23. *J. Environ. Sci. Health.* 40 (8): 1605–1617.

Dávila-Jiménez, M. M., M. P. Elizalde-González, E. García-Díaz, V. Marín-Cevada, and J. Zequineli-Pérez. 2015. Photodegradation of the anthraquinonic dye acid green 25 by TiO$_2$ immobilized on carbonized avocado kernels: Intermediates and toxicity. *Appl. Catal. B: Environ.* 166–167: 241–250.

Delnavaz, M., B. A. H. Ganjidoust, and S. Sanjabi. 2015. Application of concrete surfaces as novel substrate for immobilization of TiO$_2$ nano powder in photocatalytic treatment of phenolic water. *J. Environ. Health Sci. Eng.* 13. doi:10.1186/s40201-015-0214-y.

Ding, Z., X. Hu, P. L. Yue, G. Q. Lu, and P. F. Greenfield. 2001. Synthesis of anatase TiO$_2$ supported on porous solids by chemical vapor deposition. *Nanomater. Catal.* 68 (1–3): 173–182.

Doll, T. E. and F. H. Frimmel. 2004. Development of easy and reproducible immobilization techniques using TiO$_2$ for photocatalytic degradation of aquatic pollutants. *Acta Hydrochim. Hydrobiol.* 32 (3): 201–213.

El-Roz, M., Z. Haidar, L. Lakiss, J. Toufaily, and F. Thibault-Starzyk. 2013. Immobilization of TiO$_2$ nanoparticles on natural *Luffa cylindrical* fibers for photocatalytic applications. *RSC Adv.* 3: 3438–3445.

Feitz, A. J., B. H. Boyden, and T. D. Waite. 2000. Evaluation of two solar pilot scale fixed-bed photocatalytic reactors. *Water Res.* 34: 3927–3932.

Fernández, A., G. Lassaletta, V. M. Jiménez, A. Justo, A. R. González-Elipe, J.-M. Herrmann, H. Tahiri, and Y. Ait-Ichou. 1995. Preparation and characterization of TiO$_2$ photocatalysts supported on various rigid supports (glass, quartz and stainless steel). Comparative studies of photocatalytic activity in water purification. *Appl. Catal. B: Environ.* 7 (1–2): 49–63.

Gadiyar, C., B. Boruaha, C. Mascarenhasa, and K. V. Shetty. 2013. Immobilized nano TiO$_2$ for photocatalysis of acid yellow-17 dye in fluidized bed reactor. Proceedings of National Conference on Women in Science and Engineering Dharwad. *Int. J. Curr. Eng. Technol.* 84–87.

Goetz, V., J. P. Cambon, D. Sacco, and G. Plantard. 2009. Modeling aqueous heterogeneous photocatalytic degradation of organic pollutants with immobilized TiO$_2$. *Chem. Eng. Process.: Process Intens.* 48: 532–537.

Gumy, D., A. G. Rincon, R. Hajdu, and C. Pulgarin. 2006. Solar photocatalysis for detoxification and disinfection of water: Different types of suspended and fixed TiO$_2$ catalysts study. *Solar Energy*. 80: 1376–1381.

He, Y., N. B. Sutton, H. Rijnaarts, and A. A. M. Langenhoff. 2016. Degradation of pharmaceuticals in wastewater using immobilized TiO$_2$ photocatalysis under simulated solar irradiation. *Appl. Catal. B: Environ*. 182: 132–141.

Iguchi, Y., H. Ichiura, T. Kitaoka, and H. Tanaka. 2003. Preparation and characteristics of high performance paper containing titanium dioxide photocatalyst supported on inorganic fiber matrix. *Chemosphere*. 53 (10): 1193–1199.

Ishikawa, Y. and Y. Matsumoto. 2001. Electrodeposition of TiO$_2$ photocatalyst into nano-pores of hard alumite. *Electrochim. Acta*. 46 (18): 2819–2824.

Khalilian, H., M. Behpour, V. Atouf, and S. N. Hosseini. 2015. Immobilization of S, N-codoped TiO$_2$ nanoparticles on glass beads for photocatalytic degradation of methyl orange by fixed bed photoreactor under visible and sunlight irradiation. *Solar Energy*. 112: 239–245.

Khataee, A. R., M. Safarpour, M. Zarei, and S. Aber. 2012. Combined heterogeneous and homogeneous photodegradation of a dye using immobilized TiO$_2$ nanophotocatalyst and modified graphite electrode with carbon nanotubes. *J. Mol. Catal. A: Chem*. 363–364: 58–68.

Lee, J. M., M. S. Kim, and B. W. Kim. 2004. Photodegradation of bisphenol-A with TiO$_2$ immobilized on the glass tubes including the UV light lamps. *Water Res*. 38 (16): 3605–3613.

Lee, S. M., C. Laldawngliana, and D. Tiwari. 2012. Iron oxide nano-particles-immobilized-sand material in the treatment of Cu(II), Cd(II) and Pb(II) contaminated waste waters. *Chem. Eng. J*. 195–196: 103–111.

Lei, P., F. Wang, X. Gao, Y. Ding, S. Zhang, J. Zhao, S. Liu, and M. Yang. 2012. Immobilization of TiO$_2$ nanoparticles in polymeric substrates by chemical bonding for multi-cycle photodegradation of organic pollutants. *J. Hazard. Mater*. 227–228: 185–194.

Lu, M. C., G. D. Roam, J. N. Chen, and C. P. Huang. 1993. Factors affecting the photocatalytic degradation of dichlorvos over titanium dioxide supported on glass. *J. Photochem. Photobiol. A: Chem*. 76 (1–2): 103–110.

Ma, Y., J. B. Qiu, Y. A. Cao, Z. S. Guan, and J. N. Yao. 2001. Photocatalytic activity of TiO$_2$ films grown on different substrates. *Chemosphere*. 44 (5): 1087–1092.

Malato, S., P. Fernandez-Ibanez, M. I. Maldonado, J. Blanco, and W. Gernjak. 2009. Decontamination and disinfection of water by solar photocatalysis: Recent overview and trends. *Catal. Today*. 147: 1–59.

Matsuzawa, S., C. Maneerat, Y. Hayata, T. Hirakawa, N. Negishi, and T. Sano. 2008. Immobilization of TiO$_2$ nanoparticles on polymeric substrates by using electrostatic interaction in the aqueous phase. *Appl. Catal. B: Environ*. 83 (1–2): 39–45.

McCullagh, C., N. Skillen, M. Adams, and P. K. J. Robertson. 2011. Photocatalytic reactors for environmental remediation: A review. *J. Chem. Technol. Biotechnol*. 86: 1002–1017.

Mehrotra, K., G. S. Yablonsky, and A. K. Ray. 2003. Kinetic studies of photocatalytic degradation in a TiO$_2$ slurry system: Distinguishing working regimes and determining rate dependences. *Ind. Eng. Chem. Res*. 42: 2273–2281.

Mozia, S., P. Brożek, J. Przepiórski, B. Tryba, and A. W. Morawski. 2012. Immobilized TiO$_2$ for phenol degradation in a pilot-scale photocatalytic reactor. *J. Nanomater*. doi: 10.1155/2012/949764.

Mozia, S., M. Tomaszewska, and A. W. Morawski. 2007. Photodegradation of azo dye acid red 18 in a quartz labyrinth flow reactor with immobilized TiO$_2$ bed. *Dyes Pigments*. 75 (1): 60–66.

Na, Y., S. Song, and Y. Park. 2005. Photocatalytic decolorization of rhodamine B by immobilized TiO$_2$/UV in a fluidized-bed reactor. *Korean J. Chem. Eng*. 22 (2): 196–200.

Nakajima, A., Y. Akiyama, S. Yanagida, and K. Okada. 2009. Preparation and properties of Cu-grafted transparent TiO_2-nanosheet thin films. *Mater. Lett.* 63(20): 1699–1701.

Nogueira, R. F. P. and W. F. Jardim. 1996. TiO_2-fixed-bed reactor for water decontamination using solar light. *Solar Energy.* 56 (5): 471–477.

Pan, G. T., C. M. Huang, L. C. Chen, and W. T. Shiu. 2006. Immobilization of TiO_2 onto nonwoven fiber textile by silica sol: Photocatalytic activity and durability studies. *J. Environ. Eng. Manage.* 16(6): 413–420.

Pant, H. R., S. P. Adhikari, B. Pant, M. K. Joshi, H. J. Kim, C. H. Park, and C. S. Kim. 2015. Immobilization of TiO_2 nanofibers on reduced graphene sheets: Novel strategy in electrospinning. *J. Coll. Interface Sci.* 457: 174–179.

Parra, S., S. E. Stanca, I. Guasaquillo, and K. R. Thampi. 2004. Photocatalytic degradation of atrazine using suspended and supported TiO_2. *Appl. Catal. B: Environ.* 51: 107–116.

Pozzo, R. L., M. A. Baltanás, and A. E. Cassano. 1997. Supported titanium oxide as photocatalyst in water decontamination: State of the art. *Photocatal. Solar Energy Conver.* 39 (3): 219–231.

Pozzo, R. L., J. L Giombi, M. A. Baltanás, and A. E. Cassano. 2000. The performance in a fluidized bed reactor of photocatalysts immobilized onto inert supports. *Catal. Today.* 62 (2–3): 175–187.

Priya, D. N., J. M. Modak, and A. M. Raichur. 2009. LbL fabricated poly(styrene sulfonate)/ TiO_2 multilayer thin films for environmental applications. *ACS Appl. Mater. Interfaces.* 1 (11): 2684–2693.

Ryu, C. S., M. S. Kim, and B. W. Kim. 2003. Photodegradation of alachlor with the TiO_2 film immobilised on the glass tube in aqueous solution. *Chemosphere.* 53 (7): 765–771.

Shan, A. Y., T. Idaty, M. Ghazi, and S. A. Rashid. 2010. Immobilisation of titanium dioxide onto supporting materials in heterogeneous photocatalysis: A review. *Appl. Catal. A: Gen.* 389 (1–2): 1–8.

Shavisi, Y., S. Sharifnia, S. N. Hosseini, and M. A. Khadivi. 2014. Application of TiO_2/perlite photocatalysis for degradation of ammonia in wastewater. *J. Ind. Eng. Chem.* 20 (1): 278–283.

Subba Rao, K. V., A. Rachel, M. Subrahmanyam, and P. Boule. 2003. Immobilization of TiO_2 on pumice stone for the photocatalytic degradation of dyes and dye industry pollutants. *Appl. Catal. B: Environ.* 46: 77–85.

Tennakone, K., C. T. K. Tilakaratne, and I. R. M. Kottegoda. 1995. Photocatalytic degradation of organic contaminants in water with TiO_2 supported on polythene films. *J. Photochem. Photobiol. A: Chem.* 87: 177–179.

Thiruvenkatachari, R., V. Saravanamuth, and I. S. Moon. 2008. A review on UV/TiO_2 photocatalytic oxidation process. *Korean J. Chem. Eng.* 25: 64–72.

Tian, X., W. Wang, N. Tian, C. Zhou, C. Yang, and S. Komarneni. 2016. Cr(VI) reduction and immobilization by novel carbonaceous modified magnetic Fe_3O_4/Halloysite nanohybrid. *J. Hazard. Mater.* 309: 151–156.

Tryba, B. 2008. Immobilization of TiO_2 and Fe-C-TiO_2 photocatalysts on the cotton material for application in a flow photocatalytic reactor for decomposition of phenol in water. *J. Hazard. Mater.* 151 (2–3): 623–627.

Tytgat, T., M. Smits, S. Lenaerts, and S. W. Verbruggen. 2014. Immobilization of TiO_2 into self-supporting photocatalytic foam: Influence of calcination temperature. *Appl. Ceram. Technol.* 11 (4): 714–722.

Wang, X. Z., W. W. Yong, W. Q. Yin, K. Feng, and R. Guo. 2011. Immobilization of nano-TiO_2 on expanded perlite for photocatalytic degradation of rhodamine B. *Appl. Mech. Mater.* 110–116: 3795–3800.

Yanagida, S., A. Nakajima, Y. Kameshima, N. Yoshida, T. Watanabe, and K. Okada. 2005. Preparation of a crack-free rough titania coating on stainless steel mesh by electrophoretic deposition. *Mater. Res. Bull.* 40: 1335–1344.

Yao, S., J. Li, and Z. Shi. 2010. Immobilization of TiO$_2$ nanoparticles on activated carbon fiber and its photodegradation performance for organic pollutants. Special Section: Structure and Properties of Layered Inorganic Materials. *Particuology.* 8 (3): 272–278.

Yousef, A., N. A. M. Barakat, S. S. Al-Deyab, R. Nirmala, B. Pant, and H. Y. Kim. 2012. Encapsulation of CdO/ZnO NPs in PU electrospun nanofibers as novel strategy for effective immobilization of the photocatalysts. *Coll. Surf. A.* 401: 8–16.

Zhang, H., F. Li, and H. Zhu. 2013. Immobilization of TiO$_2$ nanoparticles on PET fabric modified with silane coupling agent by low temperature hydrothermal method. *Fibers Polym.* 14 (1): 43–51.

Zhang, H. and L. Yang. 2012. Immobilization of nanoparticle titanium dioxide membrane on polyamide fabric by low temperature hydrothermal method. *Thin Solid Films.* 520 (18): 5922–5927.

Zhang, S., Z., Shi, and L. Wu. 2011. *International Conference on Computer Distributed Control and Intelligent Environmental Monitoring (CDCIEM).* Changra: IEEE Computer Society, pp. 1449–1452.

Zhang, T., Y. Luo, B. Jia, Y. Li, L. Yuan, and J. Yu. 2015. Immobilization of self-assembled pre-dispersed nano-TiO$_2$ onto montmorillonite and its photocatalytic activity. *J. Environ. Sci.* 32: 108–117.

11 Hydrogen Generation

11.1 INTRODUCTION

Energy can be defined as the "capacity to do work" or "usable power." It is in the form of both kinetic energy and potential energy. The free energy is associated with chemical transformation and electromagnetic radiation energy in the form of photons, that is, quantized packets of energy. Energy is available in the form of heat, electricity, chemical, hydropower, wind, geothermal, tidal, biomass, liquid and gas fuels, sound, and many other forms. Energy is a basic necessity for life as it provides comfort to society in its different forms. Energy can help us to cool down during summer and keeps us warm during winter. The global demand for energy has increased in the last few decades as a result of industrial revolution (development) and population growth, which is increasing day by day. Energy is helpful for transportation, construction, manufacturing, heat and electricity production, mining, and so on.

An energy crisis occurs when a country has a huge drop in the supply of energy available or a large rise in the price of energy but does not have enough to fulfill consumer demand. As a result, energy becomes very expensive and is not available for everyone. More often, there is a shortage of crude oil and electricity, as well as other natural sources of nonrenewable energy.

The world energy crisis has arrived due to a shortage of existing nonrenewable energy sources. Primitive energy resources, such as crude oil, charcoal, gas, and so on, are limited and are being exhausted at a rapid pace. The advancement in technologies, industrialization, and increased dependence of users on different forms of energy are the main causes of the present energy crisis. The energy demand of any country is decisively linked with its economic progress and development. A developing country has huge energy consumption demand so that energy prices jump to the higher side and probably will still rise in future.

The world's economic growth relies on renewable energy sources. It is quite clear that the energy crisis is basically a fossil fuel crisis. The only solution to resolve this energy crisis is to develop renewable energy resources because these are the biggest nominees for energy production and they can serve to fulfill the energy demands of the world for a longer time. One can produce much more clean energy using renewable energy sources than burning fossil fuels and it could reduce carbon emissions also by 60%–80%. These renewable energy technologies provide energy far more efficiently than fossil fuels.

In forthcoming decades, expected global energy consumption demand may be increased dramatically, due to rising standards of living and a growing population worldwide. The increased energy demand will require enormous growth in energy production capacity, more reliable and diversified energy sources, and, therefore, there is a need to develop successful strategy for it. It was estimated by the International Energy Agency (IEA) that the energy needs of the world will be 50% more in 2030 than at present. In the same context, in coming years, worldwide energy demand could increase almost threefold by the year 2050. Renewable energy is a rising energy source in recent years, and it has been estimated that renewable sources produced 16.5% of the primary energy requirements of the world in 2005. The proportion of renewable energy resources increases each day with the result that dependence on nonrenewable energy resources is regularly decreasing.

All the conventional energy sources, such as oil, gas, coal, and so on, are nonrenewable energy sources and their availability on the earth is being exhausted rapidly. Therefore, the focus is now shifted toward renewable energy sources and these have become the apparent choice over conventional energy sources. The future constraints of carbon emissions, energy security, downfall in existing energy sources, rising prices, resulting changes in climate and environment, high energy demands of industrial and economic development, abundant and free renewable energy sources, awareness of consumers, financial risk mitigation, flexibility, and resilience are the prime reasons behind the development of renewable energy sources. Renewable energy will surpass nuclear energy and other traditional energy sources as it cracks the code of sustainable economic growth by reducing energy demands. A futuristic energy scenario can be considered an alternative image, and the future might unfold with an appropriate tool to analyze how driving forces may influence future outcomes and also to assess the associated uncertainties.

Hydrogen is one of the most widely occurring elements of the universe. It is the third-most abundant element on the earth's surface. It exists in multiple chemical forms such as hydrocarbons, hydrides, prebiotic organic compounds, water, and so on. Among the various alternative energy strategies, hydrogen is the primary source that connects a host of energy sources with end users as a secure and clean energy source for the future of any nation. Hydrogen is not an energy source but an energy carrier. It is the simplest element present on the earth, and consists of only one electron and one proton. Usable energy can be stored and delivered by hydrogen, but it is not easily available in a free form in nature. It must be produced from compounds that contain it. Hydrogen can be utilized in fuel cells to produce electrical power using a chemical reaction and generate only water and heat as byproducts. It can be used in cars and in houses, and has many more possible applications.

Hydrogen can be easily generated as a clean fuel and it is a promising candidate to resolve our improving energy problems as a renewable source. During the combustion of hydrogen, only water vapors are produced without any traces of noxious carbon emission in the form of carbon dioxide, carbon monoxide, or unburnt hydrocarbons. It is a big temptation for us to generate and utilize hydrogen as an energy source or energy carrier because 90% of the materials available on the earth contain hydrogen. Hydrogen has attracted much attention from scientists as the fuel of the future due to its qualities, but the hydrogen must be produced in a cost-effective, benign, and reliable way. Nowadays, hydrogen can be produced by cleavage of natural gas such as methane, but in this process, harsh conditions of high temperature and pressure are required. Not only this, but a large amount of greenhouse gases are emitted and as a result, the world is facing ozone depletion and global warming. Therefore, this method is not considered eco-friendly.

Evolution of hydrogen in its purest form using eco-friendly methods seems to be a very difficult task. Solar energy is used as a renewable energy source to produce environmentally benign hydrogen by the cleavage of water. Photocatalytic or photoelectrochemical water-splitting techniques are used for hydrogen generation. Moreover, there are many other water splitting methods such as thermochemical, biophotolytic, mechanocatalytic, plasmolytic, magnetolytic, radiolytic, and so on. Photocatalytic water splitting is a simple method and it will prove to be advantageous for the large-scale production of hydrogen using solar energy. Hydrogen has been advocated as the fuel of the future also.

11.2 WATER SPLITTING

Water splitting occurs naturally when photon energy is absorbed and it is converted into chemical energy (hydrogen) in the photosynthesis process through a complex biological pathway. A huge amount of incident energy is required to produce hydrogen from water. The water-splitting process is a highly endothermic process ($\Delta H > 0$).

- A semiconductor should have a minimum 1.23 eV band gap for successful water splitting at pH = 0 and 1008 nm light irradiation.
- This process is initiated by absorbing photon energy by a semiconductor, with a band gap equal to or greater than the energy gap.
- The electron is promoted from the valence band (VB) of the photocatalyst to the conduction band (CB) and as a result, an electron (e_{CB}^-)—hole (h_{VB}^+) pair is generated.
- As produced, charge carriers can either recombine (consequent loss of energy as heat or emission of photons), or begin some electron transfer reactions at the semiconductor surface (oxidation and/or reduction).
- On the semiconductor surface both reactions may take place: the reduction of an electron acceptor species, when the reduction potential is lower in energy than its CB level, and the oxidation of electron donor species, when the potential is higher in energy than the VB level.
- In the case of water splitting, the electron acceptor species would be the H^+ ion, whereas water, or hydroxyl anions, would be the electron donor species, according to the following reactions:

$$TiO_2 + h\nu \longrightarrow e_{CB}^- + h_{VB}^+ \tag{11.1}$$

Oxidation $\quad H_2O + 2\,h_{VB}^+ \longrightarrow 1/2\,O_2 + 2\,H^+ \tag{11.2}$

Reduction $\quad 2\,H^+ + 2\,e_{CB}^- \longrightarrow H_2 \tag{11.3}$

Net reaction $\quad H_2O \xrightarrow{\ TiO_2/h\nu\ } H_2 + \tfrac{1}{2}\,O_2 \tag{11.4}$

Generally, TiO_2 is used to increase the rate of H_2 production with a cocatalyst such as platinum (Pt) (Figure 11.1).

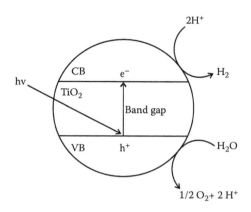

FIGURE 11.1 The mechanism of photocatalytic water splitting over an illuminated TiO_2 semiconductor particle.

11.3 PHOTOCATALYTIC HYDROGEN GENERATION

The pioneering work has been done by Fujishima and Honda (1972) in the field of photocatalysis. They developed a photocatalytic process with improved efficiencies for hydrogen production from water using solar energy, but it still faces some major challenges, although the field has come out of its infancy and is regarded now as a developed subject. Photocatalytic water splitting is regarded as an important pathway to produce hydrogen as renewable energy (Zhao and Liu 2014). The conversion of photon energy to hydrogen via water-splitting process, assisted by photocatalysts, is one of the most promising approaches for the future. Large quantities of hydrogen can potentially be generated in a clean and sustainable manner from this technique (Navarro et al. 2009).

For more than the last four decades, a large number of semiconductor materials have been investigated for the photocatalytic hydrogen production process. Generally, semiconductor systems used in this process are as follows:

- Efficient solar energy converter
- An optimized band gap so as to make maximum utilization of solar radiation
- Sufficient chemical stability against photo or other corrosion processes

Inorganic oxide semiconductors have been used in the last few decades for the photoassisted generation of hydrogen from water. Various types of semiconductors have been used for water photosplitting. Ideal photoelectrolysis systems, such as titanium dioxide, tungsten trioxide and other binary metal oxides, perovskites and other ternary oxides, tantalates and niobates, miscellaneous multinary oxides, semiconductor alloys and mixed semiconductors, composites, twin-photosystem configurations, and so on, have been reported from time to time for water splitting (Rajeshwar 2007).

Among them, titania (TiO_2) becomes a prominent material as a photocatalyst and was extensively studied for photocatalytic hydrogen production from water splitting due to its unique qualities such as relatively high efficiency, high photostability, abundance in nature, low cost, and nontoxicity. The chemical and physical properties of titania have been investigated including crystal phase, crystallinity, particle size, and surface area, to study its photoactivity toward hydrogen generation. But it has some limitations as well such as high overpotential for hydrogen generation, rapid recombination of photogenerated electrons and holes, rapid reverse reaction of molecular hydrogen and oxygen, and inability to absorb visible light. These factors restricted the photoactivity of titania, and some strategies have been developed to resolve these barriers (Leung et al. 2010).

Photocatalytic activities of various tantalates such as $LiTaO_3$, $NaTaO_3$, $KTaO_3$, $MgTa_2O_6$, and $BaTa_2O_6$ for the decomposition of distilled water into H_2 and O_2 were also investigated (Kudo et al. 1999). These alkali and alkaline earth tantalates showed photocatalytic activities for water decomposition without cocatalysts. $BaTa_2O_6$ showed the highest efficiency among all and it was the most active in its orthorhombic phase. The photocatalytic reaction was enhanced with the addition of a small amount of $Ba(OH)_2$ in the water and supporting NiO on the $BaTa_2O_6$ catalyst. The transition metal tantalate, $NiTa_2O_6$, produced both H_2 and O_2 without cocatalysts.

Sayama et al. (2002) reported water splitting into H_2 and O_2 using two different semiconductor photocatalysts and a redox mediator, mimicking the Z-scheme mechanism of photosynthesis. The Pt–$SrTiO_3$ (Cr–Ta-doped) system was found to evolve H_2 using an I^- electron donor, whereas Pt–WO_3 photocatalyst exhibited excellent activity for O_2 evolution using an IO_3^- electron acceptor under visible light irradiation. A mixture of the Pt–WO_3 and the Pt–$SrTiO_3$ (Cr–Ta-doped) powders suspended in NaI aqueous solution was used for both these catalysts. H_2 and O_2 gases evolved in the stoichiometric ratio ($H_2/O_2 = 2$) for more than 250 hours under visible light. The stoichiometric water splitting occurred over oxide semiconductor photocatalysts under visible light irradiation. They also proposed a two-step photoexcitation mechanism using a pair of I^-/IO_3^- redox mediators and the quantum efficiency of the stoichiometric water splitting was approximately 0.1% at 420.7 nm.

Ji et al. (2005) reported the visible light active nitrogen-doped perovskite-type photocatalysts, $Sr_2Nb_2O_{7-x}N_x$ (0, 1.5 < x < 2.8), for hydrogen production from methanol–water mixtures. The doping of nitrogen in $Sr_2Nb_2O_7$ affects light absorption of red-shift edge toward the visible light range and as a result, induced visible light photocatalytic activity. As the N doping increased, the original orthorhombic structure of the layered perovskite was transformed to an unlayered cubic oxynitride structure. The intermediate phase remained as the original layered perovskite structure, whereas its oxygen was replaced by nitrogen and oxygen vacancy to adjust the charge difference between oxygen and doped nitrogen. Nitrogen doping changes the top VB of the $Sr_2Nb_2O_{7-x}N_x$, N2p orbital by causing band gap narrowing, while the bottom of the CB remained almost unchanged due to the Nb 4d orbital.

A novel composite photocatalyst system Cr-doped $Ba_2In_2O_5/In_2O_3$ (C–BIO) oxide was synthesized by a solid-state reaction method and was found to have enhanced activity for photocatalytic water splitting (Wang et al. 2005). Hydrogen evolution was evolved using Pt- or NiO-loaded C–BIO and individual precursor materials from aqueous solution under visible light and also by pure water splitting under UV light irradiation. The C–BIO composite showed higher H_2 evolution rate as compared to its counterpart. The photophysical and photocatalytic properties such as the overall band structure, charge carrier excitation, separation, transportation and redox reactions for H_2 and O_2 evolution in the C–BIO system were discussed (Figure 11.2).

Korzhak et al. (2008) designed a new template-Na^+ complex with dibenzo-18-crown-6 for the synthesis of titanium dioxide and used it for photocatalytic hydrogen generation. These metal nanocomposites were produced on the photoreduction of copper (II), nickel (II) chlorides, and silver (I) nitrate on the surface of mesoporous TiO_2. Remarkable photocatalytic activity for hydrogen production from water-alcohol mixtures was obtained, which was 50%–60% more efficient than metal containing nanocomposites based on TiO_2 (Degussa P25).

Wang et al. (2008) introduced a new crystal structure (body centric cubic [bcc]) for nanostructured vanadium dioxide (VO_2) with a large optical band gap of approximately 2.7 eV. It showed surprisingly excellent photocatalytic activity in hydrogen production. The VO_2 nanorods exhibited a high quantum efficiency of approximately 38.7%. The hydrogen production rate can be tuned by varying the incident angle of UV light on the films of the aligned VO_2 nanorods, and it reached a high rate of 800 mmol/m²/h from a mixture of water and ethanol under UV light at approximately 27 mW/cm² power density. Thus, this material showed potential for commercial application as photoassisted hydrogen generators.

Z-scheme photocatalysis systems consisting of $SrTiO_3$:Rh for H_2 evolution, $BiVO_4$ for O_2 evolution, and Fe^{3+}/Fe^{2+} as an electron mediator were investigated (Sasaki et al. 2008). Visible light driven $SrTiO_3$:Rh was loaded with cocatalysts. The overall water splitting using the system Ru cocatalyst was as high as that of the Pt cocatalyst system. The photocatalytic activity of Pt cocatalyst decreased as the partial pressures of evolved H_2 and O_2 increased, but such deactivation was not observed for the system using the Ru cocatalyst. The formation of water from H_2 and O_2, where reduction of

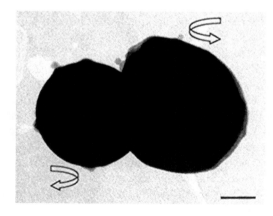

FIGURE 11.2 Cr-doped $Ba_2In_2O_5/In_2O_3$ for hydrogen generation. (Adapted from Wang, D. et al., *Chem. Mater.*, 17, 3255–3261, 2005. With permission.)

Fe^{3+} by H_2 and oxidation of Fe^{2+} by O_2 take place, revealed back reactions. These were found to be significantly suppressed in the system using Ru cocatalyst, resulting in good photocatalytic performance for water splitting. The $(Ru/SrTiO_3:Rh)-(BiVO_4)-(Fe^{3+}/Fe^{2+})$ photocatalysis system gave 0.3% quantum yield and showed stability for more than 70 hours. Hence, this system was considered suitable for water splitting using a solar simulator.

Polymer-supported CdS/ZnS nanocomposites have been synthesized by Deshpande et al. (2009). These nanocomposite CdS/ZnS moieties coated over polyester strip were found to be better in visible–light mediated photoactivity as compared to corresponding pure CdS- or ZnS-containing coupons for the splitting of water. The increase in activity depends on the molar ratio of two component sulfides in a particular sample. It has a guest–host system, where 1–3 nm CdS particles were embedded over larger size ZnS clusters. The average crystallite size was found to be around 5 and 15 nm for CdS and ZnS, respectively, which is dispersed individually over polyester. When ZnS was added to CdS, a blue shift was observed in the UV–vis absorption spectrum, in conformation with the quantum size effects. The nanocomposites have face-centered cubic (alpha) phases for both CdS and ZnS in close contact with each other. A solid solution of $Cd_{1-x}Zn_xS$ was generated at the interfaces of these two semiconductors. This is due to the smaller size of CdS particles and the microstructural properties associated with the nanostructured CdS or CdS/ZnS interfaces. An increase in the number of reaction sites may play a vital role together with the augmented catalytic activity of CdS/ZnS composite photocatalysts.

Cu_2O nanosystems and CuO nanowires (NWs) were synthesized by chemical vapor deposition (CVD) and used by Barreca et al. (2009) in the photocatalytic splitting of methanol/water solutions to produce hydrogen. These systems were used for the clean conversion of sunlight into storable chemical energy.

A nanosized core-shell (CdS)/(ZnS) structure was synthesized via a microemulsion technique by Wang et al. (2010). The average particle size of the photocatalysts was around 10–100 nm and band gaps of the catalysts ranged from 2.25 to 2.46 eV. The photocatalytic H_2 generation from $(CdS_x)/(ZnS_{1-x})$ with x ranging from 0.1 to 1 was able to produce hydrogen from water photolysis under visible light, and the highest amount of hydrogen was produced with $x = 0.9$ catalyst. The photohydrogen production rate was influenced by catalyst loading density and the best catalyst concentration in water was 1 g/L. The highest hydrogen production rate was 2.38 mmol/g/L/h with 16.1% quantum yield under visible light. These results revealed that the (CdS)/(ZnS) core/shell nanoparticles is a novel photocatalyst for renewable hydrogen generation from water under visible light. It is due to the large band gap of the ZnS shell that separates the electron–hole pairs generated by the CdS core and hence reduces their recombinations.

A novel nitrogen-doped tantalite $(Sr_2Ta_2O_{7-x}N_x)$ photocatalyst was used in hydrogen generation by Mukherji et al. (2011). The nitrogen-doped tantalite showed significantly increased visible light absorption as well as 87% improved photocatalytic hydrogen production under solar irradiation, when it was compared with its undoped counterpart $(Sr_2Ta_2O_7)$. This photocatalyst also showed a strong capability of photoinduced reduction toward exfoliated graphene oxide (GO) to graphene sheets. A new type of composite was designed that consisted of graphene-Pt and $Sr_2Ta_2O_{7-x}N_x$.

It demonstrated an additional ~80% increase in hydrogen production and a quantum efficiency of 6.45% (~177% increase from pristine undoped $Sr_2Ta_2O_7$) due to the efficient charge carrier separation on the photocatalyst. So it was suggested that the graphene can play an important role as an electron-transfer highway, which facilitates the charge carrier collection onto Pt cocatalysts. Thus, this method can be considered as an excellent strategy to increase photocatalytic hydrogen production in addition to a commonly applied doping method.

N-doped TiO_2 nanofibers have been synthesized and their photocatalytic efficiency for hydrogen generation from ethanol–water mixtures under UV-A and UV-B irradiation was reported by Wu et al. (2011). These nanofibers were synthesized by the hydrothermal method and annealed in air and/or ammonia to achieve N-doped anatase fibers. Then N-doped TiO_2 was used as support for Pd and Pt nanoparticles deposited with wet impregnation followed by calcination and reduction. These N-doped TiO_2 nanofibers performed clearly much better than their undoped counterparts, as well as exhibited remarkable efficiency for both UV-B and UV-A illumination. The H_2 evolution rate was observed for 100 mg of catalyst (N-doped TiO_2 nanofiber decorated with Pt nanoparticles) in 1 L of water–ethanol mixture. It was observed that the H_2 evolution rates were as high as 700 µmol/h (UV-A) and 2250 µmol/h (UV-B) corresponding to photoenergy conversion percentages of ~3.6% and ~12.3%, respectively.

The p-type (F) doped Co_3O_4 nanostructured films were synthesized by a plasma-assisted process and tested for photocatalytic production of H_2 from water/ethanol solutions by Gasparotto et al. (2011) under both near-UV and solar irradiation. $F-Co_3O_4$ nanostructures demonstrated remarkable improvement with respect to the corresponding undoped oxide. F-doped Co_3O_4 film was suggested as a highly promising system for hydrogen generation. H_2 generated from this semiconductor was the best ever reported for similar semiconductor-based photocatalytic processes.

A novel visible light–driven CuS/ZnS porous nanosheet has been used as a photocatalyst in photocatalytic hydrogen production through water splitting (Zhang et al. 2011). CuS/ZnS porous nanosheet photocatalysts were synthesized via a simple hydrothermal and cation exchange reaction between ZnS(en)(0.5) nanosheets and $Cu(NO_3)_2$. The as-prepared CuS/ZnS porous nanosheets without a Pt cocatalyst reached a high H_2 production rate of 4147 µmol/h^{-1}g^{-1} at 2 mol% content loading of CuS and an apparent quantum efficiency of 20% at 420 nm. This showed high visible light photocatalytic H_2 production activity due to the interfacial charge transfer (IFCT) from the VB of ZnS to CuS, which caused the reduction of CuS to Cu_2S. Hence, it was proposed that low-cost CuS noble metals show a possibility for photocatalytic H_2 production. Thus, a facile method for enhancing H_2 production activity by photoinduced IFCT was exhibited.

Townsend et al. (2011) introduced layered $K_4Nb_6O_{17}$ with a 3.5 eV band gap as a UV light–driven photocatalyst for overall water splitting. The photochemical deposition of Pt and IrO_x (x = 1.5–2) nanoparticles onto the surface of the nanoscrolls produced two- and three-component photocatalysts. Hydrogen generation obtained from pure water and aqueous methanol using these nanostructures under UV irradiation ranged from 2.3 to 18.5 turnover over a 5-hour period. The photocatalytic activity of nanostructures for hydrogen evolution was directly correlated with variation in overpotentials for water reduction (210–325 mV). Instead of oxygen formation, a peroxide surface bond with hydrogen was observed in 1:1 stoichiometry.

Gallo et al. (2012) presented highly active bimetallic Pt–Au nanoparticles supported on reduced anatase nanocrystals as photocatalysts for hydrogen production by photoreforming of aqueous solutions of renewable feedstocks, such as ethanol and glycerol. This catalyst was easily obtained by metal impregnation of commercial TiO_2, followed by a reductive treatment resulting in an enhanced photoactivity on pre-reduction. More H_2 is produced in the case of using ethanol as a sacrificial agent, whereas in the case of glycerol, significant amounts of CO_2 have also been detected under both UV-A or simulated sunlight irradiation that indicated more efficient oxidation of the organic sacrificial agent. The key parameters maximizing light absorption is the presence of bimetallic Pt–Au nanoparticles and Ti^{3+} sites/O^{2-} vacancies in the bulk structure, resulting in good photocatalytic performances.

Thornton and Raftery (2012) synthesized undoped and carbon-doped cadmium indate ($CdIn_2O_4$) powders using a sol–gel pyrolysis method and utilized it for hydrogen generation under the UV–visible irradiation without using any sacrificial reagent. Each catalyst was loaded with platinum cocatalyst, which increased electron–hole pair separation and promoted surface reaction. Carbon-doped indium oxide and cadmium oxide were also prepared and analyzed for comparison. The band gap for $C–CdIn_2O_4$ was observed to be 2.3 eV. The hydrogen generation rate for C-doped In_2O_4 was approximately double than that of undoped material. It was also compared with platinized TiO_2 in methanol, which showed a fourfold increase in hydrogen production. The quantum efficiency was found to be 8.7% at 420–440 nm. It was observed that this material was capable of producing hydrogen using only visible light only and showed good efficiency even at 510 nm.

Shen et al. (2012) reported preparation of $CdS/ZnS/In_2S_3$ microspheres for photocatalytic H_2 production. The microspheres were prepared by embedded ZnS and CdS nanocrystals (5–10 nm) on $CdS/ZnS/In_2S_3$ via a sonochemical method at room temperature and normal pressure without using any templates or surfactants. The photocatalytic activity of $CdS/ZnS/In_2S_3$ for water splitting was investigated under visible light irradiation ($\lambda > 400$ nm) and an especially high photocatalytic activity yield of 40.9% at 420 nm was achieved in the absence of cocatalysts.

$SrTiO_3$ (STO) semiconductor, with a large band gap 3.2 eV, was used for a water-splitting reaction under UV light irradiation in the presence of NiO cocatalyst. The bulk STO, 30 ± 5 nm STO, and 6.5 ± 1 nm STO were synthesized by three different methods. Water splitting splitting into stoichiometric mixtures of H_2 and O_2, for all the samples in connection with NiO, was measured, and it was observed that the activity decreased for bulk STO, 30 nm STO, and 6.5 nm STO from 28, 19.4, and 3.0 µmol H_2/g/h, respectively. It was so observed as oxidation overpotential of the water was increased for the smaller particles, and due to the quantum size effect light absorption was reduced (Townsend et al. 2012).

Ouyang et al. (2012) reported a strategy of surface-alkalinization to enhance photocatalytic performance of a semiconductor photocatalyst. The surface alkalinization of $SrTiO_3$ photocatalyst induced a high alkalinity of the solution environment and it significantly shifted the surface energy band of $SrTiO_3$ to a more negative level. It also affected the H_2O reduction and consequently promoted the photocatalytic efficiency of H_2 evolution. The visible light–sensitive La, Cr-codoped $SrTiO_3$ photocatalyst also follows this mechanism, and 25.6% highly apparent quantum efficiency was achieved for H_2 evolution in CH_3OH aqueous solution containing 5 M NaOH at an incident wavelength of 425 ± 12 nm.

Core/shell nanofibers CdS/ZnO were synthesized by Yang et al. (2013) using one-pot single-spinneret electrospinning. These fabricated nanofibers as nanoheterojunction photocatalysts presented excellent visible light photocatalytic activity and stability for hydrogen production. Fluorine-doped α-Fe_2O_3 nanomaterials were synthesized by plasma-enhanced CVD (PE-CVD) at 300°C–500°C temperature by Carraro et al. (2013). For this purpose, fluorinated iron(II) diketonate-diamine compound was used as a single source precursor for both Fe and F. The photocatalytic H_2 production from water/ethanol solutions under simulated solar irradiation evidenced promising gas evolution rates. It was suggested that the PE-CVD approach is a valuable strategy to fabricate highly active supported materials.

Hybrid photocatalysts such as ZrO_2, TiO_2, and CdS have been synthesized and their photocatalytic activity for hydrogen generation was reported by Sasikala et al. (2013). Photocatalytic activity of these hybrid photocatalysts was in decreasing order as

$$ZrO_2\text{–}TiO_2\text{–}CdS > TiO_2\text{–}CdS > ZrO_2\text{–}CdS > CdS > ZrO_2\text{–}TiO_2 \approx TiO_2 > ZrO_2$$

Quantum efficiency of ZrO_2–TiO_2–CdS was obtained as 11.5% with Pd as cocatalyst.

Wang et al. (2013) synthesized calcium tantalate composite and used it for photocatalytic hydrogen production. The phase composites such as cubic α-$CaTa_2O_6$/hexagonal $Ca_2Ta_2O_7$, cubic $CaTa_2O_6$/hexagonal $Ca_2Ta_2O_7$/orthorhombic β-$CaTa_2O_6$, or cubic α-$CaTa_2O_6$/orthorhombic β-$CaTa_2O_6$ provided very high photocatalytic H_2 production without any cocatalysts and in the presence of methanol. The photoexcited charge carrier separation rate was significantly improved due to the presence of junctions and interfaces in the composites. The photocatalytic activity of these composites was greatly promoted for H_2 production by *in situ* photodeposition of noble metal nanoparticles (Pt or Rh) as cocatalysts.

The fly ash–based mesoporous CdS/Al-MCM-41 nanocomposites were synthesized and used for hydrogen production by Zhang et al. (2013a). The CdS/Al-MCM-41 nanocomposites generated 3.3 mL/g of H_2 in 6 hours by photocatalytic water splitting under visible light irradiation. These nanocomposites show enhanced performance due to the synergistic effect between CdS clusters and mesoporous Al-MCM-41 matrix.

CdS/g-C_3N_4 core/shell nanowires (NWs) with different g-C_3N_4 contents were fabricated by Zhang et al. (2013b) via combined solvothermal and chemisorption methods. Photocatalytic hydrogen generation of these samples was evaluated under visible light illumination ($\lambda \geq 420$ nm) with Na_2S and Na_2SO_3 as sacrificial reagents in water. CdS NWs coated with g-C_3N_4 showed significantly enhanced photocatalytic hydrogen production rate, that is, up to 4152 µmol/h/g. It was observed that g-C_3N_4 coating can substantially reinforce the photostability of CdS NWs even in a nonsacrificial system. Hence, the synergistic effect between g-C_3N_4 and CdS was considered responsible for the enhancement of the photocatalytic activity and photostability. The charge separation and transfer of corrosive holes from CdS to robust C_3N_4 can effectively accelerate this process.

Fang et al. (2014) introduced dye-sensitized Pt at TiO_2 core-shell nanostructures for the efficient photocatalytic generation of hydrogen. These nanostructures were prepared through a hydrothermal method. The Pt at TiO_2 core-shell structures sensitized with a dye have a high photocatalytic activity for hydrogen generation from proton reduction under visible light irradiation. Yield of H_2 generation was found to be enhanced, when the dyes and TiO_2 coexcited through the combination of two irradiation beams of different wavelengths. At this time, they showed a synergistic effect. They also developed an Au at TiO_2–CdS ternary nanostructure by decorating CdS nanoparticles onto Au at TiO_2 core-shell structures for photocatalytic hydrogen generation under UV light irradiation (Fang et al. 2013). This unique ternary design is responsible for increased photocatalytic activity. The photoexcited electron transfer path was built up from CdS to the core Au particles via the TiO_2 nanocrystal bridge (CdS→TiO_2→Au) that effectively suppressed electron–hole recombination on a CdS photocatalyst. The core Au nanoparticles can act as the interior active catalyst for proton reduction toward hydrogen evolution so that there is no need of postdeposition of the metal cocatalyst. The metal–semiconductor hybrid photocatalysts with high photocatalytic efficiency are promising candidates for use in production of solar fuels.

$CaFe_2O_4/TiO_2$ composite photocatalysts consisting of $CaFe_2O_4$ and TiO_2 hierarchical spheres of nanosheets were prepared by the solid-state dispersion (SSD) method by Police et al. (2014). The photocatalytic activity of the composites was studied for hydrogen production using methanol–water mixtures. The activity of $CaFe_2O_4/TiO_2$ composite was responsible for possible charge transfer processes under visible and solar light irradiation. The highly conductive nature of TiO_2 spheres of nanosheets was considered responsible for improved charge separation and charge mobility. The $CaFe_2O_4/TiO_2$ composite was highly photocatalytically active under the sunlight for H_2 production.

Hakamizadeh et al. (2014) synthesized mesoporous TiO_2/AC, Pt/TiO_2, and $Pt/TiO_2/AC$ (AC = activated carbon) nanocomposites by functionalizing the AC using acid treatment and the sol–gel method. Pt was loaded by the photochemical deposition method. Both AC and Pt were found to improve hydrogen production via water splitting and methanol. Methanol also acted as a good hole scavenger. Mesoporous $Pt/TiO_2/AC$ nanocomposite was highly efficient for photocatalytic hydrogen production as compared to TiO_2/AC, Pt/TiO_2, and the commercial photocatalyst P25. The hydrogen production rate of $Pt/TiO_2/AC$ was found to be 7490 µmol (h g cat)$^{-1}$, which was about 75 times higher than that with P25 photocatalyst.

Aslan et al. (2014) investigated photocatalytic hydrogen generation using oleic acid-capped CdS, CdSe, and $CdS_{0.75}Se_{0.25}$ alloy nanocrystals (quantum dots, QDs). Hydrogen generated by these photocatalysts under visible light irradiation was observed and Na_2S and Na_2SO_3 were employed as hole scavengers. $CdS_{0.75}Se_{0.25}$ was found to be highly photostable and gave the highest hydrogen evolution rate (1466 µmol/h/g), which was about three times higher than that of CdS and seven times higher than that of CdSe.

The hydrogen generation from a double heterojunction system, CdSe, CdSe/CdS core/shell NWs, and their Pt nanoparticle-decorated counterparts was studied by Tongying et al. (2014). Pt nanoparticle-decorated CdSe/CdS NWs, a double heterojunction system, showed a H_2 generation rate of ~434.29 ± 27.40 µmol/h/g under the UV/Visible irradiation. Wang et al. (2014) investigated visible light–driven H_2 evolution on nickel-hybrid CdS QDs (Ni_h–CdS QDs) from glycerol and water. A hydrogen quantity of 403.2 µmol was obtained from 20-hour visible light irradiation with a high H_2 evolution rate of approximately 74.6 µmol/h/mg. The modified CdS QDs exhibited the greatest affinity toward Ni^{2+} ions and the highest activity for H_2 evolution compared to CdTe QDs and CdSe QDs.

Recently, a novel TiO_2–In_2O_3 at g-C_3N_4 hybrid material was synthesized via facile solvothermal method and evaluated for degradation of RhB and hydrogen production by Jiang et al. (2015). The RhB degradation rate on TiO_2–In_2O_3 at g-C_3N ternary composites was 6.6 times higher than that of pure g-C_3N_4. The H_2 evolution rate with this ternary material was found to be 48 times that of pure g-C_3N_4. The enhanced activities of TiO_2, In_2O_3, and g-C_3N_4 were mainly attributed to the low recombination rate of photogenerated carriers as well as the high surface area of ternary composites.

Huerta-Flores et al. (2015) synthesized pervoskite strontium zirconate ($SrZrO_3$) powders with an orthorhombic phase by three methods: solid-state reaction, molten state method, and ultrasound-assisted synthesis. These catalysts were used for photocatalytic hydrogen evolution from water splitting. $SrZrO_3$ was prepared using a solid-state reaction, and exhibited the highest hydrogen evolution rate, that is, 49 µmol/g/h, whereas powder synthesized using ultrasound and molten salt methods showed 40 and 34 µmol/g/h, respectively. It was confirmed that $SrZrO_3$ is a suitable photocatalyst for clean hydrogen generation from water splitting under UV light irradiation, and the crystallinity is one of the primary determining factors of the catalytic activity.

Hu et al. (2015) synthesized a heterojunction-type photocatalyst g-C_3N_4/nano-$InVO_4$ by in situ growth of $InVO_4$ nanoparticles on the surface of g-C_3N_4 sheets using a hydrothermal process. It was observed that g-C_3N_4/nano-$InVO_4$ nanocomposites were effective in charge–hole pair separation and also have stronger reducing power. Its use also improved H_2 evolution from water splitting as compared to bare g-C_3N_4 sheets and g-C_3N_4/micro-$InVO_4$ composites. The results showed that the g-C_3N_4/nano-$InVO_4$ nanocomposite with a mass ratio of 80:20 possessed the maximum photocatalytic activity for hydrogen production under visible light radiation.

A novel ZnO at PbS/GO (graphene oxide) nanostructure photocatalyst was constructed by Shi et al. (2015) using a simple assembly method. The multiple exciton generation (MEG) application in a photocatalytic hydrogen evolution system was used for the first time by them. In the MEG process, absorption of a single photon produces more than one electron–hole pair; this is also known as carrier multiplication (CM). PbS QDs with the MEG property were used to construct a novel ZnO at PbS/GO structured photocatalyst, which was found beneficial for photocatalytic H_2 evolution from water. This strategy was suggested as a newer technique for the design of photocatalysts to achieve high efficiency QD materials with the MEG property.

Yuan et al. (2015) introduced a novel CdS nanorods/g-C$_3$N$_4$ heterojunctions loaded metal-free NiS as a cocatalyst by an *in situ* hydrothermal method. The photocatalytic hydrogen was evaluated using an aqueous solution containing triethanolamine under visible light ($\lambda \geq 420$ nm). Visible light–driven 1D CdS nanorods and 2D g-C$_3$N$_4$ nanosheets are a promising strategy to achieve highly efficient photocatalytic H$_2$ evolution. The ternary hybrid g-C$_3$N$_4$-CdS-9%NiS composite material exhibited the highest photocatalytic performance with an H$_2$ production rate of 2563 µmol/h/g, which is 1582 times higher than that with pristine g-C$_3$N$_4$ alone. The improved photocatalytic activity was achieved due to the combined effects of NiS cocatalyst loading and the formation of the intimate nanoheterojunctions between 1D CdS nanorods and 2D g-C$_3$N$_4$ nanosheets. These were also favorable for promoting charge transfer, improving the separation efficiency of photoinduced electron–hole pairs from the bulk to the interfaces, and accelerating the surface H$_2$ evolution kinetics.

Liang et al. (2015) reported novel CdS/triptycene-based polymer (NTP) nanocomposites that were fabricated via a facile precipitation process using Cd(OAc)$_2$, Na$_2$S, and prefabricated NTP as raw materials. As-prepared CdS-NTP nanocomposites were used in photocatalytic hydrogen generation in the presence of a sacrificial reagent. This visible–light driven CdS-NTP gave a 10 times higher rate of hydrogen production than that of pure CdS. The photocorrosion of CdS was simultaneously suppressed and hence the composites show high stability. Additionally, the CdS QDs of NTP have a high surface area and stable covalent structure that prevented aggregation, thus its catalytic activity was increased. It is clearly reflected in enhanced photocatalytic performance of the hybrid nanocomposites. Hence, there is a high potential of using these porous triptycene-based materials to develop multifunctional porous materials for semiconductor-based photocatalytic hydrogen evolution.

The NiSe nanowire film on nickel foam (NiSe/NF) was used as a 3D bifunctional electrode for both the oxygen and hydrogen evolution reaction in a strongly alkaline electrolyte (Tang et al. 2015). *In situ* hydrothermal treatment was used to synthesize NiSe/NF by NF using NaHSe as the Se source. NiSe/NF showed high photocatalytic activity for hydrogen generation, whereas NiOOH species formed at the NiSe surface served as the actual catalytic site. This system was highly efficient in basic media for catalyzing hydrogen evolution reaction and also showed superior stability. This bifunctional electrode enhanced the performance of alkaline water electrolyzer with 10 mA/cm^2 at a cell voltage of 1.63 V.

Dong et al. (2015) designed a novel CdSe/NiO heteroarchitecture and utilized it as a photocathode for hydrogen production from water. Enhanced photoelectrochemical properties and light harvesting in the visible light region were observed for the CdSe/NiO heteroarchitecture due to CdSe deposited on the NiO film. The CdSe/NiO photoelectrode exhibited superior stability in both the neutral environments in presence of nitrogen and saturated air. The average hydrogen evolution rate of the MoS$_2$/CdSe/NiO photocathode was obtained as 0.52 µmol/h/cm^2 at –0.131 V at pH 6, with almost 100% Faradaic efficiency.

Plascencia-Hernández et al. (2016) synthesized mesoporous TiO$_2$ hollow shells via a conventional templating method that combines sol–gel coating and selective etching of the silica cores. Pt nanocatalysts were loaded as a supporter on these

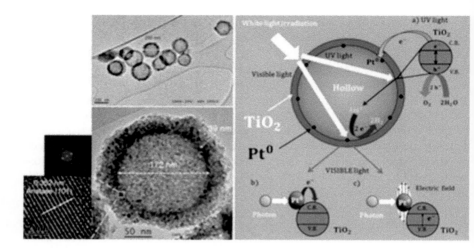

FIGURE 11.3 Use of Pt nanocatalysts supported on mesoporous TiO$_2$ hollow shells. (Adapted from Plascencia-Hernández, F. et al., *J. Sol-Gel Sci. Technol.*, 77, 39–47, 2016. With permission.)

mesoporous TiO$_2$ hollow shells by varying the metal loading: 0%, 1%, 3%, 5%, and 7% and then calcined at 500°C or 900°C. It was observed that mesostructures were smaller than 300 nm and the TiO$_2$ shells had 40 nm average wall thickness. The sample calcined at 900°C was pure anatase phase, whereas those calcined at 500°C were amorphous in nature. The photocatalytic hydrogen production was measured under the white light illumination (UV and visible). As-prepared samples were suspended in an aqueous solution of methanol and compared to TiO$_2$ (P25, Degussa) as a reference. The highest hydrogen yield was obtained from the crystalline TiO$_2$ hollow shells annealed at 900°C containing 1 or 7 wt.% Pt. It was found that the amorphous samples were inactive at all metal loadings (Figure 11.3).

The existing sources of energy such as coal, biomass, petrol, diesel, gas, and so on are being exhausted very rapidly on one hand and their resources are limited to a few more decades on the other. Therefore, there is an urgent need to find an alternate source of energy that is renewable. Hydrogen has been advocated as the fuel of the future, as it is generated by photosplitting of water and has the highest storage capacity. Thus, this area will attract many more chemists to find newer photocatalytic systems for splitting of water.

REFERENCES

Aslan, E., O. Birinci, A. Aljabour, F. Ozel, I. Akın, I. Hatay Patir, M. Kus, and M. Ersoz. 2014. Photocatalytic hydrogen evolution by oleic acid-capped CdS, CdSe, and CdS$_{0.75}$Se$_{0.25}$ alloy nanocrystals. *ChemPhysChem.* 15 (13): 2668–2671.

Barreca, D., P. Fornasiero, A. Gasparotto, V. Gombac, C. Maccato, T. Montini, and E. Tondello. 2009. The potential of supported Cu$_2$O and CuO nanosystems in photocatalytic H$_2$ production. *ChemSusChem.* 2 (3): 230–233.

Carraro, G., D. Barreca, D. Bekermann, T. Montini, A. Gasparotto, V. Gombac, C. Maccato, and P. Fornasiero. 2013. Supported F-doped alpha-Fe$_2$O$_3$ nanomaterials: Synthesis, characterization and photo-assisted H$_2$ production. *J. Nanosci. Nanotechnol.* 13 (7): 4962–4968.

Deshpande, A., P. Shah, R. S. Gholap, and N. M. Gupta. 2009. Interfacial and physico-chemical properties of polymer-supported CdSZnS nanocomposites and their role in the visible-light mediated photocatalytic splitting of water. *J. Colloid Interface Sci.* 333 (1): 263–268.

Dong, Y., Y. Chen, P. Jiang, G. Wang, X. Wu, R. Wu, and C. Zhang. 2015. Efficient and stable MoS$_2$/CdSe/NiO photocathode for photoelectrochemical hydrogen generation from water. *Chem. Asian J.* 10 (8): 1660–1667.

Fang J., L. Xu, Z. Zhang, Y. Yuan, S. Cao, Z. Wang, L. Yin, Y. Liao, and C. Xue. 2013. Au@TiO$_2$-CdS ternary nanostructures for efficient visible-light-driven hydrogen generation. *ACS Appl. Mater. Interfaces.* 5 (16): 8088–8092.

Fang, J., L. Yin, S. Cao, Y. Liao, and C. Xue. 2014. Dye-sensitized Pt@TiO$_2$ core-shell nanostructures for the efficient photocatalytic generation of hydrogen. *Beilstein J. Nanotechnol.* 5: 360–364.

Fujishima, A. and K. Honda. 1972. Electrochemical photolysis of water at a semiconductor electrode. *Nature.* 238: 37–38.

Gallo, A., T. Montini, M. Marelli, A. Minguzzi, V. Gombac, R. Psaro, P. Fornasiero, and V. Dal Santo. 2012. H$_2$ production by renewables photoreforming on Pt-Au/TiO$_2$ catalysts activated by reduction. *ChemSusChem.* 5 (9): 1800–1811.

Gasparotto, A., D. Barreca, D. Bekermann, A. Devi, R. A. Fischer, P. Fornasiero et al. 2011. F-doped Co$_3$O$_4$ photocatalysts for sustainable H$_2$ generation from water/ethanol. *J. Am. Chem. Soc.* 133 (48): 19362–19365.

Hakamizadeh, M., S. Afshar, A. Tadjarodi, R. Khajavian, M. R. Fadaie, and B. Bozorgi. 2014. Improving hydrogen production via water splitting over Pt/TiO$_2$/activated carbon nano-composite. *Int. J. Hydrogen Energy.* 39 (14): 7262–7269.

Hu, B., F. Cai, T. Chen, M. Fan, C. Song, X. Yan, and W. Shi. 2015. Hydrothermal synthesis g-C$_3$N$_4$/Nano-InVO$_4$ nanocomposite and enhanced photocatalytic activity for hydrogen production under visible light radiation. *ACS Appl. Mater. Interfaces.* 7 (33): 18247–18256.

Huerta-Flores, A. M., L. M. Torres-Martínez, D. Sánchez-Martínez, and M. E. Zarazúa-Morín. 2015. SrZrO$_3$ powders: Alternative synthesis, characterization and application as photocatalysts for hydrogen evolution from water splitting. *Fuel.* 158: 66–71.

Ji, S. M., P. H. Borse, H. G. Kim, D. W. Hwang, J. S. Jang, S. W. Bae, and J. S. Lee. 2005. Photocatalytic hydrogen production from water-methanol mixtures using N-doped Sr$_2$Nb$_2$O$_7$ under visible light irradiation: Effects of catalyst structure. *Phys. Chem. Chem. Phys.* 7 (6): 1315–1321.

Jiang, Z., D. Jiang, Z. Yan, D. Liu, K. Qian, and J. Xie. 2015. A new visible light active multifunctional ternary composite based on TiO$_2$–In$_2$O$_3$ nanocrystals heterojunction decorated porous graphitic carbon nitride for photocatalytic treatment of hazardous pollutant and H$_2$ evolution. *Appl. Catal. B: Environ.* 170–171: 195–205.

Korzhak, A. V., N. I. Ermokhina, A. L. Stroyuk, V. K. Bukhtiyarov, A. E. Raevskaya, V. I. Litvin, S. Y. Kuchmiy, V. G. Ilyin, and P. A. Manorik. 2008. Photocatalytic hydrogen evolution over mesoporous TiO$_2$/metal nanocomposites. *J. Photochem. Photobiol. A: Chem.* 198: 126–134.

Kudo, A., H. Kato, and S. Nakagawa. 1999. Water splitting into H$_2$ and O$_2$ on new Sr$_2$M$_2$O$_7$ (M = Nb and Ta) photocatalysts with layered perovskite structures: Factors affecting the photocatalytic activity. *J. Phys. Chem. B.* 104: 571–575.

Leung, D. Y., X. Fu, C. Wang, M. Ni, M. K. Leung, X. Wang, and X. Fu. 2010. Hydrogen production over titania-based photocatalysts. *ChemSusChem.* 3 (6): 681–694.

Liang, Q., G. Jiang, Z. Zhao, Z. Li, and M. J. Maclachlan. 2015. CdS-decorated triptycene-based polymer: Durable photocatalysts for hydrogenproduction under visible-light irradiation. *Catal. Sci. Technol.* 5 (6): 3368–3374.

Mukherji, A., B. Seger, G. Q. Lu, and L. Wang. 2011. Nitrogen doped $Sr_2Ta_2O_7$ coupled with graphene sheets as photocatalysts for increased photocatalytic hydrogen production. *ACS Nano.* 5 (5): 3483–3492.

Navarro, R. M. Y., M. C. Alvarez Galván, F. del Valle, J. A. Villoria de la Mano, and J. L. Fierro. 2009. Water splitting on semiconductor catalysts under visible-light irradiation. *ChemSusChem.* 2 (6): 471–485.

Ouyang, S., H. Tong, N. Umezawa, J. Cao, P. Li, Y. Bi, Y. Zhang, and J. Ye. 2012. Surface-alkalinization-induced enhancement of photocatalytic H_2 evolution over $SrTiO_3$-based photocatalysts. *J. Am. Chem. Soc.* 134 (4): 1974–1977.

Plascencia-Hernández, F., G. Valverde-Aguilar, N. Singh, A. R. Derk, E. W. McFarland, F. Zaera, N. Cayetano-Castro, R. Vázquez-Arreguín, and M. A. Valenzuela. 2016. Photocatalytic hydrogen production from aqueous methanol solution using Pt nanocatalysts supported on mesoporous TiO_2 hollow shells. *J. Sol-Gel Sci. Technol.* 77 (1): 39–47.

Police, A. K. R., S. Basavaraju, D. K. Valluri, V. S. Muthukonda, S. Machiraju, and J. S. Lee. 2014. $CaFe_2O_4$ sensitized hierarchical TiO_2 photo composite for hydrogen production under solar light irradiation. *Chem. Eng. J.* 247: 152–160.

Rajeshwar, K. 2007. Hydrogen generation at irradiated oxide semiconductor–solution interfaces. *J. Appl. Electrochem.* 37 (7): 765–787.

Sasaki, Y., A. Iwase, H. Kato, and A. Kudo. 2008. The effect of co-catalyst for Z-scheme photocatalysis systems with an Fe^{3+}/Fe^{2+} electron mediator on overall water splitting under visible light irradiation. *J. Catal.* 259: 133–137.

Sasikala, R., A. R. Shirole, and S. R. Bharadwaj. 2013. Enhanced photocatalytic hydrogen generation over ZrO_2-TiO_2-CdS hybrid structure. *J. Colloid Interface Sci.* 409: 135–140.

Sayama, K., K. Mukasa, R. Abe, Y. Abe, and H. Arakawa. 2002. A new photocatalytic water splitting system under visible light irradiation mimicking a Z-scheme mechanism in photosynthesis. *J. Photochem. Photobiol.* 148: 71–77.

Shen, Z., G. Chen, Q. Wang, Y. Yu, C. Zhou, and Y. Wang. 2012. Sonochemistry synthesis and enhanced photocatalytic H_2-production activity of nanocrystals embedded in CdS/ZnS/In_2S_3 microspheres. *Nanoscale.* 4(6): 2010–2017.

Shi, X. F., X. Y. Xia, G. W. Cui, N. Deng, Y. Q. Zhao, L. H. Zhuo, and B. Tang. 2015. Multiple exciton generation application of PbS quantum dots in ZnO@PbS/graphene oxide for enhanced photocatalytic activity. *Appl. Catal. B: Environ.* 163: 123–128.

Tang, C., N. Cheng, Z. Pu, W. Xing, and X. Sun. 2015. NiSe nanowire film supported on nickel foam: An efficient and stable 3D bifunctional electrode for full water splitting. *Angew. Chem. Int. Ed.* 54 (32): 9351–9355.

Thornton, J. M. and D. Raftery. 2012. Efficient photocatalytic hydrogen production by platinum-loaded carbon-doped cadmium indate nanoparticles. *ACS Appl. Mater. Interfaces.* 4 (5): 2426–2431.

Tongying, P., F. Vietmeyer, D. Aleksiuk, G. J. Ferraudi, G. Krylova, and M. Kuno. 2014. Double heterojunction nanowire photocatalysts for hydrogen generation. *Nanoscale.* 6 (8): 4117–4124.

Townsend, T. K., N. D. Browning, and F. E. Osterloh. 2012. Nanoscale strontium titanate photocatalysts for overall water splitting. *ACS Nano.* 6 (8): 7420–7426.

Townsend, T. K., E. M Sabio, N. D. Browning, and F. E. Osterloh. 2011. Improved niobat-enanoscroll photocatalysts for partial water splitting. *ChemSusChem.* 4 (2): 185–190.

Wang, D., Z. Zou, and J. Ye. 2005. Photocatalytic water splitting with the Cr-doped $Ba_2In_2O_5$/In_2O_3 composite oxide semiconductors. *Chem. Mater.* 17 (12): 3255–3261.

Wang, J. J., Z. J. Li, X. B. Li, X. B. Fan, Q. Y. Meng, S. Yu, C. B. Li, J. X. Li, C. H. Tung, and L. Z. Wu. 2014. Photocatalytic hydrogen evolution from glycerol and water over nickel-hybrid cadmium sulfide quantum dots under visible-light irradiation. *ChemSusChem.* 7 (5): 1468–1475.

Wang, P., P. Chen, A. Kostka, R. Marschall, and M. Wark. 2013. Control of phase coexistence in calcium tantalate composite photocatalysts for highly efficient hydrogen production. *Chem. Mater.* 25 (23): 4739–4745.

Wang, X., K. Shih, and X. Y. Li. 2010. Photocatalytic hydrogen generation from water under visible light using core/shell nano-catalysts. *Water Sci Technol.* 61 (9): 2303–2308.

Wang, Y., Z. Zhang, Y. Zhu, Z. Li, R. Vajtai, L. Ci, and P. M. Ajayan. 2008. Nanostructured VO_2 photocatalysts for hydrogen production. *ACS Nano.* 2 (7): 1492–1496.

Wu, M. C., J. Hiltunen, A. Sápi, A. Avila, W. Larsson, H. C. Liao et al. 2011. Nitrogen-doped anatase nanofibers decorated with noble metal nanoparticles for photocatalytic production of hydrogen. *ACS Nano.* 5 (6): 5025–5030.

Yang, G., W. Yan, Q. Zhang, S. Shen, and S. Ding. 2013. One-dimensional CdS/ZnO core/shell nanofibers via single-spinneret electrospinning: Tunable morphology and efficient photocatalytic hydrogen production. *Nanoscale.* 5 (24): 12432–12439.

Yuan, J., J. Wen, Y. Zhong, X. Li, Y. Fang, S. Zhang, and W. Liu. 2015. Enhanced photocatalytic H_2 evolution over noble-metal-free NiS cocatalyst modified CdS nanorods/g-C_3N_4 heterojunctions. *J. Mater. Chem. A.* 3 (35): 18244–18255.

Zhang, Y., L. Kang, J. Shang, and H. Gao. 2013a. A low cost synthesis of fly ash-based mesoporous nanocomposites for production of hydrogen by photocatalytic water-splitting. *J. Mater. Sci.* 48 (16): 5571–5578.

Zhang, J., Y. Wang, J. Jin, J. Zhang, Z. Lin, F. Huang, and J. Yu. 2013b. Efficient visible-light photocatalytic hydrogen evolution and enhanced photostability of core/shell CdS/g-C_3N_4 nanowires. *ACS Appl. Mater. Interfaces.* 5(20): 10317–10324.

Zhang J., J. Yu, Y. Zhang, Q. Li, and J. R. Gong. 2011. Visible light photocatalytic H_2-production activity of CuS/ZnS porous nanosheets based on photoinduced interfacial charge transfer. *Nano Lett.* 11 (11): 4774–4779.

Zhao, W.-N. and Z.-P. Liu. 2014. Mechanism and active site of photocatalytic water splitting on titania in aqueous surroundings. *Chem. Sci.* 5: 2256–2264.

12 Solar Cells

12.1 INTRODUCTION

Energy is the basic necessity for life and it is an essential resource for fulfilling many human needs. It is required to do all kinds of works such as driving vehicles, cooking food, running industries, and so on. Fast depletion of existing world's energy resources has become a major problem and this has put the whole globe in the grip of an energy crisis. An energy crisis is defined in terms of rapid fall in the supply of available energy or escalating prices of energy sources.

Over the years, this energy crisis has increased due to increasing world population and development of global industries. The only solution to resolve such energy scarcity is to develop some renewable energy resources. Here, solar cells are major candidates for energy production as they are key factors for the future of energy, food, and security.

Solar energy offers a clean, eco-friendly, abundant, and inexhaustible energy resource. Many types of solar cells have been developed such as organophotovoltaic, photoelectrochemical, dye-sensitized, and hybrid solar cells to harness solar energy. Photocatalytic materials in various forms are used to either generate electricity in these devices or chemical energy in the form of hydrogen at the cost of solar energy.

12.2 PHOTOELECTROCHEMICAL CELLS

Photoelectrochemical cells (PECs) are the most efficient cells for converting solar energy into electrical or chemical energy. These cells are quite simple and they consist of a photoactive semiconductor electrode (either n- or p-type) and a metal counter electrode. These electrodes are immersed in a suitable redox electrolyte. The basic difference is that photovoltaic (PV) solar cells have a solid–solid interface while PEC have a solid–liquid interface. Various efforts have been made such as electrolyte modification, surface modification of the semiconductors, photoetching of layered semiconductors, a semiconductor-septum-based PEC solar cell, and so on, to make these PEC cells more efficient. Irradiation of semiconductor/electrolyte junction with light > Eg resulted in generation and separation of charge carriers. Here, the majority carriers are electrons in an n-type semiconductor, which move to the counter electrode through an external circuit and take part in the counter reaction while holes are the minority charge carriers, which in turn, migrate to the electrolyte and participate in electrochemical reactions.

PEC cells can be categorized in two major classes on the basis of change in Gibbs free energy. These are as follows:

- Regenerative PEC solar cells with $\Delta G = 0$. Here, the photoenergy is converted into electric energy.
- Photoelectrosynthetic cells with $\Delta G \neq 0$. Here, the photoenergy is used to affect chemical reactions. These cells can be further categorized into the following two types of cells:

  1. Photoelectrolytic cells with $\Delta G < 0$, where the photoenergy is stored as chemical energy in endergonic reactions, for example $H_2O \rightarrow H_2 + \frac{1}{2} O_2$
  2. Photocatalytic cells with $\Delta G < 0$, where photoenergy provides activation energy for exergonic reactions, for example $N_2 + 3 H_2 \rightarrow 2 NH_3$

Regenerative PEC solar cells are based on a narrow band gap semiconductor and a redox couple, which convert light energy into electrical energy without bringing any change in the free energy of the redox electrolyte ($\Delta G = 0$). An opposite electrochemical reaction occurred at the counter electrode to that of a semiconductor working electrode. These are also called electrochemical PV cells.

Park et al. (2008) prepared doubly β-functionalized porphyrin sensitizers and studied the photoelectrochemical properties of dye-sensitized nanocrystalline-TiO_2 solar cells. These porphyrin sensitizers were functionalized at meso- and β-positions with different carboxylic acid groups. Multiple pathways through olefinic side chains at two β-positions improved the overall electron injection efficiency, and the moderate distance between the porphyrin sensitizer and the TiO_2 semiconductor layer was responsible for retardation of the charge recombination processes.

Some new compounds were incorporated in sandwich-type regenerative PECs by Stergiopoulos et al. (2005). A transition metal complex with two terpy ligands [(2,2':6',2''-terpyridine-4'-iodophenyl)(2, 2':6',2''-terpyridine-4'-phenylphosphonic acid)-ruthenium(II)]dichloride was used to sensitize thin nanostructured SnO_2 film electrodes. A high molecular mass poly(ethylene) oxide electrolyte was filled with titania, while LiI and I_2 were used to transport the current of the cell at the counter electrode. A continuous photocurrent (0.63 mA/cm^2) and a photovoltage (290 mV) were produced by this cell. Incident photon-to-current conversion efficiencies (IPCEs) (16%) and energy conversion values (0.1%) were similar to that with the standard N_3 dye under identical conditions.

Lee et al. (2009) prepared selenide (Se^{2-}) and deposited CdSe quantum dots (QDs) over mesopore TiO_2 photoanodes by the successive ionic layer adsorption and reaction (SILAR) process in ethanol. QD-sensitized TiO_2 films were optimized using a cobalt redox couple [$Co(o$-phen$)_3$]$^{2+/3+}$ in PECs. They achieved over 4% efficiency at 100 W/m^2 with about 50% IPCE at its maximum on addition of a final layer of CdTe. CdTe-terminated CdSe QD cells gave better charge collection efficiencies compared with CdSe QD cells. They also prepared multilayered semiconductor (CdS/CdSe/ZnS)-sensitized TiO_2 mesoporous solar cells by the SILAR process (Lee et al. 2010). This multicomponent sensitizer (CdS/CdSe/ZnS) was used in a polysulfide electrolyte solution as a redox mediator in regenerative PECs.

The photoelectrosynthetic (photoelectrolytic or photocatalytic) cells utilize photon energy input ($E \geq Eg$) to produce a net chemical change in the electrolyte solution

($\Delta G \neq 0$), when the reaction at the counter electrode is not exactly opposite of the hole transfer reaction at the illuminated semiconductor–liquid interface.

In photoelectrocatalytic cells, the rate of reaction will increase if $\Delta G < 0$. Aqueous suspensions comprising of irradiated semiconductor particles may be considered to be an assemblage of short-circuited microelectrochemical cells operating in the photocatalytic mode.

A novel PEC was proposed for generation of hydrogen via photocatalytic water splitting (Jeng et al. 2010). This PEC consists of a membrane electrode assembly integrated with Degussa P25 TiO_2 powder as a photocatalyst for the photoanode and Pt catalyst powder as the dark cathode. This serves as an effective separator for the generated hydrogen and oxygen as well as a compact photocatalytic reactor for water splitting. This PEC can be operated without adding water to the cathode compartment and showed improved photoconversion efficiency. It was observed that Degussa P25/$BiVO_4$ mixed photocatalyst enhanced the hydrogen generation significantly.

A PEC was designed by Li et al. (2011) for hydrogen generation via photoelectrocatalytic water splitting. It consisted of a TiO_2 nanotube (TNT) photoanode, a Pt/C cathode, and a commercial asbestos diaphragm. This PEC could generate hydrogen under ultraviolet (UV) light irradiation with applied bias in KOH solution. The Ti mesh was used as the substrate to synthesize the self-organized TiO_2 nanotubular array layers, and the effect of the morphology of these layers on the PV performances was also investigated. When TiO_2 photocatalyst was irradiated with UV light, it prompted the water splitting. Photocurrent generation of 0.58 mA/cm^2 was obtained, which showed good performance on hydrogen production.

Sun et al. (2011) reported an efficient and economical technology to produce hydrogen from solar energy by splitting water in a two-compartment PEC without any external applied voltage. Highly ordered TNT arrays of 4 µm in length were synthesized by them via a rapid anodization process in ethylene glycol electrolyte to enhance solar conversion efficiency. The photocatalytic activity of this PEC cell was evaluated by hydrogen production. The crystal phase and morphology of TNTs showed no major changes at low annealing temperatures. The crystallization transformation from anatase to rutile phase was observed with increase in temperature along with the destruction of tubular structures. TNTs annealed at 450°C exhibited the highest photoconversion efficiency of 4.49% and maximum hydrogen production rate of 122 µmol/h/cm^2 and this was attributed to the excellent crystallization and the maintenance of tubular structures.

Rahman et al. (2012) observed the effect of doping (C or N) and codoping (C + N) on the titania coating for solar water splitting. Efficient materials in PEC cells should have a smaller band gap (approximately 2.4 eV). They reported that the type and amount of doping affected the coating growth rate, structure, surface morphology, and roughness. The photocurrent density of the C-doped photoanode was approximately 26% higher than undoped photoanode.

Reduced graphene oxide and $BiVO_4$ (rGO–$BiVO_4$) composite was reported by Gao et al. (2013) for PEC. The working electrode was prepared by a doctor blade method onto fluorine-doped tin oxide (FTO) coated glass. Graphene oxide–$BiVO_4$ composite was synthesized by hydrothermal reaction and then rGO-$BiVO_4$ was obtained with annealing in N_2 atmosphere. The rGO–$BiVO_4$ films were found to have enhanced photoelectrochemical properties under the visible light compared to pure BiVO$_4$ film. A high photocurrent response of 160 µA/cm^2 and quantum efficiency of over 1.81% were achieved. This enhanced PEC activity of rGO–BiVO$_4$ was attributed to larger recombination resistance and longer electron lifetime.

A photostable p-type NiO photocathode based on a bifunctional cyclometalated ruthenium sensitizer and a cobaloxime catalyst was developed for visible light–driven reduction of water to produce H$_2$ (Ji et al. 2013). The ruthenium sensitizer was anchored firmly on the surface of NiO, and this binding was resistant to the hydrolytic cleavage. The bifunctional sensitizer can also immobilize the water reduction catalyst. This photoelectrode exhibited superior stability in aqueous solutions and as a result stable photocurrents have been observed over a longer time. This can be helpful in cases of degradation in dye-sensitized PECs due to desorption of dyes and catalysts. Such a high stability of photocathodes is important for the practical application of these devices for solar fuel production.

n-type ZnO:Cu photoanodes were fabricated by Dom et al. (2013) using a simple spray pyrolysis deposition technique. The influence of low concentration (range ~ 10^{-4} to 10^{-1}%) of Cu doping in hexagonal ZnO lattice on its photoelectrochemical performance was evaluated. The doped photoanodes showed sevenfold higher conversion efficiencies compared with their undoped counterpart, as evident from the photocurrents generated under simulated solar radiation. This enhanced performance was attributed to the red shift in the band gap of the Cu-doped films and it was in agreement with the IPCE measurements. Electrochemical studies revealed that Cu-doped ZnO was n-type in nature.

TiO$_2$ branched nanorod arrays (TiO$_2$ BNRs) were synthesized by Su et al. (2013) with plasmonic Au nanoparticles attached on the surface. These Au/TiO$_2$ BNR composites exhibited high photocatalytic activity in photoelectrochemical water splitting. The unique structure of Au/TiO$_2$ BNRs showed enhanced activity with a photocurrent of 0.125 mA/cm^2 under visible light and 2.32 ± 0.1 mA/cm^2 under sunlight irradiation. Au/TiO$_2$ BNRs achieved the highest efficiency of ~1.27% at a very low bias of 0.50 V versus RHE, which indicated an elevated charge separation and transportation efficiencies. This high PEC performance was mainly attributed to the plasmonic effect of Au nanoparticles due to visible light absorption, large surface area, efficient charge separation, and high carrier mobility of the TiO$_2$ BNRs. The carrier density of Au/TiO$_2$ BNRs was found to be six times higher than that of the pristine TiO$_2$ BNRs.

TiO$_2$ nanorod arrays (TiO$_2$ NRAs) were synthesized by Ho and Chen (2014) via a hydrothermal method. The maximum current density and conversion efficiency were 2.5 mA/cm^2 at 1.23 V versus RHE and 0.95% with length of 6–7 µm under AM 1.5G 1 Sun conditions, respectively. Cu$_2$ZnSnS$_4$ (CZTS) has a direct band gap of 1.5 eV and suitable band positions, and therefore it was employed to extend the absorption range. CZTS nanoparticles were decorated on TiO$_2$ NRAs with an adhesion layer using a solvothermal method. The current density increased from 2.92 mA/cm^2 to 6.91 mA/cm^2 and the conversion efficiency increased from 1.44% to 3.50% using ZnS as a passivation layer. They showed that the ZnS layer could reduce electron–hole recombination and as a result, the conversion efficiency was improved.

Guo et al. (2014) prepared ZnO/Cu_2S core/shell NRs from ZnO NRs by a versatile hydrothermal chemical conversion method ($H-ZnO/Cu_2S$ core/shell NRs) and SILAR method ($S-ZnO/Cu_2S$ core/shell NRs). They used these for photoelectrochemical water splitting. The photoelectrode was composed of a core/shell structure. The core portion was ZnO NRs and the shell portion was Cu_2S nanoparticles sequentially located on the surface. It was found that the ZnO NR array provides a fast electron transport pathway due to its high electron mobility properties. Such a PEC system produced very high photocurrent density and has a higher photoconversion efficiency for hydrogen generation. It was also reported that $H-ZnO/Cu_2S$ core/shell NRs exhibited much higher photocatalytic activity than $S-ZnO/Cu_2S$ core/shell NRs. The photocurrent density and photoconversion efficiency of $H-ZnO/Cu_2S$ core/shell NRs were found to be 20.12 mA/cm^2 at 0.85 V versus SCE and 12.81% at 0.40 V versus SCE, respectively.

Wei et al. (2014) reported a highly efficient all-vanadium photoelectrochemical storage cell. This storage cell takes advantage of fast electrochemical kinetics of vanadium redox couples of VO_2^+/VO^{2+} and V^{3+}/V^{2+}, and therefore it appeared as a possible alternative to photoproduction of hydrogen from water. A VO^{2+} conversion rate of 0.0042 $\mu mol/h$ and Faradaic efficiency of 95% were obtained with continuous photocharging for 25 hours without an external voltage bias. The IPCE was calculated to be around 12%.

A novel bio-PEC consisting of a MoS_3 modified p-type Si nanowire (NW) photocathode and a microbially catalyzed bioanode has been reported by Zang et al. (2014) for hydrogen production under visible light illumination. Here, microbial pollutant oxidation occurred spontaneously in the bioanode providing sufficient electrons for the photocathode reaction without any external bias. The recombination of the photogenerated hole and electron pairs at the photocathode was found to be retarded by the supply of electrons from the bioanode, thus leaving enough photogenerated electrons for hydrogen evolution. Hydrogen was continuously produced from the bio-PEC, with a maximum power density of 71 mW/m^2 and an average hydrogen-producing rate of 7.5 ± 0.3 $\mu mol/h/cm^2$ under light illumination.

Li et al. (2014) fabricated a hybrid heterojunction and solid-state photoelectrochemical solar cell based on graphene woven fabrics (GWFs) and silicon. The GWFs were transferred onto n-Si to form a Schottky junction with embedded polyvinyl alcohol based solid electrolyte. Solid electrolyte simultaneously serves three purposes in this hybrid solar cell: (1) it is an antireflection layer, (2) a chemical modification carrier, and (3) a photoelectrochemical channel. The open-circuit voltage, short circuit current density, and fill factor of the cell were all significantly improved giving an impressive power conversion efficiency (PCE) of 11%. Such solar cell models were constructed to confirm the hybrid working mechanism, in which the heterojunction junction and photoelectrochemical effect may work synergistically.

Hematite, $\alpha-Fe_2O_3$, satisfies many requirements for a good PEC photoanode, but its efficiency is insufficient in its pristine form. A promising strategy for enhancing photocurrent density may be using photosynthetic proteins. Ihssen et al. (2014) concluded that electrode surfaces, particularly hematite photoanodes, can be functionalized with light-harvesting proteins. Low-cost biomaterials such as cyanobacterial phycocyanin and enzymatically produced melanin are likely to increase the overall performance of low-cost metal oxide photoanodes in a PEC system. The use of such biomaterials will change the overall nature of the photoanode, so that aggressive alkaline electrolytes such as concentrated KOH are no longer required.

A microbial electrolysis cell (MEC) is one of the promising techniques for converting organic matter to hydrogen, but it lacks efficient and cost-effective cathode catalysts and also needs additional electricity input. He et al. (2014) reported a light driven microbial photoelectrochemical cell (MPC) system consisting of a TiO_2 photocathode and a microbial anode, which utilizes light energy and harvest electrons respectively. Continuous hydrogen production was achieved without external applied voltage under UV irradiation in this MPC system. This system worked well continuously over 200 hours in a batch-fed mode under light irradiation. A hydrogen production rate of about 3.5 μmol/h was achieved.

Li et al. (2015a) constructed ternary CdS/rGO/TNT array hybrids for enhancing visible light–driven photoelectrochemical and photocatalytic activity. They used a coupling technique of electrophoretic deposition (EPD) for the synthesis of ternary nanocomposite photoelectrodes that are composed of CdS nanocrystallites, rGO and TNT arrays. This ternary CdS/rGO/TNTs hybrid showed more activity because the outer layer of CdS acts as a sensitizer for trapping photons from visible light and the middle layer of rGO serves as a transporter for suppressing the recombination of photogenerated carriers and as a green sensitizer for increasing visible light absorption. The inner TNTs with narrowed band gap collected the hot electrons from the visible light absorption, while CdS and rGO participated in subsequent redox reaction for hydrogen production and degradation of organic pollutants.

Xu et al. (2015) studied the role of carbon nitride (C_3N_4) as an absorber in a PEC. They prepared C_3N_4-sensitized TiO_2 mesoporous film via in situ, vapor transport growth with direct Ti–O–C bonding. Material hybridization showed a unique electronic transition at the interface, which leads to strong visible light absorption and photoactivity.

Liu et al. (2015) designed and prepared Ag/Cu_2O/ZnO tandem triple-junction photoelectrode. An eleven-fold increase in photocurrent was achieved using Ag/Cu_2O/ZnO photoelectrode compared with the Cu_2O film. This high performance was attributed to the optimized design of the tandem triple-junction structure, where efficient absorption of solar energy was due to localized surface plasmon resonance of Ag and the heterojunctions in production and separation of electron–hole pairs in the photocathode.

Wang et al. (2015) reported the feasibility of simultaneous production of hydrogen and electricity with removal of contaminants from actual urban wastewater using a dye-sensitized photoelectrochemical cell (DSPC). The photoanode in this DSPC was nanostructured plasmonic Ag/AgCl at chiral TiO_2 nanofibers. The electrolyte used was actual wastewater with added estrogen (17-β-ethynylestradiol, EE2) and heavy metal (Cu^{2+}). I^-/I_3^- acted as electron bridges for the stabilization of charges. Nearly total removal of total organic carbon (TOC), Cu^{2+}, and 70% removal of total nitrogen (TN) were achieved under visible light irradiation. They also reported relatively high solar energy conversion efficiency (PCE 3.09%) and about 98% of the electricity was converted to H_2 after the consumption of dissolved oxygen (DO), Cu^{2+}, and TN. This was attributed to the symbiotic relationship between the TiO_2 chiral nanofibers and the plasmonic effect of Ag nanoparticles at the photoanode. Silver may act as a recombination site hindering the generation of electricity. The dye N719 exhibited a temporary sensitization effect.

Hydrogen evolution was obtained during electrochemical anodization by self-organized TNT arrays (TiO_2 NTA) in ethylene glycol based electrolytes using different NH_4F concentrations (Xue et al. 2015). Hydrogen production by photocatalytic water splitting was performed in a two-compartment PEC without any applied voltage. The highest hydrogen evolution was observed with TiO_2 anodized with 0.50 wt% of NH_4F concentration for 60 minutes, that is 2.53 mL/h/cm², and the maximum photoconversion efficiency was 4.39%. Another series of TNT array samples with equal charge consumption (designated as TiO_2 NTA-EC) were also synthesized by them via controlling the anodization time in electrolytes containing different NH_4F concentrations.

Overall solar water splitting without external bias was also demonstrated by Xu et al. (2016) using a photoelectrochemical tandem device composed of a $BiVO_4$ photoanode and Si nanoarray photocathode. An unassisted photocurrent density of 0.6 mA/cm² was determined under AM 1.5G illumination with current density–voltage curves of respective photoelectrodes and the operating point in the three-electrode system. Unassisted two-electrode operation of the tandem cell and a solar-to-hydrogen efficiency of 0.57% were also achieved. The solar photocurrent density of this tandem cell decreased during stability testing, possibly due to the dissolution of the Co–Pi electrocatalyst rather than the instability of the p-Si photocathode.

It was observed that photocurrent density of a WO_3-based PEC decreases as a function of increasing photoactive area (Park et al. 2016). This was caused by a non-linear decrease in resistance of the interface between the conducting FTO electrode and the photoactive material (WO_3). It was observed that the relatively high interfacial resistance of large area electrodes can be reduced by introduction of conductive layers. Enhanced photocurrents were obtained for relatively large, graphitized C_{70} modified WO_3 photoelectrodes.

12.3 DYE-SENSITIZED SOLAR CELLS

A heterojunction three-dimensional dye-sensitized solar cell (DSSC) was fabricated by O'Regan and Grätzel (1991). More than 7% efficiency was reported in the presence of direct sunlight using n-type semiconductor material such as TiO_2. Current was generated when dye molecules absorbed photons from light and injected electrons into the conduction band. DSSCs are also called "Gratzel cells." These cells may prove to be reliable alternatives of existing p–n junction PV devices. However, further improvements are still required for future development.

Light absorption and charge transfer processes separately occur in DSSCs. Basically, incident light is absorbed by organic dyes, or sensitizer molecules, adsorbed on the surface of nanocrystalline metal oxide. A typical DSSC is constructed by two glass sheets coated with a transparent conductive oxide (TCO) layer. One is used as a working electrode that is covered with a film of dye-sensitized semiconductor particles, while the other is the counter electrode coated with a catalyst. Both these electrodes are kept together and an electrolyte is filled in the gap between them, which is commonly a redox couple in an organic solvent.

The photo-to-electric conversion efficiency of DSSC was limited due to electron–hole charge recombination at the electrode–electrolyte interface. Therefore, present research trends are based on alternative nanomaterials, such as nanofibers, NWs, NRs, NTs, and so on. These can be used as photoanodes having large surface areas, low recombination rates, and high energy conversion efficiencies.

There are some basic steps for generation of electrical energy in the presence of light in DSSC. These steps are as follows:

- Excitation process
- Injection process
- Energy generation
- Regeneration of dye
- e^- recapture reaction

DSSC consists of the following components:

- Working electrode—Porous semiconducting nanostructures (naïve and modified semiconductors, nanomaterials) attached to a conducting FTO glass.
- Counter electrode—Platinized conducting substrate.
- Light-absorbing layer—Adsorbed dye as a sensitizer.
- Redox system—Some redox couple systems as electrolyte. These may be liquid electrolytes, polymer gel-electrolytes, and so on.
- Sensitizers such as metal complexes, dyes, and natural pigments.

Other types of DSSCs include solid-state DSSC, quasi-solid state DSSC, QD SSC, and so on.

Onicha and Castellano (2010) observed the effects of electrolyte composition, particularly the role of Li^+ and I^- ions, on the PV performance of DSSCs based on an Os(II) polypyridine complex, $[Os(^tBu_3tpy)(dcbpyH_2)(NCS)]PF_6$. The suitability of the dye to serve as a sensitizer on mesoporous titania films in regenerative cells was confirmed. The performance of Os(II)-based DSSCs may be enhanced by simple modification of the composition of redox electrolytes used in such sandwich cells. Abundance of I^- played an important role for the effective regeneration of oxidized surface-bound osmium sensitizers. The PCE for an Os(II)-based DSSC was achieved as 4.7%.

Johansson et al. (2011) synthesized and characterized three ruthenium compounds, that is, cis-Ru(dcbq)$_2$(NCS)$_2$, cis-Ru(dcbq)(bpy)(NCS)$_2$, and cis-Ru(dcb)(bq)(NCS)$_2$ (where bpy is 2,2'-bipyridine, dcb is 4,4'-(CO$_2$H)$_2$-2,2'-bipyridine, bq is 2,2'-biquinoline, and dcbq is 4,4'-(CO$_2$H)$_2$-2,2'-biquinoline), with the well-known N3 compound (i.e., cis-Ru(dcb)$_2$(NCS)$_2$) in DSSCs. The lowered π^* orbitals resulted in enhanced red absorption compared with N3. Sensitization from 400 to 900 nm was observed with cis-Ru(dcb)(bq)(NCS)$_2$ with HCl-pretreated TiO$_2$ in regenerative solar cells. PCEs as high as 6.5% were obtained.

Chung et al. (2012) reported that the solution-processable p-type direct band gap semiconductor CsSnI$_3$ can be used for hole conduction in place of a liquid electrolyte.

The solid-state DSSCs consisting of $CsSnI_{2.95}F_{0.05}$ doped with SnF_2 and nanoporous TiO_2 sensitized with N719 exhibited up to 10.2% conversion efficiencies. $CsSnI_3$ had a band gap of 1.3 eV and enhanced visible light absorption on the red side of the spectrum.

ZnO photoelectrode sensitized with rhodamine B and polyaniline base (EB), single-wall carbon NT (EB-SWCNT)-coated ITO-based DSSCs was fabricated by Abdel-Fattah et al. (2013). ZnO photoanode sensitized with a mixture of rhodamine B and riboflavin dyes showed improvement in short circuit current density as well as energy conversion efficiency compared with ITO/EB-SWCNTs/ZnO-rhodamine B/ITO DSSC.

Chen et al. (2013) synthesized SnS nanosheets (NSs), SnS NWs, and SnS_2 NSs as counter electrode catalysts in I_3^-/I^--based DSSCs. PCE of SnS NS-based DSSCs was 6.56% compared with DSSCs based on Pt 7.56%, while SnS NW and SnS_2 NS-based DSSCs exhibited 5.00% and 5.14% cell efficiencies, respectively, indicating an excellent catalytic activity of SnS_x for the reduction of triiodide to iodide.

The highly crystalline perovskite $BaSnO_3$ nanoparticles were synthesized by Kim et al. (2013a) and used as photoanode materials in DSSCs. They showed remarkably rapid charge collection in DSSC with $BaSnO_3$ (~43 µm film thickness) and a high energy conversion efficiency of 5.2%. Thus, $BaSnO_3$ cells showed superior charge collection in nanoparticle films compared with TiO_2 cells and could offer a breakthrough in the efficiencies of DSSCs.

Coral-like TiO_2 nanostructured films were chemically synthesized by Bahramian (2013) through the sol–gel method and used in the fabrication of DSSC. The effect of operational parameters such as precursor hydrolysis rate, reaction time, type and concentration of acid, and annealing temperature was studied. The coral-like TiO_2 film had excellent light-scattering properties and the mesoporous structure had a fairly large specific surface area of 164 m²/g. DSSCs consisting of a dense, coral-like TiO_2 nanostructured film, dye N719 with an electrolyte showed better performance compared with DSSCs using TiO_2. A photocurrent of 16.1 mA/cm², fill factor of 77.6%, and conversion efficiency of 9.4% were obtained (Figure 12.1).

Magnesium tetrapyrrole is the natural choice of metal tetrapyrrole in photosynthesis and its use as a photosensitizer in DSSCs was conducted by constructing solar

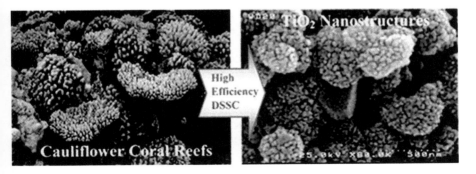

FIGURE 12.1 Coral-like nanostructured titania film. (Adapted from Bahramian, A., *Ind. Eng. Chem. Res.*, 52, 14837–14846, 2013. With permission.)

cells using a metal-ligand axial coordination approach on TiO_2 surface modified with 4-carboxyphenyl imidazole and was compared with cells constructed using zinc porphyrin as a sensitizer (Subbaiyan and D'souza 2013). It was suggested that magnesium tetraphenylporphyrin (MgTPP) is a better photosensitizer compared with zinc tetraphenylporphyrin (ZnTPP) for DSSC applications. The open-circuit potential, short circuit current, fill factor, IPCE, and overall efficiency of the solar cell were also found to be better in the cells using the MgTPP photosensitizer. It was observed that slight addition of acetonitrile improved the performance of the solar cells, but the performance of cell constructed using pure acetonitrile was poor as acetonitrile competitively binds with MgTPP instead of imidazolde on the TiO_2 surface.

Wang et al. (2014a) developed a new visible light photocatalyst $Bi_{24}O_{31}Cl_{10}$ with high dye-sensitized photocatalytic activity. It showed excellent visible light photocatalytic activity toward rhodamine B degradation, which was initiated by dye sensitization due to compatible energy levels and high electronic mobility. $Bi_{24}O_{31}Cl_{10}$ was also considered as a photoanode material for DSSC and exhibited a PCE of 1.5%. Titanium nitride/titanium oxide (TiN/TiO_2) composite photoanodes were utilized in DSSCs and also for water splitting by Li et al. (2015b). The DSSC with TiN/TiO_2 composite photoanode annealed for 1 hour, and showed 7.27% PCE, while the cell without TiN showed only 7.02% efficiency.

Wang et al. (2014b) synthesized ternary semiconductor oxide Zn_2SnO_4 (ZTO) via a hydrothermal method and utilized it as a photoanode for DSSCs. An improved PCE of 5.72% was achieved for Zn_2SnO_4–DSSCs through surface modification and structural optimization. The electron diffusion coefficient of Zn_2SnO_4 was found to be higher than TiO_2 and the lifetime of Zn_2SnO_4-based DSSCs was also increased after surface modification. The hierarchically structured Zn_2SnO_4 beads (ZTO-Bs) were used in DSSCs by Hwang et al. (2014). The DSSCs with ZTO-B and SJ-E1 and SJ-ET1 organic dyes exhibited the highest performance for DSSCs with ternary metal oxide-based photoelectrodes. The morphology of the ZTO-Bs permits enhancement in dye absorption, light scattering, electrolyte penetration, and the charge recombination lifetime. A PCE of 6.3% was achieved for ZTO-Bs DSSCs.

A molecularly engineered porphyrin dye, coded as SM315, features the prototypical structure of a donor–π-bridge–acceptor, maximizes electrolyte compatibility, and improves light harvesting properties. A high open-circuit voltage of 0.91 V, short circuit current density of 18.1 mA/cm², fill factor of 0.78, and PCE of 13% were achieved by Mathew et al. (2014) using this porphyrin dye with the cobalt(II/III) redox shuttle in DSSCs.

Hierarchical mesoporous SnO_2 (HM-SnO_2) spheres were synthesized by Park et al. (2014) with large surface area (85.3 m²g⁻¹) through a one-pot controlled solvothermal process. Tin chloride pentahydrate and graft copolymer, that is, poly(vinyl chloride)-g-poly(oxyethylene methacrylate) (PVC-g-POEM), were used as Sn precursor and structure directing agent, respectively. Solid-state DSSCs fabricated with HM-SnO_2 spheres on an organized mesoporous SnO_2 interfacial (om-SnO_2 IF) layer as the photoanode had a long-term efficiency of 3.4%, which was only 1.9% with photoanode comprising nonporous SnO_2(NP-SnO_2) spheres. This enhancement was attributed to well-organized hierarchical structure with dual pores (23.5 and 162.3 nm) providing larger surface area, improved light scattering, and decreased charge recombination compared with the nonporous SnO_2 (NP-SnO_2) photoanode. Introduction of an om-SnO_2 IF layer between the HM-SnO_2 spheres and FTO substrate enhanced light harvesting, increased electron transport, and reduced charge recombination and interfacial/internal resistance. HM-SnO_2 spheres also showed high activity with good recyclability for photodegradation of methyl orange under UV light irradiation.

As mangosteen peel is an inedible portion of a fruit, carbonized mangosteen peel was used with a mangosteen peel dye as a natural counter electrode and a natural photosensitizer, respectively, in DSSCs (Maiaugree et al. 2015). It was found that carbonized mangosteen peels had mesoporous honeycomb-like carbon structure with rough nanoscale surface. The efficiency of a DSSC using this carbonized mangosteen peel was compared with that of DSSCs with Pt and poly(3,4-ethylenedioxythiophene)polystyrene sulfonate (PEDOT-PSS) counter electrodes. The highest solar conversion efficiency (2.63%) was obtained using carbonized mangosteen peel along with an organic disulfide/thiolate electrolyte.

DSSCs were fabricated by Abodunrin et al. (2015) with mango leaf dye extracts as natural dye sensitizers at pH 5.20 and temperature 18.1°C. They used methanol as the dye-extracting solvent. DSSCs with dye extract of *Mangifera indica*. L. with $KMnO_4$ electrolyte had the highest photocurrent density of 1.3 mA/cm^2 and fill factor of 0.46. Potassium iodide, potassium bromide, and mercury chloride electrolytes showed only 0.2, 0.08, and 0.02 mA/cm^2 photocurrent densities, respectively.

Use of TiO_2 mesoporous beads as photocatalyst and photoelectrode for DSSCs has been investigated by Wu and Tsou (2015). Sol–gel and microwave-assisted hydrothermal (MH) techniques were used for the synthesis. MH-synthesized TiO_2 beads showed much higher values of specific surface areas, pore volumes, and porosities compared with the commercial P25 TiO_2 powders and hydrothermal synthesized TiO_2 beads.

Fatima et al. (2015) used α-bismuth oxide (α-Bi_2O_3) as a photoanode material in DSSC. They observed that electron transfer and transport behavior of the photoanode could be controlled by its morphology, size, and surface properties. Two different dyes, N719 (organometallic dye) and eosin Y (organic dye), were used as light harvesters in the cell structure with an iodine/iodide electrolyte. The efficiency values of 0.09% and 0.05% were obtained for dyes N719 and eosin Y, respectively under 1 sun illumination. Nano α-Bi_2O_3 have good charge carrier concentration, mobility, and conductivity, but the efficiency of the cell was quite low due to the poor dye attachment and back recombination at the photoanode–electrolyte interface.

Rajamanickam et al. (2016) reported a DSSC based on $BaSnO_3$/$TiCl_4$ treated and BSO/scattering layer photoelectrodes. BSO/$TiCl_4$-treated photoanode showed 3.88% PCE under optimum conditions. The highest PCEs of 5.68% was achieved using BSO/$TiCl_4$-treated/TiO_2 scattering layer photoanode, while BSO/$TiCl_4$-treated/ZnO scattering layer exhibited 4.28% PCE. It was suggested that BSO may be one of the most important future technological materials.

Sun et al. (2016) used polyethylene terephthalate (PET), a commonly used textile fiber, in the form of a wet-laid nonwoven fabric as a matrix for electrolytes in DSSCs. It also functions as a separator between the photoanode and cathode of a DSSC. They prepared this nonwoven membrane by a wet-laid manufacturing process followed by calendaring to reduce the thickness and increase the uniformity. This membrane can absorb the electrolyte, turning into a quasi-solid, which provides an interfacia contact between both electrodes of the DSSC, thus preventing short circuiting. Th quasi-solid-state DSSC assembled with an optimized membrane exhibited 10.248% PCE. The membrane was also plasma-treated with argon and oxygen separately resulting in retention of the electrolyte and avoiding its evaporation. Here, a 15% longer lifetime of the DSSC was observed compared with liquid electrolytes.

TiO_2 nanofibers with different phases were successfully prepared by Zheng et al (2016) using electrospinning followed by annealing at different temperatures. The DSSCs were fabricated using these TiO_2 nanofiber mats as the photoanode. A PCE of 6.12% and a fill factor of 0.65 were obtained by optimizing the annealing tem perature and the thickness of the photoanode. A treatment with $TiCl_4$ was found to further improve the photoanodes. DSSC with $TiCl_4$-treated anodes exhibited ar improved efficiency of 7.06% and a fill factor of 0.73. Improvements in cell perfor mance were attributed to the different crystalline structures of the mesoporous com posite anatase/rutile TiO_2 nanofibers, which has stair-step energy levels and the high surface area of close grain packing. It was suggested that this electrospun nanofibe photoanode may provide efficient charge transport for application in DSSCs.

Novel backside-illumination DSSCs were fabricated by Chang et al. (2016 utilizing high aspect ratio microstructures created by a colloidal lithography proces involving reactive ion etching. They observed that the charge collection efficiency and light-harvesting efficiency of the photoanode were dramatically improved and a high solar cell efficiency of 11% was achieved.

Electricity is a form of energy that has many-fold applications as it can generate heat, mechanical power, sound, light, and so on, and society has benefited from elec tricity in its different forms. The world has faced an energy crisis for the last three to four decades. The development of solar cells such as PECs and DSSCs has helped rescue us as these can convert the light energy of the Sun either directly into electri cal energy or to a form of chemical energy such as hydrogen. The efficiency of these cells achieved so far is appreciable, but we have many more miles to go to depend solely on these solar cells.

REFERENCES

Abdel-Fattah, T. M., S. Ebrahim, M. Soliman, and M. Hafez. 2013. Dye-sensitized solar cells based on polyaniline-single wall carbon nanotubes composite. *ECS J. Solid State Sci. Technol.* 2 (6): M13–M16.

Abodunrin, T. J., O. Obafemi, A. O. Boyo, T. Adebayo, and R. Jimoh. 2015. The effect of elec trolyte on dye sensitized solar cells using natural dye from Mango (*M. indica* L.) leaf as sensitizer. *Adv. Mater. Phys. Chem.* 5: 205–213.

Bahramian, A. 2013. High conversion efficiency of dye-sensitized solar cells based on coral-like TiO_2 nanostructured films: Synthesis and physical characterization. *Ind. Eng. Chem. Res.* 52 (42): 14837–14846.

Chang, C.-H., H.-H. Lin, C.-C. Chen, and F. C.-N. Hong. 2016. Fabrication of high-efficiency (11%) dye-sensitized solar cells in backside illumination mode with microstructures. *Adv. Mater. Interface.* doi: 10.1002/admi.201500769.

Chen, X., Y. Hou, B. Zhang, X. H. Yang, and H. G. Yang. 2013. Low-cost SnS_x counter elec trodes for dye-sensitized solar cells. *Chem. Commun.* 49: 5793–5795.

Chung, I., B. Lee, J. He, R. P. H. Chang, and M. G. Kanatzidis. 2012. All-solid-state dye-sensitized solar cells with high efficiency. *Nature*. 485: 486–489.

Dom, R., L. R. Baby, H. G. Kim, and P. H. Borse. 2013. Enhanced solar photoelectrochemical conversion efficiency of ZnO: Cu electrodes for water-splitting application. *Int. J. Photoenergy*. doi. org/10.1155/2013/928321.

Fatima, M. J. J., C. V. Niveditha and S. Sindhu. 2015. α-Bi_2O_3 photoanode in DSSC and study of the electrode–electrolyte interface. *RSC Adv.* 5: 78299–78305.

Gao, L., F. Qu, and X. Wu. 2013. Reduced graphene oxide-$BiVO_4$ composite for enhanced photoelectrochemical cell and photocatalysis. *Sci. Adv. Mater.* 5: 1485–1492.

Guo, K., X. Chen, J. Han, and Z. Liu. 2014. Synthesis of ZnO/Cu2S core/shell nanorods and their enhanced photoelectric performance. *J. Sol-Gel Sci. Tech.* 72: 92–99.

He, Y.-R., F.-F. Yan, H.-Q. Yu, S.-J. Yuan, Z.-H. Tong, and G.-P. Sheng. 2014. Hydrogen production in a light-driven photoelectrochemical cell. *Appl. Energy.* 113 (C): 164–168.

Ho, T.-Y., and L.-Y. Chen. 2014. The study of Cu_2ZnSnS_4 nanocrystal/TiO_2 nanorod heterojuction photoelectrochemical cell for hydrogen generation. *Energy Proc.* 61: 2050–2053. *International Conference on Applied Energy, ICAE2014.*

Hwang, D., J.-S. Jin, H. Lee, H.-J. Kim, H. Chung, D. Y. Kim et al. 2014. Hierarchically structured Zn_2SnO_4 nanobeads for high-efficiency dye-sensitized solar cells. *Sci. Rep.* 4. doi: 10.1038/srep07353.

Ihssen, J., A. Braun, G. Faccio, K. Gajda-Schrantz, and L. Thöny-Meyer. 2014. Light harvesting proteins for solar fuel generation in bioengineered photoelectrochemical cells. *Curr. Protein Pept. Sci.* 15(4): 374–384.

Jeng, Ki.-T., Y.-C. Liu, Y.-F. Leu, Y.-Z. Zeng, J.-C. Chung, and T.-Y. Wei. 2010. Membrane electrode assembly-based photoelectrochemical cell for hydrogen generation. *Int. J. Hydrogen Energy.* 35: 10890–10897.

Ji, Z., M. He, Z. Huang, U. Ozkan, and Y. Wu. 2013. Photostable p-type dye-sensitized photoelectrochemical cells for water reduction. *J. Am. Chem. Soc.* 135 (32): 11696–11699.

Kim, D. W., S. S. Shin, S. Lee, I. S. Cho, D. H. Kim, C. W. Lee et al. 2013. $BaSnO_3$ Perovskite nanoparticles for high efficiency dye-sensitized solar cells. *ChemSusChem.* 6 (3): 449–454.

Lee, H. J., J. Bang, J. Park, S. Kim, and S.-M. Park. 2010. Multilayered semiconductor (CdS/CdSe/ZnS)-sensitized TiO_2 mesoporous solar cells: All prepared by successive ionic layer adsorption and reaction processes. *Chem. Mater.* 22: 5636–5643.

Lee, H., M. Wang, P. Chen, D. R. Gamelin, S. M. Zakeeruddin, M. Grätzel et al. 2009. Efficient CdSe quantum dot-sensitized solar cells prepared by an improved successive ionic layer adsorption and reaction process. *Nano Lett.* 9: 4221–4227.

Li, C.-T., S.-R. Li, L.-Y. Chang, C.-P. Lee, P.-Y. Chen, S.-S. Sun et al. 2015b. Efficient titanium nitride/titanium oxide composite photoanodes for dye-sensitized solar cells and water splitting, *J. Mater. Chem. A.* 3: 4695–4705.

Li, H., Z. Xia, J. Chen, L. Lei, J. Xing. 2015a. Constructing ternary CdS/reduced graphene oxide/TiO_2 nanotube arrays hybrids for enhanced visible-light-driven photoelectrochemical and photocatalytic activity. *J. Appl. Catal. B: Environ.* 168–169: 105–113.

Li, X., X. Zang, X. Li, M. Zhu, Q. Chen, K. Wang et al. 2014. Hybrid heterojunction and solid-state photoelectrochemical solar cells. *Adv. Energy Mater.* 4 (14): doi:10.1002/aenm.201400224.

Li, Y., H. Yu, W. Song, G. Li, B. Yi, and Z. Shao. 2011. A novel photoelectrochemical cell with self-organized TiO_2 nanotubes as photoanodes for hydrogen generation. *Int. J. Hydrogen Energy.* 36 (22): 14374–14380.

Liu, Y., F. Ren, S. Shen, Y. Fu, C. Chen, C. Liu, Z. Xing et al. 2015. Efficient enhancement of hydrogen production by Ag/Cu_2O/ZnO tandem triple-junction photoelectrochemical cell. *Appl. Phys. Lett.* 106. doi.org/10.1063/1.4916224.

Maiaugree, W., S. Lowpa, M. Towannang, P. Rutphonsan, A. Tangtrakarn, S. Pimanpang et al. 2015. A dye sensitized solar cell using natural counter electrode and natural dye derived from mangosteen peel waste. *Sci. Rep.* 5. doi:10.1038/srep15230.

Mathew, S., A. Yella, P. Gao, R. Humphry-Baker, B. F. E. Curchod, N. Ashari-Astani et al. 2014. Dye-sensitized solar cells with 13% efficiency achieved through the molecular engineering of porphyrin sensitizers. *Nat. Chem.* 6: 242–247.

O'Regan, B. and M. Gratzel. 1991. A low-cost, high-efficiency solar cell based on dye-sensitized colloidal TiO$_2$ films. *Nature.* 353: 737–739.

Onicha, A. C. and F. N. Castellano. 2010. Electrolyte-dependent photovoltaic responses in dye-sensitized solar cells based on an osmium(II) dye of mixed denticity. *J. Phys. Chem. C.* 114: 6831–6840.

Park, J. K., H. R. Lee, J. Chen, H. Shinokubo, A. Osuka, and D. Kim. 2008. Photoelectrochemical properties of doubly β-functionalized porphyrin sensitizers for dye-sensitized nanocrystalline-TiO$_2$ solar cells. *J. Phys. Chem. C.* 112: 16691–16699.

Park, J. T., C. S. Lee, and J. H. Kim. 2014. One-pot synthesis of hierarchical mesoporous SnO$_2$ spheres using a graft copolymer: Enhanced photovoltaic and photocatalytic performance. *RSC Adv.* 4: 31452–31461.

Park, S.-Y., E. M. Hong, D. C. Lim, J.-Y. Lee, and G. Mul. 2016. Photoactive area dependent electrochemical characteristics of photoelectrochemical cells. *J. Electrochem. Soc.* 163 (2): H105–H109.

Rahman, M., B. H. Q. Dang, K. McDonnell, J. M. D. MacElroy, and D. P. Dowling. 2012. Effect of doping (C or N) and co-doping (C + N) on the photoactive properties of magnetron sputtered titania coatings for the application of solar water-splitting. *J. Nanosci. Nanotechnol.* 12: 4729–4735.

Rajamanickam, N., P. Soundarrajan, V. K. Vendra, J. B. Jasinski, M. K. Sunkara and K. Ramachandran. 2016. Efficiency enhancement of cubic perovskite BaSnO$_3$ nanostructures based dye sensitized solar cells. *Phys. Chem. Chem. Phys.* 18: 8468–8478.

Stergiopoulos, T., I. M. Arabatzis, M. Kalbac, I. Lukes, and P. Falaras. 2005. Incorporation of innovative compounds in nanostructured photoelectrochemical cells. *J. Mater. Proc. Technol.* 161: 107–112.

Su, F., T. Wang, R. Lv, J. Zhang, P. Zhang, J. Lu et al. 2013. Dendritic Au/TiO$_2$ nanorod arrays for visible-light driven photoelectrochemical water splitting. *Nanoscale.* 5: 9001–9009.

Subbaiyan, N. K. and F. D'Souza. 2013. Studies of a supramolecular photoelectrochemical cell using magnesium tetraphenylporphyrin as photosensitizer. *J. Porphyrins Phthalocyanines* 17. doi:10.1142/S1088424613500156.

Sun, K. C., I. A. Sahito, J. W. Noh, S. Y. Yeo, J. N. Im, S. C. Yi et al. 2016. Highly efficient and durable dye-sensitized solar cells based on a wet-laid PET membrane electrolyte. *J. Mater. Chem. A.* 4: 458–465.

Sun, Y., K. Yan, G. Wang, W. Guo, and T. Ma. 2011. Effect of annealing temperature on the hydrogen production of TiO$_2$ nanotube arrays in a two-compartment photoelectrochemical cell. *J. Phys. Chem. C.* 115: 12844–12849.

Wang, D., Y. Li, G. L. Puma, C. Wang, P. Wang, W. Zhang, and Q. Wang. 2015. Dye-sensitized photoelectrochemical cell on plasmonic Ag/AgCl @ chiral TiO$_2$ nanofibers for treatment of urban wastewater effluents, with simultaneous production of hydrogen and electricity. *Appl. Catal. B: Environ.* 168–169: 25–32.

Wang, K., Y. Shi, W. Guo, X. Yu, and T. Ma. 2014b. Zn$_2$SnO$_4$-Based dye-sensitized solar cells: Insight into dye-selectivity and photoelectric behaviors. *Electrochim. Acta.* 135: 242–248.

Wang, L., J. Shang, W. Hao, S. Jiang, S. Huang, T. Wang et al. 2014a. A dye-sensitized visible light photocatalyst-Bi$_{24}$O$_{31}$Cl$_{10}$. *Sci. Rep.* 4. doi:10.1038/srep07384.

Wei, Z., D. Liu, C. Hsu, and F. Liu. 2014. All-vanadium redox photoelectrochemical cell: An approach to store solar energy. *Electrochem. Commun.* 45: 79–82.

Wu, W.-Y. and Y.-Y. Tsou. 2015. TiO$_2$ beads as photocatalyst and photoelectrode for dye-sensitized solar cells synthesized by a microwave-assisted hydrothermal method. *Int. J. Energy Res.* 39 (10): 1420–1429.

Xu, J., I. Herraiz-Cardona, X. Yang, S. Gimenez, M. Antonietti, and M. Shalom. 2015. The complex role of carbon nitride as a sensitizer in photoelectrochemical cells. *Adv. Optic. Mater.* 3 (8): 1052–1058.

Xu, P., J. Feng, T. Fang, X. Zhao, Z. Li, and Z. Zou. 2016. Photoelectrochemical cell for unassisted overall solar water splitting using a BiVO$_4$ photoanode and Si nanoarray photocathode. *RSC Adv.* 6: 9905–9910.

Xue, Y., Y. Sun, G. Wang, K. Yan, and J. Zhao. 2015. Effect of NH$_4$F concentration and controlled-charge consumption on the photocatalytic hydrogen generation of TiO$_2$ nanotube arrays. *Electrochim. Acta.* 155: 312–320.

Zang, G.-L., G.-P. Sheng, C. Shi, Y.-K. Wang, W.-W. Li, and H.-Q. Yu. 2014. A bio-photoelectrochemical cell with a MoS$_3$-modified silicon nanowire photocathode for hydrogen and electricity production. *Energy Environ. Sci.* 7: 3033–3039.

Zheng, D., J. Xiong, P. Guo, Y. Li, and H. Gu. 2016. Fabrication of improved dye-sensitized solar cells with anatase/rutile TiO$_2$ nanofibers. *J. Nanosci. Nanotechnol.* 16 (1): 613–618.

13 Wastewater Treatment

13.1 INTRODUCTION

Our environment is composed of atmosphere, earth, water, and space. Under normal circumstances, it remains clean and, therefore, enjoyable. However, with increasing world population and with limited natural resources, the composition and complex nature of our environment has changed. Our world is beautiful, but the increasing use and improper disposal of the effluents from various industries are creating pollution in the environment. Human activities such as rapidly growing industrialization, new construction, and increase in transportation lead to the generation of objectionable materials into the environment, thus making it polluted. There are various types of pollution, for example, air pollution, water pollution, soil pollution, and noise pollution. Out of all these, water pollution is the most dangerous for mankind, aquatic plants and animals, and the environment as a whole.

In the world, life is not possible without water. In general, water accounts for almost 70%–90% of the weight of living organisms. One cannot imagine life without clean water. No doubt, there is no raw material in the world which is more important than water. Thus, the quality of this valuable resource will directly influence the normal life of human beings.

Water pollution occurs when pollutants are directly or indirectly discharged into water bodies without adequate treatment to remove harmful compounds. In almost all cases, the effect is damaging not only to individual species and populations, but also to the natural biological communities. The effluents from dyeing, textile, and printing industries pollute the water bodies. Dye effluents originating from dye industries pose a major threat to surrounding ecosystems, because of their toxicity and potentially carcinogenic nature. The situation, if not controlled in a timely manner, would become a malignant problem for the survival of mankind on this planet.

Many chemicals undergo reactive decay or chemical change especially over long periods of time in ground water reservoirs. A noteworthy class of such chemicals is the chlorinated hydrocarbons such as trichloroethylene (used in industrial metal degreasing and electronics manufacturing) and tetrachloroethylene (used in the dry cleaning industry). Both of these chemicals, which are carcinogens themselves, undergo partial decomposition reactions, leading to new hazardous chemicals including dichloroethylene and vinyl chloride. Ground water pollution is much more difficult to abate than surface pollution, because ground water can move great distances through unseen aquifers. The recycling of wastewater is associated with the presence of suspended solids, health-threat coliforms, and soluble organic compounds, and it is also quite expensive to treat wastewater.

Dyes are widely used in the textile, leather, paper, printing inks, plastics, cosmetics, paints, pharmaceutical, and food industries. About 15% of the dyes are lost in the synthesis and processing of colorants, dyeing, printing, and finishing. Thi invariably corresponds to a release of large amount of dyes (about 615 tons per day into the environment and ecological system. It is known that dyes are hazardous t the environment and even when they are present in very low concentrations, the can cause serious carcinogenic effects. Degradation of dye molecules sometime results in the formation of noncolored dye fragments that, although satisfying the requirement of decolorization, leads to the formation of environmentally unfriendl degradation products, such as aromatic amines, which are sometimes much more toxic than the parent dyes. Color removal from textile dyeing effluents has been the target of great attention in the last few decades due to its aesthetic effect, even a lower concentrations.

13.2   METHODS OF WASTEWATER TREATMENT

It is the prime objective of environmental education to make people aware about the importance of protection and conservation of our environment, because indiscrim inate release of various pollutants in the environment leads to serious health haz ards. Water treatment is the process of removing undesirable chemicals, materials and biological contaminants from raw water. Drinking water quality is defined in terms of physical, chemical, biological, and radiological parameters, and limiting values are set either as national regulations within countries or as recommenda tions from international organizations, especially the World Health Organization (WHO). Organic waste represents one of the most problematic groups of pollutant because it can be easily identified by the human eye and is not easily biodegrad able. Treatment of wastewater includes biological treatment, catalytic oxidation membrane filtration, sorption process, ion-exchange and coagulation–flocculation and so on. Each one of these is associated with various demerits.

13.3   DYES

Zhao et al. (2008) prepared aluminum(III)-modified TiO_2 by the sol–gel process via a sudden gelating method. They observed the effect of the aluminum modification on interaction between the dye and photocatalyst, the interfacial electron transfer process, and the degradation of dye pollutants under visible irradiation. It was found that the aluminum forms an overlayer of Al_2O_3 on the surface of TiO_2, interfaced with Ti–O–Al bonds rather than incorporating into its crystal lattice. It was interest ing to note that the carboxylate-containing dyes such as rhodamine B (RhB) adsorb preferentially on the alumina, rather than the Ti(IV) sites on the surface of TiO_2 A five times faster photodegradation rate was observed for RhB as compared to a pristine TiO_2 system. The photodegradation of dyes on the aluminum(III)-modified photocatalyst depends on the structure as well as anchoring group of the dyes. It was observed that dye with carboxylate as the anchoring group and the amino group as electron donor was favorably degraded.

The photocatalytic degradation of methylene blue (MB) was studied by Gandhi et al. (2008) using Nb_2O_5 as semiconductor and visible light as the source of energy

The effect of various parameters, such as amount of semiconductor, pH, light intensity, dye concentration, and so forth, on the photodegradation was investigated. An aqueous solution of MB was photocatalytically degraded by UV radiation in the presence of TiO_2 photocatalyst with different concentrations of hydrogen peroxide (Madhu et al. 2009). The effect of different dye concentrations (12 and 20 ppm), catalyst loading, pH, and H_2O_2 dosage (1–10 mL/L) was observed. The decolorization of dye was almost complete when H_2O_2 was used. The optimum degradation was observed with 0.1 wt.% of catalyst loading at pH 2 for a TiO_2/UV system. Optimum concentration of H_2O_2 for 12 and 20 ppm of the dye was found to be 2 mL/L for the UV/H_2O_2 system. The degradation of the dye followed pseudo-first-order kinetics.

Kansal et al. (2009) observed the photocatalytic degradation of reactive black 5 (RB5) and reactive orange 4 (RO4) dyes in a heterogeneous system. Photocatalytic activity of different semiconductors such as titanium dioxide and zinc oxide has been examined. The effect of various reaction parameters such as amount of catalyst, concentration of dye, and pH on photocatalytic degradation of RB5 and RO4 was monitored. The optimum dose of catalyst was found to be 1.25 and 1 g/L for RB5 and RO4, respectively. Maximum rate of decolorization was observed at pH 4 in the case of RB5, whereas the decolorization of RO4 reached a maximum at pH 11. The performance of a photocatalytic system employing ZnO/UV light was found to be better than a TiO_2/UV system. Complete decolorization of RB5 was observed after 7 minutes with ZnO, whereas only 75% of the dye degraded in 7 minutes with TiO_2. The decolorization noticed after 7 minutes was 92% and 62% for ZnO and TiO_2, respectively, in the case of RO4.

The photocatalytic degradation of two azo dyes, monoazo dye acid orange 10 (AO10) and diazo dye acid red 114 (AR114), present in wastewater were studied by Abo-Farha (2010). It was found that the rate of decolorization increased by increasing the initial dosage of H_2O_2 up to a certain limit and thereafter it inhibited. The reactions followed pseudo-first-order kinetics. The photocatalytic degradation rate was found to be dependent on structure of dye, its concentration, amount of TiO_2, and pH of the medium. The photodegradation process under UV–visible light illumination involves an electron excitation into the conduction band of the TiO_2 semiconductor leading to the generation of active oxygenated species. The photocatalytic decomposition of the two azo dyes was enhanced in the presence of electron scavengers such as H_2O_2. The observation that monoazo dye (AO10) decolorized more rapidly than diazo dye (AR114) indicates that the number of azo and sulfonate groups in the dye molecule may be a deciding factor for the degradation rates.

Kumar and Bansal (2010) reported that immobilization of nanosized titanium dioxide on cotton fabric enhanced the dye degradation efficiency of the photocatalyst, which may be due to the dual effects of adsorption and photodegradation. They immobilized titania particles on cotton fabric to overcome the problem of TiO_2 suspensions in aqueous solution. The sol–gel coated cotton fabric was used for degradation of amaranth dye in aqueous solution. As-prepared fabric showed better dye degradation capabilities.

TiO_2 and ZnO were used by Joshi and Shrivastava (2011) to remove alizarin red-S, a hazardous textile dye, from aqueous solution. The factors affecting the degradation process were mainly initial dye concentration, contact time, dose of catalyst, and pH. The adsorption rate data were analyzed using pseudo-first-order kinetics. The optimum contact time was fixed at 120 minutes for both semiconductors, TiO_2 and ZnO. Freundlich and Langmuir isotherm equations were applied for the equilibrium.

The photocatalytic degradation of two commercial textile azo dyes, RB5 and reactive red 239, has been observed by Saggioro et al. (2011). TiO_2 P25 Degussa was used as semiconductor and photodegradation was carried out in aqueous solution under a 125 W mercury vapor lamp. The effects of different parameters such as the amount of TiO_2 used, UV-light irradiation time, pH of different concentrations of the azo dye, and hydrogen peroxide were investigated. They also investigated simultaneous photodegradation of these two azo dyes and observed that the degradation rates achieved in mono and bicomponent systems were almost identical. The recyclability of the photocatalyst was also tested. It was observed that after using TiO_2 for five cycles, the rate of discoloration was 77% of the initial rate. Adequate selection of optimal operational parameters led to a complete decolorization of the aqueous solutions of both azo dyes.

Jafari et al. (2012) observed the decolorization and degradation of RB5 azo dye by biological, photocatalytic (UV/TiO_2), and combined processes. Complete decolorization of the dye with 200 mg/L RB5 was recorded in the presence of *Candida tropicalis* JKS2 in less than 24 hours, but degradation of the aromatic rings did not occur during this treatment. On the other hand, mineralization of 50 mg/L RB5 solution was obtained after 80 minutes by photocatalytic process in the presence of 0.2 g/L TiO_2 and chemical oxygen demand (COD) was not detectable at the end. However, the photocatalytic process was also not effective in the removal of the dye with high concentrations (≥200 mg/L) and only 74.9% decolorization was achieved after 4 hours of illumination. A two-step treatment process was attempted, namely, biological treatment by the yeast followed by photocatalytic degradation in this combined process (with 200 mg/L RB5). They reported that the absorbance peak due to the aromatic ring in the UV region disappeared after 2 hours of illumination and nearly 60% COD removal was achieved in the biological step. It was suggested that the combined process is more effective in the destruction of aromatic rings than the biological and photocatalytic treatments individually.

Photocatalytic degradation of toxic organic chemicals is considered to be the most efficient green method for surface water treatment. The sol–gel synthesis of Ga-doped anatase TiO_2 nanoparticles was reported by Banerjee et al. (2012) and they used it for the photocatalytic oxidation of organic dye into nontoxic inorganic products. Very good photocatalytic activity of Ga-doped TiO_2 nanoparticles was observed with almost 90% degradation efficiency within 3 hours of UV irradiation. Doping of Gd created additional levels within the band gap of TiO_2, which act as trapping centers for photogenerated electron–hole recombination. This will help in timely utilization of charge carriers for the generation of strong oxidizing radicals for degradation of the dye. Photocatalytic degradation was found to follow pseudo-first-order kinetics. These cost-effective, sol–gel derived Gd-TiO_2 nanoparticles can be used effectively for light-assisted oxidation of toxic organic molecules for environmental remediation.

Photocatalytic degradation of MB over ferric tungstate in the presence of light has been carried out by Ameta et al. (2013). They observed the effect of some parameters affecting the rate of reaction such as pH, dye concentration, amount of semiconductor, light intensity, and so on. This reaction also followed pseudo-first-order kinetics. Nanogold-doped TiO_2 catalysts were synthesized and applied in the photodegradation of dye pollutants (Padikkaparambil et al. 2013). They revealed a strong interaction between the metallic gold nanoparticles and the TiO_2 (anatase) support. Au-doped systems showed appreciable photoactivity in the degradation of pollutants (dye) under UV irradiation as well as in the sunlight. The photocatalyst can give more than 98% degradation of the dye even after 10 cycles.

Roy and Mondal (2013) observed the efficacy of thermally activated ZnO for photocatalytic degradation of Congo red (CR) dye. The effect of various parameters such as catalyst loading, pH, oxidant dose, intensity variations, and initial concentration of the dye on degradation was also investigated. The pseudo-second-order kinetic model fit best for the CR degradation process. Thermodynamic parameters such as ΔH, ΔS, and ΔG were also determined, which indicated the process to be spontaneous and exothermic in nature.

Photocatalytic degradation of azo dyes using nano-strontium titanate was reported by Karimi et al. (2014). The influence of the different variables such as photocatalyst concentration, dye concentration, temperature, pH, and the presence of hydrogen peroxide was observed on dye removal. They revealed that nano-strontium titanate has quite high and significant photocatalytic activity as compared to that of nano-titanium dioxide.

Alahiane et al. (2014) observed the photocatalytic degradation of the synthetic textile dye reactive yellow 145 (RY145) in aqueous solution using TiO_2-coated nonwoven fibers as a photocatalyst under UV irradiation. The effects of the different operational parameters such as initial dye concentration, pH, and addition of oxidant hydrogen peroxide and ethanol on the reaction rate were investigated. The effect of some inorganic ions such as SO_4^{2-}, Cl^-, NO_3^-, CH_3COO^-, HCO_3^-, and HPO_4^- (commonly present in real effluents) on the photodegradation of RY145 was also reported. The maximum rate of complete decolorization of RY145 was observed in the acidic medium at pH 3. It was observed that the presence of SO_4^{2-} and Cl^- ions led to an increase in the effectiveness of the photocatalytic degradation, while the presence of CH_3COO^-, HCO_3^-, and $H_2PO_4^-$ ions adversely affect it.

The photocatalytic degradation of reactive blue 4 (RB4), an anthraquinone dye, has been investigated by Samsudin et al. (2015), using pure anatase nano-titanium(IV) oxide. The dye molecules were fully degraded in this process and the addition of hydrogen peroxide enhanced the photodegradation efficiency of the photocatalyst. Hydroxyl radicals were considered responsible for degradation in the bulk solution of dye leading to complete mineralization. The disappearance of the dye followed pseudo-first-order kinetics. The effect of pH, amount of photocatalyst, UV-light intensity, light source, and concentration of hydrogen peroxide was also observed.

Boruah et al. (2016) synthesized Fe_3O_4/reduced graphene oxide (rGO) nanocomposite photocatalyst by an eco-friendly solution chemistry approach. Fe_3O_4/rGO nanocomposite was found effective in photocatalytic degradation of carcinogenic and mutagenic cationic as well as anionic dye molecules, namely methyl green, methyl blue, and RhB under direct sunlight irradiation. Excellent photocatalytic reduction of aqueous Cr(VI) solution to nontoxic aqueous Cr(III) solution (more than 96%) was observed within 25 minutes in the presence of Fe_3O_4/rGO nanocomposite under sunlight irradiation. The reusability of the magnetically recovered photocatalyst was studied and it was also reported that this photocatalyst can retain its efficiency up to 10 cycles. The particle size of the Fe_3O_4 nanoparticles on the rGO sheets remained unchanged after photocatalytic degradation. They focused on the importance of the use of Fe_3O_4/rGO nanocomposite toward photocatalytic degradation of wastewater containing organic dye pollutants and toxic Cr(VI), and suggested it as an easily recoverable and reusable photocatalyst with possibilities for many environmental remediation applications.

Photocatalytic degradation of an azo dye, direct red 23 (DR23), was carried out by Sobana et al. (2016) with activated carbon-loaded ZnO (AC-ZnO) as a photocatalyst under solar light irradiation. The effects of operational parameters such as pH of the solution, amount of catalyst, initial dye concentration, and effect of grinding on photodegradation of DR23 by AC-loaded ZnO were analyzed. It was observed that the degradation of DR23 follows pseudo-first-order kinetics according to the Langmuir–Hinshelwood model. The AC–ZnO showed higher dye degradation efficiency under sunlight than with UV light.

13.4 CHLORO COMPOUNDS

A fluidized-bed-type flow reactor was designed by Kometani et al. (2008) for the photocatalytic treatment of the suspension of model soil under high temperature and high pressure conditions. An aqueous suspension containing hydrogen peroxide (an oxidizer) and inorganic oxides as a model soil (titania, silica, or kaoline) was continuously fed into the reactor with the temperature 20°C–400°C and the pressure 30 MPa. The photocatalytic degradation of chlorobenzene (CB) in aqueous solution was chosen as a model system. It was observed that most of the soils were not so harmful to the super critical water oxidation (SCWO) treatment of CB in solutions, but when the titania suspension containing H_2O_2 was exposed to near-UV light, the degradation of CB was observed at all temperatures. Persistence of such photocatalytic activity in the oxidation reaction at high temperature and high pressure water opened new possibilities of a hybrid process based on the combination of these two processes (SCWO and TiO_2 photocatalysis) for the treatment of environmental pollutants in soil and water, which are ordinarily difficult to perform by the conventional SCWO process or catalytic SCWO process alone.

The photocatalytic degradation of 4-chlorophenol (4-CP) in aqueous solution was studied by Ghorai (2011) using Cu–MoO_4-doped TiO_2 nanoparticles under visible light radiation. The photocatalysts were synthesized by a chemical route from TiO_2 using different concentrations of $CuMoO_4$ ($Cu_xMo_xTi_{1-x}O_6$; 0.05–0.5). The $Cu_xMo_xTi_{1-x}O_6$ ($x = 0.05$) was found to exhibit high activity for degradation of 4-CP under visible light. The surface area of the catalyst was found to be 101 m^2/g. The photodegradation process was optimized using $Cu_xMo_xTi_{1-x}O_6$ ($x = 0.05$) catalyst at a concentration level of 1 g/L. Maximum photocatalytic efficiency of 96.9% was btained at pH 9 after irradiation for 3 hours. The effect of different parameters such s catalyst loading, concentration of the catalyst and the dopant concentration, solution pH, and concentration of 4-CP was observed.

Kansal and Chopra (2012) observed the photocatalytic degradation of 2, 6-dichloophenol (2,6-DCP) in aqueous phase using titania (PC-105) as a photocatalyst. The queous suspensions of the 2,6-DCP was irradiated in the presence of photocatalysts under UV light. Various parameters affecting the degradation process such s catalyst dose, pH, initial substrate concentration, and time were investigated, nd optimum conditions were obtained. The maximum degradation of 2,6-DCP vas achieved with 1.25 g/L catalyst dose at pH 4. The disappearance of 2,6-DCP ollowed pseudo-first-order kinetics and the rate constant was 4.78×10^{-4} per second.

The photochemical oxidation of 4-CP in aqueous solutions was studied by Alimoradzadeh et al. (2012) using UV irradiation, hydrogen peroxide, and nickel oxide. The efficiency of the system was evaluated with respect to reaction time, pH, eed concentration of reactants, catalyst load, and light intensity. The concentrations of 4-CP and chloride ions were monitored by high-performance liquid chromatography (LC) and ion chromatography, respectively. The optimum conditions for the complete 4-CP removal (100%) were obtained at neutral pH, 0.2 mol/L H_2O_2, and 0.05 g/L of nickel oxide. The degradation rate was found to increase with increasing JV light intensity and decreasing initial concentration of 4-CP.

Verma et al. (2013) made a comparison of photocatalytic, sonolytic, and sonophotocatalytic degradation of 4-chloro-2-nitrophenol using titania. The degradation of compound followed first-order kinetics. The optimum conditions obtained were catalyst concentration 1.5 g/L, pH 7, and oxidant concentration 1.5 g/L. Degradation vas obtained for photocatalytic treatment in 120 minutes, whereas use of ultrasound tad a synergistic effect as 96% degradation was achieved in 90 minutes during sonophotocatalysis. The degradation follows the trend:

Sonophotocatalysis > Photocatalysis > Sonocatalytic > Sonolysis

Mesoporous nanocrystalline anatase was prepared by Avilés-García (2014) via EISA employing cetyltrimethylammonium bromide (CTAB) as a structure-directing agent. As-prepared mesoporous crystalline phases exhibited specific surface area 55–150 m²/g), average unimodal pore size (3.4–5.6 nm), and average crystallite size 7–13 nm). These mesophases were used as photocatalysts for the degradation of 4-CP with UV light. The mesoporous anatase degraded 100% 4-CP, which was two imes faster than Degussa P25. They also achieved 57% reduction of COD value.

The efficiency of W-doped TiO_2 was evaluated by Oseghe et al. (2015) in the photocatalytic degradation of 4-chloro-2-methylphenoxyacetic acid (MCPA). They synthesized W-doped TiO_2 via the sol–gel method. Pore size distribution indicated that the materials were mesoporous in nature with Brunauer–Emmett–Teller BET) surface area ranging from 86.08 to 91.71 m²/g. All materials had polycrystalline anatase phase with a decreasing crystal size on increasing the percentage of dopant. Optimal photocatalytic activity was obtained with 0.1 wt.% W-doped TiO_2 (0.1 wt.%) at pH 5, which may be due to reduced band gap, crystal size, and esser amount of oxygen vacancy. The intermediates/products were analyzed by iquid chromatography–mass spectrometry (LC-MS) as 2-hydroxybuta-1, 3-diene-1,

4-diyl-bis (oxy) dimethanol, and 2-(4-hydroxy-2-methylphenoxy) acetic acid, whicl are relatively safer than the starting material, MCPA.

Castañeda et al. (2016) synthesized phosphated TiO_2 photocatalysts with differer phosphate content by *in situ* phosphatation of sol–gel TiO_2. Phosphoric acid wa used as a hydrolysis catalyst and as an anionic precursor. The photoactivity o as-prepared photocatalyst was tested in the degradation of 4-CP and 2,4-DCP unde UV irradiation. It was observed that the presence of phosphate anions decrease the crystallite size without any modification of the anatase phase. Specific surfac areas on the phosphated samples were higher than the pristine TiO_2. Its photoactivit improved in case of phosphated TiO_2. Degradation rate of the chlorophenolic com pounds was found to increase with the degree of substitution in the aromatic ring a the degradation rate of 2,4-DCP was higher than that of 4-CP. The photocatalyst can be reused with good efficiency.

The adsorption of 1-chloro-4-nitrobenzene (1C4NB) on carbon nanofibers (CNFs was investigated by Mehrizad and Gharbani (2016) in a batch system. They ana lyzed the combined effects of operating parameters such as contact time, pH, initia 1C4NB concentration, and CNFs dosage on the adsorption of 1C4NB by CNFs. High efficiency removal (>90%) of 1C4NB was obtained under optimal values in the firs 6 minutes. The isotherm data were well fitted by the Freundlich isotherm equation.

13.5 PHENOLS

Solid-state nuclear magnetic resonance (NMR) spectroscopy was used to investigat the local structure of the TiO_2 surface modified with electron-donating bidentat ligands such as catechols (Tachikawa et al. 2006). The adsorption and degradatio processes of catechols at the TiO_2 surface were also observed. They interpreted pho tocatalytic degradation of catechols in terms of the interfacial charge recombinatio reaction with conduction band electrons. Photocatalytic degradation of catechol anc resorcinol with TiO_2 in the presence of acetic acid has been studied (Araña et al 2006). Low concentrations of the acetic acid (30–50 ppm) affect the adsorption anc degradation of the dihydroxybenzenes. Catechol and resorcinol molecules adsorl on different TiO_2 surface centers. Acetic acid molecules are able to displace Ti^4 coordinated water molecules. As a result, the catalyst surface becomes restructured favoring the adsorption of the dihydroxybenzenes. Photogenerated electrons giv $O_2^{\bullet -}$ radicals that react with OH groups with acidic character to give some new radi cals that improve catechol and resorcinol degradation.

Performance of a photocatalytic process for degrading catechol with titaniun dioxide and zinc oxide catalysts has been evaluated by Kansal et al. (2007). Th effect of various factors was investigated including catalyst dose, pH, and amoun of oxidant. Residual catechol concentration was monitored and the performance o the photocatalytic process was observed in terms of percentage degradation. The showed that catalyst dose and amount of oxidant have a significant effect on the process of degradation.

Photocatalytic oxidation of phenol was carried out in aquatic solutions using UV TiO_2, and the combination of them (Rahmani et al. 2008). It was observed that the combination of UV and TiO_2 had higher efficiencies of phenol removal. The remova efficiencies of UV, TiO_2, and a UV/TiO_2 photocatalytic oxidation system with various operation conditions were 1.8%–19.64%, 2.38%–17.8%, and 34.65%–82.91%, respectively. Maximum removal efficiency was obtained at pH 11 in 9 hours contact time and with 0.2 grams of TiO_2.

The photocatalytic degradation of an endocrine disruptor (resorcinol) and two fungicides (pyrimethanil and triadimenol) has been studied and a comparison has been made by Araña et al. (2008). The effect of pH, oxygen, and H_2O_2 on the photocatalytic degradation of these compounds has been observed. Mineralization of these three organics was analyzed by total organic concentration (TOC) measurements. The resorcinol and pyrimethanil solutions were detoxified after 30 minutes of reaction, while 92% detoxification was achieved in the case of triadimenol after 60 minutes of reaction.

Photocatalytic degradation of resorcinol, a potent endocrine-disrupting chemical, in aqueous medium was investigated by Pardeshi and Patil (2009) using ZnO under sunlight irradiation in a batch photoreactor. The influence of various operating parameters such as photocatalyst amount, initial concentration of resorcinol, and pH was examined. A significant influence of pH was observed upon COD disappearance. It was observed that neutral or basic pH was favorable for COD removal of resorcinol. Two initial oxidation intermediates were detected as 1,2,4-trihydroxybenzene and 1,2,3-trihydroxybenzene.

Photocatalytic systems using TiO_2 and ZnO suspensions were utilized by Lam et al. (2013a) to evaluate the degradation of resorcinol (ReOH). The effect of catalyst concentration and solution pH was optimized. They found greater degradation and mineralization activities in the presence of ZnO as compared to TiO_2 under optimized conditions. Participation of hydroxyl radicals as well as a certain radical scavenger, a positive hole, as the oxidative species was ascertained for ReOH degradation on TiO_2 while ZnO photocatalysis occurred basically via hydroxyl radicals.

Ag_2O/ZnO heterostructure has been synthesized by Lam et al. (2013b) using a facile chemical-precipitation method. Loading of Ag_2O nanoparticles on ZnO nanorods was also confirmed. Ag_2O addition increased the visible light absorption ability of the composite, and a red shift for Ag_2O/ZnO heterostructure was observed as compared to pure ZnO. A decrease in the emission yield indicated that the fraction of the excited state Ag_2O sensitizer was involved in the charge injection process, under compact fluorescent lamp irradiation. The Ag_2O/ZnO heterostructure demonstrated higher photocatalytic activity than pure ZnO in the degradation of resorcinol. It was due to the high separation efficiency of the photogenerated electron–hole pairs based on the cooperative role of Ag_2O loading on ZnO nanarods.

Photocatalytic degradation of phenol in a batch system gave some aromatic intermediates such as catechol, hydroquinone, and resorcinol. The concentrations of these intermediates were obtained in the following order:

Catechol > Resorcinol > Hydroquinone

Resorcinol attained a steady-state concentration of 3.73 mg/L after 2 minutes of photocatalytic degradation of phenol, but thereafter there was no appreciable change in its concentration, which suggested that production of resorcinol is approximately the same as its rate of removal from solution. Catechol formed rapidly, attaining a maximum concentration of 7.00–7.11 mg/L after 6 minutes while a maximum concentration of hydroquinone was obtained as 0.96–1.66 mg/L, which then decreases steadily until all phenol is degraded.

The effect of irradiation time, initial pH, and dosage of TiO_2 on photocatalytic degradation of phenol was observed by Alalm and Tawfik (2014) under sunlight. Aromatic intermediates (catechol, benzoquinone, and hydroquinone) were determined during the reaction to know the pathways of the oxidation process. Degradation efficiency of 94.5% of phenol was achieved after 150 minutes of irradiation with initial concentration of 100 mg/L. It was found that dosage of TiO_2 significantly affected the degradation efficiency of phenol and optimum pH for the reaction was 5.2.

Zhang et al. (2015) demonstrated that resorcinol–formaldehyde resin polymers are good visible light responsive photocatalysts as their band gap energies are in the range 1.80–2.00 eV. These polymers were photoactive for decomposition of organic substrates under visible light irradiation. The photocatalytic performance of resins could be significantly improved by coupling with electron-conducting materials such as rGO. Photocatalytic water oxidation can also be achieved on this hybrid with some sacrificial agents. These widely used industrial resins exhibited various merits as photocatalysts such as low cost, high surface area, large pore size and volume, easy preparation, and scalability for development of eco-friendly commercial products with self-cleaning properties, so that organic pollutants can be oxidatively removed under visible light irradiation.

13.6   NITROGEN-CONTAINING COMPOUNDS

Photocatalytic degradation as well as adsorption of nitrobenzene (NB) in the presence and absence of phenol (Ph) over UV-illuminated arginine-modified TiO_2 colloids was studied by Cropek et al. (2008). High-performance liquid chromatography and gas chromatography/mass spectrometry techniques were used for monitoring degradation products. It was reported that photodegradation of NB and Ph was strongly dependent on the nature of the TiO_2 surface. Photocatalytic decomposition rate of NB and Ph using bare TiO_2 was almost identical. This degradation proceeds via oxidative mechanism. A threefold increase in the NB decomposition rate was observed by using arginine-modified TiO_2 nanoparticles, while no such increase was observed in the case of Ph decomposition. The degradation pathway using the arginine-modified photocatalyst was completely reverted to a reductive mechanism as compared to TiO_2. It provided a more efficient means to degrade nitro-compounds that are in a highly oxidized state, and also limited the number of byproducts.

Wang et al. (2010) prepared phosphotungstic acid ($H_3PW_{12}O_{40}$), titanium dioxide, and $H_3PW_{12}O_{40}$ supported on TiO_2 ($H_3PW_{12}O_{40}/TiO_2$). They proved that $H_3PW_{12}O_{40}$ and $H_3PW_{12}O_{40}/TiO_2$ possessed classical Keggin structure, and TiO_2 was in anatase form. Photocatalytic degradation of NB containing wastewater in the presence of $H_3PW_{12}O_{40}/TiO_2$ was observed under visible light. It was observed that reaction time, catalyst dosage, and NB concentration were the main factors affecting the degradation. The maximum degradation of NB wastewater was found to be 94.1% at the optimum conditions. They concluded that photocatalytic activity of this novel photocatalyst will provide a promising solution for the degradation of water contaminants like NB.

Graphitic carbon–TiO_2 nanocomposites with different carbon loadings were synthesized by Jo et al. (2014) via a one-pot hydrothermal method. They confirmed that the presence of graphite in the composite did not alter the TiO_2 structure. The photocatalytic efficiencies of the as-synthesized composites were determined by the degradation of aqueous NB under UV irradiation. An increase in the adsorption of NB (24%) on the composite surface was observed because of the presence of graphitic carbon in the composite. This led to a higher photocatalytic yield (about 96%) in 4 hours with 1% graphitic carbon content in TiO_2.

Titanium dioxide impregnated zeolite Y (Si/Al ratio 5.5) was modified by silver metal ion exchange by Surolia and Jasra (2016) and its photocatalytic activity was observed for the mineralization of p-nitrotoluene (PNT) in aqueous medium. Degradation intermediates and mineralization pathways were established using high-performance liquid chromatography (HPLC) and mass spectroscopy (MS). PNT was mineralized into simple products such as CO_2, H_2O, NO_3^-, and NH_4^+. The Langmuir–Hinshelwood kinetic model was proposed for this degradation. Optimization of TiO_2 loading was decided from the percentage degradation, rate constant, and mineralization values. COD study revealed that ~60% mineralization took place in 240 minutes of irradiation using TiO_2/AgY_2 catalyst.

13.7 SULFUR-CONTAINING COMPOUNDS

Zeolite–TiO_2 nanocomposite was prepared by Hossein and Shakiba (2013) and used for degradation of dibenzothiophene, a typical aromatic organosulfur compound of transportation fuels. As-synthesized TiO_2 was immobilized on the surface of clinoptilolite by a solid-state dispersion (SSD) method. They studied photodegradation of dibenzothiophene in n-hexane solution by the photocatalyst under different experimental conditions. It was reported that the photocatalyst was able to degrade 88% of dibenzothiophene under optimized conditions and the degradation products obtained were adsorbed by zeolite so that the solution was deeply desulfurized.

Hossein and Reza (2014) desulfurized a solution containing dibenzothiophene. A nanocomposite containing polyethylene glycol (PEG) and $FeTiO_2$ as photocatalyst was synthesized and used for degradation of dibenzothiophene. The presence of Fe^{3+} and PEG in the TiO_2 structure broadened the light absorption zone and changed the surface structure of the photocatalyst, respectively. The effect of different experimental parameters on the degradation efficiency of dibenzothiophene was observed. Mesoporous SBA-15 adsorbent was used to remove the sulfur-containing products. It was concluded that the as-synthesized photocatalyst was an efficient composite for degradation of dibenzothiophene and the mesoporous SBA-15 was quite efficient in removing sulfur-containing degradation products.

Photocatalytic degradation of sulfanethazine (SMT) aqueous solution was investigated using TiO_2 (Fukahori and Fujiwara 2015). The structures of seven intermediates formed in this degradation were also analyzed by LC/MS/MS. They observed the decomposition behaviors of other model compounds such as sulfanilic acid (SA) and 4-amino-2,6-dimethylpyrimidine (ADMP) using the TiO_2/UV system. The formation of p-aminophenol during SMT decomposition was reported, which was not reported earlier, while the direct substitution of the sulfonamide group with a hydroxyl group has been suggested.

13.8 PHARMACEUTICAL DRUGS

Three nonsteroidal anti-inflammatory drugs (NSAIDs) were degraded by Méndez-Arriaga et al. (2008) via heterogeneous TiO_2 photocatalysis in aqueous solution at a laboratory scale. Diclofenac (DCF), naproxen (NPX), and ibuprofen (IBP) were selected as model pharmaceutical compounds. These compounds were used in the form of their sodium salts. Influences of different operational conditions such as catalyst load, temperature, and dissolved oxygen concentration were observed. The maximum degradation for DCF or NPX was observed at a TiO_2 loading of only 0.1 g/L, but it was 1 g/L for degradation of IBP. No significant differences were observed for DCF and IBP at 20°C, 30°C, and 40°C, but temperature had a significant effect for NPX degradation. Dissolved oxygen concentration was found to affect degradation for NPX and IBP favorably. Only photocatalytic treatment of IBP had a satisfactory biodegradability index biological oxygen demand (BOD)/COD (chemical oxygen demand) (0.16–0.42) and, therefore, a postbiological treatment could be suggested in case of IBP.

The photocatalytic degradation of some sulfa drugs such as sulfanilamide, sulfacetamide, sulfathiazole, sulfamethoxazole, and sulfadiazine in aqueous solutions was carried out under UV-A irradiation in the presence of TiO_2, Fe salts, and TiO_2/$FeCl_3$ catalysts (Baran et al. 2009). It was found that sulfonamides underwent photocatalytic degradation in the presence of TiO_2, TiO_2/$FeCl_3$, and Fe^{3+} salts. A relationship between the pH of irradiated solutions, initial concentrations of sulfanilamide, and $FeCl_3$ was reported.

El-Kemary et al. (2010) prepared nanostructure ZnO semiconductor with ~2.1 nm diameter using a chemical precipitation method. The absorption spectra exhibited a sharp absorption edge at ~334 nm, which corresponds to a band gap of ~3.7 eV. A near band edge UV excitonic emission at ~410 nm and a green emission peak at ~525 nm was observed, which may be due to the transition of a photogenerated electron from the conduction band to a deeply trapped hole. The photocatalytic activity of as-prepared ZnO nanoparticles has been evaluated for the degradation of the drug ciprofloxacin under UV light irradiation. They reported that the photocatalytic degradation process was effective at pH 7 and 10, but was rather slow at pH 4. Higher degradation efficiency (~50%) of the ciprofloxacin was observed at pH 10 in 60 minutes.

Photocatalytic degradation kinetics of the antivirus drug lamivudine in aqueous titania dispersions was studied by An et al. (2011). Three variables, titania content, initial pH, and lamivudine concentration, were selected to know the dependence of degradation efficiencies of lamivudine. They indicated that degradation efficiencies of lamivudine were significantly affected by TiO_2 content as well as initial lamivudine concentration. It was also found that hydroxyl radicals were the major reactive species involved in lamivudine degradation in aqueous TiO_2. Six degradation intermediates were identified using HPLC/MS/MS.

Ciprofloxacin solution was mixed with TiO_2 nanoparticles and then irradiated with two different light sources: a UV lamp and ordinary electric bulb (Hayder et al.

(012). Ciprofloxacin degradation was observed in the presence of 0.01 mg/mL of TiO_2. Around 90% and 70% of concentration of the drug was observed in 120 minutes on using a UV lamp and ordinary electric bulb, respectively.

Karan and Jayaram (2013) reported photocatalytic degradation of DCF, an NSAID, using various photocatalysts. The efficiency of degradation was affected by several parameters such as initial concentration of the drug, catalyst loading, agitation, and addition of H_2O_2 as a cooxidant. ZnO- and Ag-doped TiO_2 exhibited a higher rate of degradation as compared to TiO_2. The degradation was found to follow first-order kinetics. It was found that Ag-doped TiO_2 (2 mol% loading) leads to almost complete degradation within 180 minutes for an initial concentration of 20 ppm of the drug and at a catalyst loading of 1 g/L. Addition of hydrogen peroxide as a cooxidant enhanced the degradation of DCF further.

Photocatalytic degradation of acetaminophen ((N-(4-hydroxyphenyl)acetamide)), an analgesic drug, was investigated by Jallouli et al. (2014) in a batch reactor using TiO_2 P25 as a photocatalyst (slurry) under UV light. Much faster photodegradation of paracetamol occurred in the presence of TiO_2 P25 nanoparticles and more than 90% of 2.65×10^{-4} M paracetamol was degraded under UV irradiation. It was found that any change in pH values affected the adsorption as well as the photodegradation of paracetamol. The optimum photodegradation of paracetamol was observed at pH 9.0. Hydroquinone, benzoquinone, p-nitrophenol, and 1,2,4-trihydroxybenzene were detected as intermediates during the TiO_2-assisted photodegradation of paracetamol. They showed that the TiO_2 suspension/UV system was more efficient than the TiO_2/cellulosic fiber mode combined with solar light for this photocatalytic degradation. The immobilization of TiO_2 was advantageous over the slurry system because it can enhance adsorption properties as well as allow easy separation of the photocatalyst after use.

Georgaki et al. (2014) observed degradation of carbamazepine (CBZ) and IBP in aqueous matrices in the presence of TiO_2 and ZnO under UV-A and visible light irradiation. The contribution of ·OH was evaluated in CBZ and IBP removal as an active oxidizing species. High performance liquid chromatography was used to monitor the photodegradation rate of these pharmaceuticals. IBP was found to have better degradation rate as compared to CBZ. Addition of isopropanol showed a significant inhibition effect on degradation of CBZ, suggesting solution phase mechanism, while in IBP degradation the hole mechanism may be operative as negligible effect was observed upon addition of isopropanol.

A facile synthesis, characterization, and solar light-driven photocatalytic degradation of TiO_2 quantum dots (QDs) was reported by Kaur et al. (2015). TiO_2 QDs were synthesized by a facile ultrasonic-assisted hydrothermal process. As-prepared QDs were well-crystalline, grown in high density, and exhibited good optical properties. These QDs were found to be an effective photocatalyst for the sunlight-driven photocatalytic degradation of ketorolac tromethamine, which is a well-known NSAID. To optimize the photocatalytic degradation conditions, various experiments were conducted with variation of dose, pH, and initial drug concentration. These experiments revealed that ~99% photodegradation of ketorolac tromethamine drug solution (10 mg/L) occurred under sunlight with TiO_2 QDs (0.5 g/L) and pH 4.4.

Nanosized perovskites $BaBiO_3$ and $BaBi_4Ti_4O_{15}$ were prepared by Jain et al. (2010 using the Pechini method. It was observed that $BaBiO_3$ was crystallized in the monc clinic structure while bismuth-based layer-structured $BaBi_4Ti_4O_{15}$ was crystallized in tetragonal structure. The band gaps were determined as 2.07 and 1.80 eV for $BaBiO_3$ an $BaBi_4Ti_4O_{15}$, respectively. Nanosized perovskites were applied in the degradation of IB via photocatalytic processes. It was found that $BaBi_4Ti_4O_{15}$ exhibited drastic enhance ment on degradation of drug under visible light irradiation as compared to $BaBiO_3$.

13.9 PESTICIDES

Navarro et al. (2009) reported photodegradation of eight pesticides in leachin water at pilot plant scale using tandem $ZnO/Na_2S_2O_8$ as the photosensitizer/oxidar under natural sunlight. The pesticides of different chemical groups were selecte such as azoxyxtrobin, kresoxim-methyl, hexaconazole, tebuconazole, triadimeno pyrimethanil (fungicides), primicarb (insecticide), and propyzamide (herbicide). Th influence of the photocatalyst (150 mg/L) on the degradation of pesticides was foun to be very significant in all these cases. The addition of photosensitizer remarkabl improved the elimination of pesticides in comparison with photolytic tests. It wa observed that the presence of $Na_2S_2O_8$ had a significant reduction in treatment tim as compared to ZnO alone while the addition of H_2O_2 to ZnO suspensions did nc enhance the rate of photooxidation. The disappearance of the pesticides followe first-order kinetics according to the Langmuir–Hinshelwood model and complet degradation was observed within 60–120 minutes.

Kitsiou et al. (2009) reported heterogeneous and homogeneous photocatalyti degradation of imidacloprid in aqueous solutions using artificial UV-A or visibl illumination. Three processes under various experimental conditions were evaluate by them, namely, TiO_2/UV-A, photo-Fenton/UV-A, and photo-Fenton/vis for thei activity toward degradation and mineralization of the substrate. The initial apparer photonic efficiency decreased in the following order:

Photo-Fenton/UVA > TiO_2/UV-A > Photo-Fenton/Vis

For the TiO_2/UV-A process, the efficiency increased considerably when TiO_2 wa combined with Fe^{3+} and H_2O_2, maybe because of the synergistic effect of homo geneous and heterogeneous photocatalytic reactions. Homogeneous photocatalyti reactions were found to be enhanced in the presence of the oxalate ions. Ammonium nitrate, and chloride ions have been detected in the liquid phase as mineralizatio products. These byproducts were less ecotoxic to marine bacteria than the insecti cide itself.

The oxidation powers of N-doped, undoped anatase TiO_2, and TiO_2 Degussa P2 suspensions were studied for photocatalytic degradation of the herbicides RS-2-(4 chloro-o-tolyloxy)propionic acid (mecoprop) and 3,6-dichloropyridine-2-carboxyli acid (clopyralid) using both visible and UV light. Undoped nanostructured TiO powder was prepared by a sol–gel route in the form of anatase. TiO_2 (anatase) pow der was used for mecoprop degradation under visible light irradiation. N-doped TiO Degussa P25 was also slightly more efficient than TiO_2 Degussa P25. No degrada tion of clopyralid was observed in the presence of any of these catalysts under th same experimental conditions. N-doped TiO_2 Degussa P25 powder was most efficient in the case of clopyralid degradation, while mecoprop degradation with UV light was more efficient with undoped powders. This may be due to differences in the molecular structure of these two herbicides.

Photocatalytic degradation of acetamiprid, a widely used pyridine-based neonicotinoid insecticide, was reported by Guzsvány et al. (2009) in the UV-irradiated aqueous suspensions of O_2/TiO_2. It was indicated that along with the main products, acetaldehyde, formic and acetic acid, and pyridine-containing intermediate (6-chloronicotinic acid) were also formed during the process. The pH of the solution changed from 5 to 2 during the photocatalytic process. The photocatalytic degradation of intermediate 6-chloronicotinic acid was also investigated to have a deeper insight into the complex photocatalytic process of acetamiprid.

TiO_2 doped with C, N, and S (TCNS photocatalyst) was prepared by Reddy et al. (2010) via a hydrolysis process using titanium isopropoxide and thiourea as the precursors. They observed that the prepared catalysts were anatase type and nanosized. A red shift in the adsorption edge was exhibited by the catalysts. The photocatalytic activity of TCNS photocatalysts was evaluated by the photocatalytic degradation of isoproturon pesticide in aqueous solution. Higher photocatalytic activity of TCNS photocatalysts was attributed to the synergetic effects of red shift in the absorption edge, higher surface area, and the inhibition of charge carrier recombination process.

Miguel et al. (2012) evaluated the effectiveness of photocatalytic treatment with titanium dioxide in the degradation of 44 organic pesticides in the Ebro river basin (Spain). Monitoring of effectiveness of photocatalytic processes was carried out by measuring quality parameters of water. An average photocatalytic degradation of the pesticides of 48% was achieved with 1 g/L of TiO_2 over 30 minutes. It was observed that chlorine demand, toxicity, and dissolved organic carbon (DOC) concentration of water were reduced. The average degradation of pesticides increases up to 57% on addition of hydrogen peroxide (10 mM). The chlorine demand and toxicity of water increased while DOC concentration remained almost the same with this treatment. It was surprising to note that the physicochemical parameters of water such as pH, conductivity, color, dissolved oxygen, and hardness do not vary after photocatalytic treatments.

The pesticides that undergo efficient degradation by this photocatalytic treatment are parathion methyl, chlorpyrifos, α-endosulfan, 3,4-dichloroaniline, 4-isopropylaniline, and dicofol, while hexachlorocyclohexane (HCHs), endosulfan-sulfate, heptachlors epoxide, and 4,4'-dichlorobenzophenone show very little degradation.

Photocatalytic degradation of chlorpyrifos in aqueous phase was studied by Verma et al. (2012) using photocatalyst TiO_2 in the presence of artificial UV light and sunlight. The effect of catalyst loading, pH, and addition of oxidant was observed on the reaction rate, and optimum conditions for maximum degradation were determined. The degradation of this insecticide was investigated in terms of reduction in COD. The optical conditions obtained were 4.0 g/L, pH 6.5, and oxidant concentration at 3.0 g/L, where degradation of the insecticide was 90% completed. The complete mineralization of pesticide from water or wastewater followed the first-order and Langmuir–Hinshelwood kinetic models.

The world is already facing a shortage of drinking water and increasing industrialization, transportation, construction, undesired use of chemicals for domestic work, and so on, have made this situation even worse. Many water-treatment methods are already available, but photocatalysis has got an edge over existing conventional methods because of its low cost green chemical nature. Photocatalysis can even degrade molecules such as dyes, pesticides, phenols, surfactants, nitro-compounds, and so on, some of which are nonbiodegradable or recalcitrant molecules.

REFERENCES

Abo-Farha, S. A. 2010. Photocatalytic degradation of monoazo and diazo dyes in wastewater on nanometer-sized TiO$_2$. *J. Am. Sci.* 6 (11): 130–134.

Alahiane, S., S. Qourzal, M. El Ouardi, A. Abaamrane, and A. Assabbane. 2014. Factors influencing the photocatalytic degradation of reactive yellow 145 by TiO$_2$-coated non-woven fibers. *Am. J. Anal. Chem.* 5:445–454.

Alalm, M. G. and A. Tawfik. 2014. Solar photocatalytic degradation of phenol in aqueous solutions using titanium dioxide. *Int. J. Chem. Mol. Nucl. Mater. Metall. Eng.* 8 (2): 136–139.

Alimoradzadeh, R., A. Assadi, S. Nasseri, and M. R. Mehrasbi. 2012. Photocatalytic degradation of 4-chlorophenol by UV/H$_2$O$_2$/NiO process in aqueous solution. *Iranian J. Environ. Health Sci. Eng.* 9 (1): 12. doi:10.1155/2014/210751.

Ameta, A., R. Ameta, and M. Ahuja. 2013. Photocatalytic degradation of methylene blue over ferric tungstate. *Sci. Revs. Chem. Commun.* 3 (3): 172–180.

An, T., J. An, H. Yang, G. Li, H. Feng, and X. Nie. 2011. Photocatalytic degradation kinetics and mechanism of antivirus drug-lamivudine in TiO$_2$ dispersion. *J. Hazard. Mater.* 197:229–236.

Araña, J., J. M. D. Rodríguez, O. G. Díaz, J. A. H. Melián, C. F. Rodríguez, and J. P. Peña. 2006. The effect of acetic acid on the photocatalytic degradation of catechol and resorcinol. *Appl. Catal. A: Gen.* 299:274–284.

Araña, J., C. F. Rodríguez, J. A. H. Melián, O. G. Díaz, and J. P. Peña. 2008. Comparative study of photocatalytic degradation mechanisms of pyrimethanil, triadimenol, and resorcinol. *J. Sol. Energy Eng.* 130 (4): 21–28. doi:10.1155/1.2969793.

Avilés-García, O., J. Espino-Valencia, R. Romero, J. L. Rico-Cerda, and R. Natividad. 2014. Oxidation of 4-chlorophenol by mesoporous titania: Effect of surface morphological characteristics. *Int. J. Photoenergy.* 2014. doi:10.1155/2014/210751.

Banerjee, A. N., S. W. Joo, and B. K. Min. 2012. Photocatalytic degradation of organic dye by sol-gel-derived gallium-doped anatase titanium oxide nanoparticles for environmental remediation. *J. Nanomater.* 2012. doi:10.1155/2012/201492.

Baran, W., E. Adamek, A. Sobczak, and A. Makowski. 2009. Photocatalytic degradation of sulfa drugs with TiO$_2$, Fe salts and TiO$_2$/FeCl$_3$ in aquatic environment-Kinetics and degradation pathway. *Appl. Catal. B: Environ.* 90 (3–4):516–525.

Boruah, P. K., P. Borthakur, G. Darabdhara, C. K. Kamaja, I. Karbhal, M. V. Shelke et al. 2016. Sunlight assisted degradation of dye molecules and reduction of toxic Cr(VI) in aqueous medium using magnetically recoverable Fe$_3$O$_4$/reduced graphene oxide nanocomposite. *RSC Adv.* 6:11049–11063.

Castañeda, C., F. Tzompantzi, R. Gómez, and H. Rojas. 2016. Enhanced photocatalytic degradation of 4-chlorophenol and 2,4-dichlorophenol on *in-situ* phosphated sol–gel TiO$_2$. *J. Chem. Technol. Biotechnol.* 91 (8): 2170–2178. doi:10.1002/jctb.4943.

Cropek, D., P. A. Kemme, O. V. Makarova, L. X. Chen, and T. Rajh. 2008. Selective photocatalytic decomposition of nitrobenzene using surface modified TiO$_2$ nanoparticles. *J. Phys. Chem. C.* 112 (22): 8311–8318.

El-Kemary, M., H. El-Shamy, and I. El-Mehasseb. 2010. Photocatalytic degradation of ciprofloxacin drug in water using ZnO nanoparticles. *J. Luminescence.* 130 (12): 2327–2331.

Fukahori, S. and T. Fujiwara. 2015. Photocatalytic decomposition behavior and reaction pathway of sulfamethazine antibiotic using TiO₂. *J. Environ. Manage.* 157:103–110.

Gandhi, J., R. Dangi, and S. Bhardwaj. 2008. Nb₂O₅ used as photocatalyst for degradation of methylene blue using solar energy. *Rasayan J. Chem.* 1 (3): 567–571.

Georgaki, I., E. Vasilaki, and N. Katsarakis. 2014. A study on the degradation of carbamazepine and ibuprofen by TiO₂ and ZnO photocatalysis upon UV/visible-light irradiation. *Am. J. Anal. Chem.* 5:518–534.

Ghorai, T. K. 2011. Photocatalytic degradation of 4-chlorophenol by CuMoO₄-doped TiO₂ nanoparticles synthesized by chemical route. *Open J. Phys. Chem.* 1 (2): 28–36.

Guzsvány, V. J., J. J. Csanádi, S. D. Lazić, and F. F. Gaál. 2009. Photocatalytic degradation of the insecticide acetamiprid on TiO₂ catalyst. *J. Braz. Chem. Soc.* 20 (1): 152–159.

Hayder, I., I. A. Qazi, M. A. Awan, M. A. Khan, and A. Turabi. 2012. Degradation and inactivation of ciprofloxacin photocatalysis using TiO₂ nanoparticles. *J. App. Pharm.* 1 (4): 487–497.

Hossein, F. and S. Reza. 2014. Photo degradation-adsorption process as a novel desulfurization method. *Adv. Chem. Eng. Res.* 3:18–26.

Hossein, F. and N. Shakiba. 2013. Application of a novel nanocomposite for desulfurization of a typical organo sulfur compound. *Iran. J. Chem. Chem. Eng.* 32 (3): 9–15.

Jafari, N., R. Kasra-Kermanshahi, M. R. Soudi, A. H. Mahvi, and S. Gharavi. 2012. Degradation of a textile reactive azo dye by a combined biological-photocatalytic process: *Candida tropicalis* JKS2-TiO₂/UV. *Iranian J. Environ. Health Sci. Eng.* 9 (1). doi:10.1186/1735-2746-9-33.

Jain, S., K. Sharma, and U. Chandrawat. 2016. Photocatalytic degradation of anti-inflammatory drug on Ti doped BaBiO₃ nanocatalyst under visible light irradiation. *Iran. J. Energy Environ.* 7 (1): 64–71.

Jallouli, N., K. Elghniji, H. Trabelsi, and M. Ksibi. 2014. Photocatalytic degradation of paracetamol on TiO₂ nanoparticles and TiO₂/cellulosic fiber under UV and sunlight irradiation. *Arab. J. Chem.* doi:10.1016/j.arabjc.2014.03.014.

Jo, W. K., Y. Won, I. Hwang, and R. J. Tayade. 2014. Enhanced photocatalytic degradation of aqueous nitrobenzene using graphitic carbon–TiO₂ composites. *Ind. Eng. Chem. Res.* 53 (9): 3455–3461.

Joshi, K. M. and V. S. Shrivastava. 2011. Degradation of alizarine red-S (a textiles dye) by photocatalysis using ZnO and TiO₂ as photocatalyst. *Int. J. Environ. Sci.* 2 (1): 8–21.

Kansal, S. K. and M. Chopra. 2012. Photocatalytic degradation of 2,6-dichlorophenol in aqueous phase using titania as a photocatalyst. *Engineering.* 4:416–420.

Kansal, S. K., N. Kaur, and S. Singh. 2009. Photocatalytic degradation of two commercial reactive dyes in aqueous phase using nanophotocatalysts. *Nanoscale Res. Lett.* 4 (7): 709–716.

Kansal, S. D. K., M. Singh, and D. Sud. 2007. Parametric optimization of photocatalytic degradation of catechol in aqueous solutions by response surface methodology. *Indian J. Chem. Technol.* 14 (2): 145–153.

Karan, D. and R. Jayaram. 2013. Photocatalytic degradation of diclofenac dewoolkar. *Int. J. Res. Chem. Environ.* 3 (3): 94–99.

Karimi, L., S. Zohoori, and M. E. Yazdanshenas. 2014. Photocatalytic degradation of azo dyes in aqueous solutions under UV irradiation using nano-strontium titanate as the nanophotocatalyst. *J. Saudi Chem. Soc.* 18 (5): 581–588.

Kaur, A., A. Umar, and S. K. Kansal. 2015. Sunlight-driven photocatalytic degradation of non-steroidal anti-inflammatory drug based on TiO₂ quantum dots. *J. Colloid Interface Sci.* 459:257–263.

Kitsiou, V., N. Filippidis, D. Mantzavinos, and I. Poulios. 2009. Heterogeneous and homogeneous photocatalytic degradation of the insecticide imidacloprid in aqueous solutions. *Appl. Catal. B: Environ.* 86:27–35.

Kometani, N., S. Inata, A. Shimokawa, and Y. Yonezawa. 2008. Photocatalytic degradation of chlorobenzene by TiO_2 in high-temperature and high-pressure water. *Int. J. Photoenergy*. 2008. doi:10.1155/2008/512170.

Kumar, J. and A. Bansal. 2010. Photocatalytic degradation of amaranth dye in aqueous solution using sol-gel coated cotton fabric. In: *Procceedings of the World Congress on Engineering and Computer Science* (II WCECS 2010), 26–28 October. San Francisco, CA.

Lam, S. M., J. C. Sin, A. Z. Abdullah, and A. R. Mohamed. 2013a. Photocatalytic degradation of resorcinol, an endocrine disrupter, by TiO_2 and ZnO suspensions. *Environ. Technol*. 34 (9): 1097–1106.

Lam, S. M., J. C. Sin, A. Z. Abdullah, and A. R. Mohamed. 2013b. Efficient photodegradation of resorcinol with Ag_2O/ZnO nanorodsheterostructure under a compact fluorescent lamp irradiation. *Chem. Papers*. 67 (10): 1277–1284.

Madhu, G. M., M. A. Raj, and K. V. Pai. 2009. Titanium oxide (TiO_2) assisted photocatalytic degradation of methylene blue. *J. Environ. Biol*. 30 (2): 259–264.

Mehrizad, A. and P. Gharbani. 2016. Study of 1-chloro-4-nitrobenzene adsorption on carbon nanofibers by experimental design. *Int. J. Nano Dimens*. 7 (1): 77–84.

Méndez-Arriaga, F., S. Esplugas, and J. Giménez. 2008. Photocatalytic degradation of non-steroidal anti-inflammatory drugs with TiO_2 and simulated solar irradiation. *Water Res*. 42 (3): 585–594.

Miguel, N., M. P. Ormad, R. Mosteo, and J. L. Ovelleiro. 2012. Photocatalytic degradation of pesticides in natural water: Effect of hydrogen peroxide. *Int. J. Photoenergy*. 2012. doi:10.1155/2012/371714.

Navarro, S., J. Fenoll, N. Velac, E. Ruiz, and G. Navarro. 2009. Photocatalytic degradation of eight pesticides in leaching water by use of ZnO under natural sunlight. *J. Hazard. Mater*. 172:1303–1310.

Oseghe, E. O., P. G. Ndungu, and S. B. Jonnalagadda. 2015. Photocatalytic degradation of 4-chloro-2-methylphenoxyacetic acid using W-doped TiO_2. *J. Photochem. Photobiol. A: Chem*. 312:96–106.

Padikkaparambil, S., B. Narayanan, Z. Yaakob, S. Viswanathan, and S. M. Tasirin. 2013. Au/TiO_2 reusable photocatalysts for dye degradation. *Int. J. Photoenergy*. 2013. doi:10.1155/2012/752605.

Pardeshi, S. K. and A. B. Patil. 2009. Solar photocatalytic degradation of resorcinol a model endocrine disrupter in water using zinc oxide. *J. Hazard. Mater*. 163 (1): 403–409.

Rahmani, A. R., M. T. Samadi, and M. A. Enayati. 2008. Investigation of photocatalytic degradation of phenol by UV/TiO_2 process in aquatic solutions. *J. Res. Health Sci*. 8 (2): 55–60.

Reddy, A. K., P. V. L. Reddy, V. M. Sharma, B. Srinivas, V. D. Kumari, and M. Subrahmanyam. 2010. Photocatalytic degradation of isoproturon pesticide on C, N and S doped TiO_2. *J. Water Resour. Prot*. 2:235–244.

Roy, T. K. and N. K. Mondal. 2014. Photocatalytic degradation of Congo red dye on thermally activated zinc oxide. *Int. J. Sci. Res. Environ. Sci*. 2 (12): 457–469.

Saggioro, E. M., A. S. Oliveira, T. Pavesi, C. G. Maia, L. F. V. Ferreira, and J. C. Moreira. 2011. Use of titanium dioxide photocatalysis on the remediation of model textile wastewaters containing azo dyes. *Molecules*. 16:10370–10386.

Samsudin, E. M., S. N. Goh, T. Y. Wu, T. T. Ling, S. B. A. Hamid, and J. C. Juan. 2015. Evaluation on the photocatalytic degradation activity of reactive blue 4 using pure anatase nano-TiO_2. *Sains Malays*. 44 (7): 1011–1019.

Sobana, N., K. Thirumalai, and M. Swaminathan. 2016. Kinetics of solar light assisted degradation of direct red 23 on activated carbon-loaded zinc oxide and influence of operational parameters. *Canad. Chem. Trans*. 4 (1): 77–89.

urolia, P. K. and R. V. Jasra. 2016. Photocatalytic degradation of *p*-nitrotoluene (PNT) using TiO$_2$-modified silver-exchanged NaY zeolite: Kinetic study and identification of mineralization pathway. *Desalin. Water Treat.* 57 (46). doi:10.1080/19443994.2015.1125798.

achikawa, T., Y. Takai, S. Tojo, M. Fujitsuka, and T. Majima. 2006. Probing the surface adsorption and photocatalytic degradation of catechols on TiO$_2$ by solid-state NMR spectroscopy. *Langmuir.* 22 (3): 893–896.

erma, A., H. Kaur, and D. Dixit. 2013. Photocatalytic, sonolytic and sonophotocatalytic degradation of 4-chloro-2-nitro phenol. *Arch. Environ. Protect.* 39 (2): 17–28.

erma, A., Poonam, and D. Dixit. 2012. Photocatalytic degradability of insecticide chlorpyrifos over UV irradiated titanium dioxide in aqueous phase. *Int. J. Environ. Sci.* 3 (2): 743–755.

Vang, W., Y. Huang, and S. Yang. 2010. Photocatalytic degradation of nitrobenzene wastewater with H$_3$PW$_{12}$O$_{40}$/TiO$_2$. In: *Proceedings of the International Conference on Mechanic Automation and Control Engineering (MACE)*, 26–28 June. 1303–1305. Wuhan, China.

hang, G., C. Ni, L. Liu, G. Zhao, F. Fina, and J. T. S. Irvine. 2015. Macro-mesoporous resorcinol–formaldehyde polymer resins as amorphous metal-free visible light photocatalysts. *J. Mater. Chem. A.* 3:15413–15419.

hao, D., C. Chen, Y. Wang, W. Ma, J. Zhao, T. Rajh, and L. Zang. 2008. Enhanced photocatalytic degradation of dye pollutants under visible irradiation on Al(III)-modified TiO$_2$: Structure, interaction, and interfacial electron transfer. *Environ. Sci. Technol.* 42:308–314.

14 Reduction of Carbon Dioxide

14.1 INTRODUCTION

Carbon dioxide is one of the principal greenhouse gases. Other greenhouse gases are methane, chlorofluorocarbons (CFCs), and nitrogen oxides (NO_x). Among these, about a 72% share of the totally emitted greenhouse gases is carbon dioxide; it is considered the main culprit of global warming.

Over the last few decades, the concentration of CO_2 has been increasing in the atmosphere due to ever-increasing human activity, which has added to the greenhouse effect. On the other hand, the consumption of fossil fuels is rapidly increasing worldwide year after year because of the demands for various activities. Apart from excessive use of fossil fuels, this increasing amount of carbon dioxide has also been supported by increasing population, deforestation, and so on. The increasing level of CO_2 concentration in the atmosphere has become the most serious environmental problem and reached an alarming situation due to global warming (Liao et al. 2015).

This problem can be solved by photoreduction of carbon dioxide. Here, photocatalysis enters the scene. Photocatalytic reduction of CO_2 to synthetic organic fuels such as formaldehyde, methanol, formic acid, acetic acid, methane, and so on, will provide a solution to the energy crisis as it will give us alternate fuels, which can be burnt into fuel cells to generate electricity. It also helps us in putting a check on the ever-increasing amount of carbon dioxide in the atmosphere.

Photocatalytic reduction of CO_2 to valuable chemicals like methanol offers a promising way for clean, low cost, and eco-friendly production of fuels using solar energy (Ola et al. 2013).

Photosynthesis is a well-established natural process, which has been doing this job from time immemorial. Efforts should be made to mimic such an exciting reaction under laboratory conditions. Photocatalytic reduction of carbon dioxide requires light energy. Solar energy is not only a harmless, low-cost, and abundantly available source of energy but is also associated with some limitations like lower radiation flux, daily and seasonal fluctuations, and limited photoperiod. Converting CO_2 to some valuable hydrocarbons is one of the best solutions to both problems: global warming and energy scarcity. Photocatalytic reduction is one such process that can produce a transportable hydrogen-based fuel and also reduce the levels of CO_2 in the atmosphere.

Here, the hydrogen atom abstraction reaction is of importance. The hydrogen atom generated from splitting of the H–OH bond in the photodissociation step can be used either to form hydrogen or to react with another molecule like carbon dioxide to yield alternative energy-rich molecules, such as formic acid, formaldehyde, methanol, and methane. Such reactions might prove to be a more feasible way of storing energy than

221

actually producing molecular hydrogen, and this will also provide better utilization of carbon dioxide to generate synthetic fuels. Photocatalytic reduction of carbon dioxide may proceed in steps of two electron reduction. Complete reduction will require the transfer of eight electrons, which is not favorable energetically. These steps are

$$CO_2 \text{ (aq.)} + 2\ H^+ + 2\ e^- \xrightarrow[\text{Photocatalyst}]{h\nu} HCOOH \qquad (14.1)$$

$$HCOOH + 2\ H^+ + 2\ e^- \xrightarrow[\text{Photocatalyst}]{h\nu} HCHO + H_2O \qquad (14.2)$$

$$HCHO + 2\ H^+ + 2\ e^- \xrightarrow[\text{Photocatalyst}]{h\nu} CH_3OH \qquad (14.3)$$

$$CH_3OH + 2\ H^+ + 2\ e^- \xrightarrow[\text{Photocatalyst}]{h\nu} CH_4 + H_2O \qquad (14.4)$$

The reaction between carbon dioxide and water to form organic compounds like formic acid, formaldehyde, methanol, and methane involves two, four, six, and eight electron transfers, respectively.

Carbon dioxide can also be reduced electrochemically, but this process has certain drawbacks like deactivation of electrodes within a short time, which is due to the deposition of poisoning species like adsorbed organic compounds on the electrode. Inoue et al. (1979) observed that rapid reduction of CO_2 needs an overpotential of at least 0.6 V. Aresta (2000) and Olah et al. (2009) were also of the opinion that electrochemical reduction was energy intensive with slow kinetics. Mikkelsen et al. (2010) have highlighted the difference between photocatalysis and electrochemical reduction. On the contrary, photocatalysis is more energy efficient as compared to other reduction methods because it is low cost and eco-friendly. It may provide energy sources by conversion of CO_2 into useful products, such as methanol and methane (Indrakanti et al. 2009; Jiang et al. 2010).

Photocatalytic CO_2 reduction reaction may also be termed artificial photosynthesis due to its resemblance to photosynthesis due to having the same reactants (CO_2 and H_2O) and the same source of energy (sunlight). Although photosynthesis looks very simple, mechanistically it is a very complex process resulting in glucose $(CH_2O)_6$. Another similarity between these two processes is that both have similar reaction steps. In both processes, first water is oxidized to O_2 with electromagnetic irradiation and second CO_2 is reduced with produced protons (in the form of nicotinamide adenine dinucleotide phosphate [NADPH] in the case of photosynthesis).

Various methods have been employed to date to reduce CO_2 into less harmful compounds. Researchers have also used nanomaterials for the purpose of carbon dioxide hydrogenation. Nanosized semiconductors play a very significant role in the depletion of CO_2, generating different hydrocarbons and liquid fuel, which in turn will also reduce or compensate our energy demands for carbon-containing energy feedstock (fossil fuels), which are limited and will be exhausted in a few more decades or so.

CO_2 can only absorb photons below 200 nm, which belong to the deep ultraviolet (UV) spectral region.

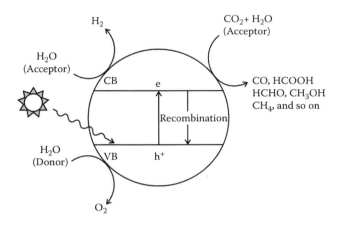

FIGURE 14.1   The role of photocatalysis.

$$2\,CO_2 + h\nu(< 200\ nm) \longrightarrow 2\,CO + O_2 \ \Delta H = +257\ kJ/mol \qquad (14.5)$$

Photocatalysis over a semiconductor is initiated by the absorption of photons in light with equal or higher energy than the band gap energy of that semiconductor. The excitation of the electron from the valence band (VB) to the conduction band (CB) results in an electron vacancy called a "hole" in the VB and thus, the electron–hole pair is generated in the semiconductor. These electrons can be utilized for the reduction of CO_2 into HCOOH, HCHO, CH_3OH, CH_4, and so on, or water into hydrogen, while the hole will oxidize water into oxygen (Figure 14.1).

The field of photocatalytic reduction of carbon dioxide has been excellently reviewed by various researchers from time to time. Some of them include Koči et al. (2008), Jacob et al. (2011), Ameta et al. (2013), Habisreutinger et al. (2013), Wang et al. (2014), Ola and Maroto-Valer (2015), Xie et al. (2016), and so on.

Woolerton et al. (2010) studied a hybrid enzyme–nanoparticle (NP) system for the reduction of CO_2 to CO using visible (VIS) light as the energy source. Carbon monoxide dehydrogenase (CODH) was attached to TiO_2 NPs, and an aqueous dispersion of modified TiO_2 NPs was sensitized using an Ru photosensitizer, which produced CO at a rate of 250 μmol CO/(g TiO_2)/h, when illuminated with VIS light at pH 6 and 20°C. Habisreutinger et al. (2013) reported the photocatalytic reduction of CO_2 using different semiconductors (metal oxide) like titanium dioxide, oxynitride, sulfide, and phosphide semiconductors.

14.2   PHOTOCATALYTIC REDUCTION OF CO_2

Many conventional semiconductors have been tested for the photoreduction of CO_2, such as TiO_2, CdS, Fe_2O_3, WO_3, ZnO, and ZrO_2 and their various combinations. A good photocatalyst should have high catalytic activity along with high stability and durability with a low tendency to photocorrode.

14.2.1 Titanium Dioxide

Titanium dioxide is particularly important due to its unique properties like low cost, nontoxicity, and fair stability as compared to other semiconductors. But it has certain limitations such that it can only be activated by UV light because of a large band gap, which represents only 2%–5% of sunlight. Three phases of TiO_2 exist in nature, that is, rutile, anatase, and brookite phases. The anatase and rutile phases have definite band gap energy of about 3.2 and 3.0 eV, respectively. Inoue et al. (1979) reported photocatalytic reduction of CO_2 to organic compounds like HCOOH, CH_3OH, and HCHO by irradiation of CO_2-saturated aqueous solution using an Xe lamp in the presence of various semiconductors such as TiO_2, ZnO, and WO_3. The best photoactivity can be achieved by combining anatase with a slight amount of rutile (Das and Daud 2014).

The photocatalytic activity of TiO_2 can be enhanced by different techniques: (1) controlling morphology of the particle by decreasing its size to nanolevel (synthesizing titania nanotube [TNT], nanorod, etc.); (2) supporting on a solid like aluminosilicates; (3) doping with some metallic or nonmetallic elements; and (4) using a photosensitizer.

Ishitani et al. (1993) observed that the deposition of a metal on TiO_2 enhanced photocatalytic reduction of carbon dioxide to methane and/or acetic acid. They also reported that product distribution was dependent on the kind of metal used on the surface of TiO_2.

The reduction of carbon dioxide in gas streams by the UV/TiO_2 process was studied by Ku et al. (2003). The effect of various parameters like retention time, humidity contents, initial carbon dioxide concentrations, and UV light intensities was observed. They used TiO_2 particles as photocatalyst and impregnated on the nonwoven fabric textile for the reduction of carbon dioxide in gas streams. The primary products from carbon dioxide reduction were detected as methane and methanol. The reduction of carbon dioxide rate was found to increase linearly with its increasing concentration. It also increased with increasing humidity content present in the gas stream up to humidity less than 55%, but then it decreased on increasing humidity further.

Dey et al. (2004) studied photocatalytic reduction of CO_2 to methane in CO_2-saturated aqueous solution in the presence of TiO_2 photocatalyst (0.1%, w/v) as a suspension using 350 nm light. The CO_2 methanation rate was very much enhanced in the presence of 2-propanol as a hole scavenger. Photocatalytic reduction of methanol in an N_2-purged system was also carried out but no methane was found as a product in the presence of TiO_2 without 2-propanol. However, the yield of methane was quite low even in the presence of 2-propanol. It was shown that the generation of CH_4 from CO_2 does not proceed via methanol as an intermediate. Methane was produced during photolysis of TiO_2 suspension in the presence of 2-propanol in an aerated system also. In an O_2-saturated system, the yield of methane was lower as compared to an aerated system, whereas the CO_2 yield was relatively higher.

Photocatalytic reduction of CO_2 by copper-doped titania catalysts has been reported by Slamet et al. (2005). The photocatalysts were prepared with different copper species [Cu(0), Cu(I), Cu(II)] by an improved impregnation method, where copper nitrate was doped into TiO_2 Degussa P25. Copper was present on the surface f the catalyst and the grain size of copper–titania catalysts was uniform (23 nm). he activation energy for Degussa P25 and 3% CuO/TiO_2 was found to be +26 and 12 kJ/mol, respectively, which suggests that the rate-limiting step was the desorpion event, while lower value for 3% CuO/TiO_2 suggested a catalytic role of the coper species enhancing methanol production.

CO_2 could be transformed into hydrocarbons when kept in contact with water apor and catalysts under UV irradiation. Tan et al. (2006) used heterogeneous phoocatalysis using the pellet form of the catalyst instead of immobilized catalysts on olid substrates. CO_2 was mixed with water vapor in the saturation state and then disharged into a quartz reactor containing porous TiO_2 pellets. It was illuminated by arious UV lamps of different wavelengths for 48 hours. The total yield of methane vas approximately 200 ppm on using UVC (253.7 nm) light. This is a good reducion yield as compared to that obtained using immobilized catalysts. CO and H_2 were lso detected as minor products. The use of UVA (365 nm) resulted in a decrease in he product yields.

Photoreduction of CO_2 was investigated by Kočí et al. (2009) using TiO_2 NPs with ize in the range 4.5–29 nm. Synthesis of methanol from CO_2 by a photocatalytic eduction process over copper-doped TiO_2 has also been investigated (Slamet et al. 009).

Efficient solar conversion of carbon dioxide and water vapor to methane and other ydrocarbons was achieved by Varghese et al. (2009). They used nitrogen-doped titaia nanotube arrays (TNAs) with a small thickness of wall to facilitate effective carier transfer to the adsorbing species, surface-loaded with nanodimensional islands of ocatalysts platinum and/or copper. Hydrogen and carbon monoxide were also detected s intermediate reaction products. Their concentrations were found to be dependent pon the nature of the cocatalysts on the nanotube array surface. The production rate f hydrocarbon was obtained as 111 ppm/cm²/h, at light intensity 100 mW/cm² on loadng nanotube array samples with both Cu and Pt NPs. This rate of CO_2 to hydrocarbon vith sunlight was about 20 times higher than reported earlier under UV illumination.

A photocatalytic system for converting carbon dioxide into carbon monoxide vas fabricated by Jacob et al. (2011). Thin films of the photocatalyst were prepared y them at low temperature using spray coating. Glass substrates were used with n active area of 100 cm². Polyethyleneterphthalate (PET), polyethylenenaphtalate PEN), and polyethylene (PE) with this area were also used as flexible substrates. 'rocesses involving photoinduced degradation of polymers, decomposition of catayst surfactant, and inactivation of the catalyst efficiency by carbonaceous residues vere also discussed.

Cybula et al. (2012) were able to photoconvert carbon dioxide and water vapor ffectively into methane in the presence of pure or modified TiO_2 under UV–VIS rradiation. This photoconversion was carried out in the gas phase with a perforated 'iO₂-coated support. The effect of different parameters of TiO_2 immobilization on hotocatalytic efficiency like time and temperature of the drying step and the phoocatalyst amount was observed. The effect of TiO_2 loading with Ag/Au NPs was lso studied. The major photoreduction product of reduction of CO_2 was found to be nethane. The highest methane production was observed over Au–TiO_2 photocatayst. About 503 ppm of methane was formed in 1 hour of UV–VIS irradiation.

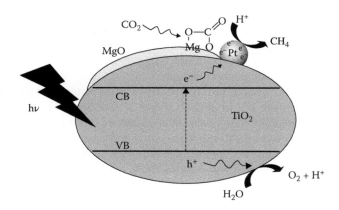

FIGURE 14.2   Photocatalytic reduction of CO_2 on Pt–TiO_2 with added MgO. (Adapted from Xie, S. et al., *Chem. Commun.*, 49, 2451–2453, 2013. With permission.)

It was interesting to note that the photocatalytic activity of Pt–TiO_2 loaded for reduction of CO_2 with H_2O to CH_4 was significantly enhanced simply by adding MgO (Xie et al. 2013). A positive correlation was observed between CH_4 formation activity and basicity. It was concluded that an interface between TiO_2, Pt, and MgO in the ternary nanocomposite has played a crucial role in the photocatalytic reduction of CO (Figure 14.2).

Nickel-based TiO_2 photocatalysts were immobilized onto quartz plate by Ola and Maroto-Valer (2012) with a high ratio of illuminated surface area of the catalyst for the reduction of CO_2 to fuels under VIS light irradiation. The improved conversion efficiency was observed in the presence of the metal nickel, which served as an electron traps suppressing electron–hole recombination. It resulted in effective charge separation and consequently CO_2 reduction.

Qin et al. (2013) used a bifunctionalized TiO_2 film containing a dye-sensitized zone and a catalysis zone for the photocatalytic reduction of CO_2 under VIS light. Charge separation was possible with the transfer of electrons to the catalysis zone and positive charge to the anode. Formic acid, formaldehyde, and methanol were obtained as the products through the transfer of electrons in the CBs of TiO_2. Reduction of CO_2 and evolution of O_2 took place in separated solutions, and this process will avoid reduction products being oxidized back by the anode.

Photocatalytic reduction of carbon dioxide was carried out using titanium(IV) oxide by Murakami et al. (2013). Methanol was detected as the main product with trace amounts of formic acid, carbon monoxide, methane, and hydrogen. As-prepared decahedral-shaped anatase TiO_2 with larger {011} and smaller {001} exposed crystal faces showed greater CH_3OH generation as compared to commercial anatase TiO_2 powder. Photodeposition of silver and gold NPs on the decahedral-shaped anatase TiO_2 induced an increase in CH_3OH production because the deposited metal particles work as reductive sites for the multielectron reduction of CO_2.

Anatase TiO_2 nanosheet porous films were prepared by the calcination of orthorhombic titanic acid films at 400°C (Li et al. 2014). They reported excellent photocatalytic activity for the photoreduction of CO_2 to methane. Pt NPs were loaded uniformly with an average size of 3–4 nm on TiO_2 porous films by the photoreduction method to further enhance the photocatalytic activity. It was found that the loading of Pt increased the light absorption ability of the porous film and improved its efficiency. The conversion yield of CO_2 to methane on Pt/TiO_2 film was obtained as 20.51 ppm/h/cm².

Self-organized V–N codoped TNAs with various doping amounts were synthesized by Lu et al. (2014) through anodizing in association with hydrothermal treatment. Effect of V–N codoping on the morphologies, phase structures, and photoelectrochemical properties of the TNAs films were also studied. The codoped TiO_2 photocatalysts show significantly enhanced photocatalytic activity for CO_2 photoreduction to methane under UV illumination.

Nhat et al. (2015) prepared molybdenum-doped hydrothermal TNTs (Mo-doped TNTs) and observed the photocatalytic activity on the reduction of NO_2 and CO_2 gases. These photocatalysts were synthesized by the hydrothermal method while Mo metal species were doped at different process steps in three fabrication methods: (1) hydrothermal, (2) precipitation, and (3) impregnation. It was observed that doping with Mo was successful by precipitation and impregnation, but not by the conventional hydrothermal method. It was also revealed that this doping did not considerably affect the morphology, microstructure, and crystalline structure. The doping of Mo sharply declined the oxidation ability of TNTs in NO_2 photocatalytic reaction but enhanced its reduction ability. In CO_2 reduction also, a higher reduction ability of Mo-doped TNTs was observed producing higher yields of methane and carbon monoxide. Mo-doped TNTs prepared by the precipitation method showed the highest reduction ability, which were attributed to the chemical oxidation states of Mo^{4+} and Mo^{5+}.

Zhang et al. (2015) observed the effect of source of nitrogen dopant on the photocatalytic properties of nitrogen-doped titanium dioxide. They prepared well-crystallized one-dimensional TiO_2 nanorod arrays via a hydrothermal treatment using hydrazine and ammonia as the source of nitrogen. Significant selectivity of the reduced products was observed with TiO_2 in carbon dioxide photocatalytic reduction under VIS light illumination. CH_4 was obtained as the main product with N_2H_4-doped TiO_2, while CO was the main product with NH_3-doped TiO_2.

Photocatalytic CO_2 reduction with H_2 over nitrogen-doped TiO_2 nanocatalyst was observed by Tahir et al. (2015) using a monolith photoreactor. N-doped TiO_2 nanocatalyst was synthesized by the sol–gel method and dip-coated over the monolith channels. Highly crystalline and anatase phase TiO_2 was obtained in the N-doped TiO_2 samples with increased surface area and comparatively reduced crystallite size. This nanocatalyst had excellent photoactivity for selective CO_2 reduction to CO. N-doped TiO_2 (3 wt. %) was found optimal and gives a rate of CO yield of 56.30 µmol(g.cat h)$^{-1}$ with selectivity of 96.3% at CO_2/H_2 feed ratio 1 and feed flow rate of 20 mL/min. Its performance for selective and continuous CO production was 4.7-fold higher than undoped TiO_2. The significant activity was considered to be due to a hindered charge recombination rate because of N doping. The N-doped TiO_2 gave prolonged stability for continuous CO and CH_4 production.

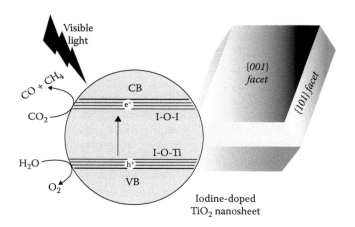

FIGURE 14.3   Role of iodine-doped TiO$_2$. (Adapted from He, Z. et al., *RSC Adv.*, 6, 23134–23140, 2016. With permission.)

Kang et al. (2015) carried out photocatalytic reduction of carbon dioxide by hydrous hydrazine (N$_2$H$_4$·H$_2$O) over Au–Cu alloy NPs supported on SrTiO$_3$/TiO$_2$ coaxial nanotube arrays. High efficiency of as-prepared NPs was attributed to the mixed catalytic effects of Au and Cu in the bimetallic alloy and rapid electron transfer in SrTiO$_3$/TiO$_2$ nanostructure.

Iodine-doped titanium dioxide nanosheets with high exposed {001} facets (IFTO) were synthesized by He et al. (2016) via a two-step hydrothermal treatment followed by calcination at 350°C. IFTO was found to be more active than surface-fluorinated TiO$_2$ nanosheets or iodine-doped TiO$_2$ NPs in the gaseous reduction of CO$_2$ with water vapor. It was observed that the photocatalytic activity was retained over five successive cycles. Coexisting I–O–Ti and I–O–I structures in the IFTO TiO$_2$ lattice were the main reason for VIS light absorption and the high percentage of exposed {001} facets that enhanced H$_2$O oxidation (Figure 14.3). This surface fluorination promoted the formation of unsaturated Ti atoms, which helped in the reduction of CO$_2$ to CO$_2^-$, and it also retards the recombination of photogenerated electron–hole pairs.

14.2.2   Other Metal Oxides

Besides TiO$_2$, other metal oxide semiconductors can be also used as photocatalysts for the reduction of CO$_2$. The photochemical reduction of CO$_2$ to formic acid can be used as a method to produce a transportable hydrogen-based fuel on one hand and it may reduce levels of CO$_2$ in the atmosphere on the other hand. There is a necessity for a sacrificial electron donor, normally a tertiary amine. A new strategy for coupling the photochemical reduction of CO$_2$ to photochemical water splitting was reported by Richardson et al. (2011). This technique makes the reducing agent recyclable. This has two potential advantages: (1) it permits two redox reactions (CO$_2$ reduction and water oxidation) to be carried and allows this (2) under mutually incompatible conditions.

The Cu_2O/SiC photocatalyst was obtained from SiC NPs modified by Cu_2O (Li et al. 2011). They observed photocatalytic activities for the reduction of CO_2 to CH_3OH under VIS light irradiation. Besides a small quantity of 6H–SiC, SiC NPs mainly consisted of 3C–SiC. The band gaps of SiC and Cu_2O were determined as 1.95 and 2.23 eV from UV–VIS spectra, respectively. It was observed that Cu_2O modification can enhance the photocatalytic performance of SiC NPs. The largest yields of methanol on SiC, Cu_2O, and Cu_2O/SiC photocatalysts were obtained under VIS light irradiation as 153, 104, and 191 μmol/g, respectively.

Khodadadi-Moghaddam (2014) reported photocatalytic reduction of carbon dioxide to formaldehyde on four different semiconductor photocatalysts (FeS, FeS/FeS_2, NiO, and TiO_2). The reaction was carried out in a continuous flow of CO_2 gas bubbled into the reactor. Sulfide ion was used as a hole scavenger. TiO_2 has greater photocatalytic activity as compared to the other photocatalysts. Addition of carbonate ion increases concentration of formaldehyde about two times. The yield of the formaldehyde was found to decrease with decreasing concentration of the sulfide ion and the pH of the reaction mixture. Carbon dioxide can be reduced to formaldehyde on semiconductor photocatalyst under UV radiation.

De Brito et al. (2014) reported photoeletrocatalytic reduction of CO_2 dissolved in slightly alkaline solution at Cu/Cu_2O electrode. This electrode was prepared by electrochemical deposition of a copper foil. The effect of a supporting electrolyte was also observed, and 80 ppm of methanol was generated in 0.1 mol/L $Na_2CO_3/NaHCO_3$. Maximum photoconversion of CO_2/CH_3OH was obtained at pH 8. They proposed that the stabilization of Cu(I) on the electrodic material seems to be the key to success in such a photoelectrochemical reduction.

Sastre et al. (2014) reported that nickel supported on silica–alumina is an efficient and reusable photocatalyst for the reduction of CO_2 to methane by H_2 with selectivity above 95% at CO_2 conversion over 90%. Although NiO behaves in a similar manner, it undergoes a gradual deactivation upon reuse. It was observed that about 26% of the photocatalytic activity of Ni/silica–alumina was derived from the VIS light photoresponse under solar light (Figure 14.4).

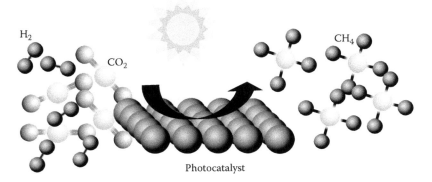

FIGURE 14.4 Reduction of CO_2 over nickel-supported silica–alumina. (Adapted from Sastre, F. et al., *J. Am. Chem. Soc.*, 136, 6798–6801, 2014. With permission.)

Alves et al. (2015) synthesized hybrid materials by combining the activity of copper and the conductivity of reduced graphene oxide (rGO) and used these for reducing carbon dioxide. These Cu NPs (Cu-NPs) coated on rGO exhibited higher current density and lower overpotential in comparison to other tested copper-based electrodes.

Yin et al. (2015) prepared amorphous copper oxide nanoclusters and used it as an efficient electrocatalyst for the reduction of carbon dioxide to carbon monoxide. Cu(II) nanoclusters act as efficient cocatalyts for CO_2 photoreduction when these are grafted onto the surface of a semiconductor, such as niobate ($Nb_3O^-_8$) nanosheets, which act as a light harvester. They reported that electrons were extracted from water to produce oxygen and then reduce CO_2. It was also confirmed that excited holes in the VB of niobate nanosheets react with water, and the excited electrons in the CB were injected into the Cu(II) nanoclusters through the interface resulting into reduction of CO_2 into CO. The Cu(II) nanocluster-grafted niobate nanosheets can be easily synthesized by a wet chemical method.

The performance of a highly efficient two-dimensional (2D) fluidized bed catalytic photoreactor with Cu/TiO_2, Ru/TiO_2, and Pd/TiO_2 was observed by Vaiano et al. (2015). Methane was detected as the main product, with very low amounts of CO of course. Out of these, Pd/TiO_2 photocatalysts showed the best performance. More than four times the CH_4 photoproduction was achieved with this as compared with TiO_2.

A facile preparation of uniform urchin-like $NiCo_2O_4$ microspheres was reported by Wang et al. (2015). They used it as an efficient and stable cocatalyst for the photocatalytic reduction of CO_2. A combined solvothermal–calcination strategy was used for synthesizing the $NiCo_2O_4$ material. The $NiCo_2O_4$ effectively promoted the deoxygenative reduction of CO_2 to CO by more than 20 times under mild reaction conditions; of course, by incorporation of a VIS light photosensitizer. Various reaction parameters were optimized and the stability and reusability of $NiCo_2O_4$ were confirmed.

Prasad et al. (2016) synthesized Bi_2S_3/CdS by hydrothermal reaction as an effective VIS light responsive photocatalyst and used it for CO_2 reduction into methanol. The yield of methanol was found to increase with increasing CdS concentration in Bi_2S_3/CdS. The highest yield was obtained for 45 wt.% of CdS.

Carbon dioxide was reduced photocatalytically to produce methanol and ethanol by Li et al. (2016) in aqueous suspension of CuO-loaded titania powders. They used Na_2SO_3 as a hole scavenger. The photocatalysts were prepared by an impregnation method using P25 (Degussa) as support. The maximum yields of methanol and ethanol were 12.5 and 27.1 µmol/g catalyst, respectively, with copper oxide loading (3 wt.%) under 6 hours of UV illumination. The redistribution of photogenerated charge carriers in CuO/TiO_2 assisted the electron trapping, which prohibited the recombination of electrons and holes. As a result, photoefficiency was found to increase significantly. The addition of Na_2SO_3 was found to promote the formation of ethanol.

14.2.3　NONOXIDES

Nonoxide photocatalysts have low band gap energy compared to oxide photocatalysts and as a result, they respond in the VIS light region and show high photocatalytic activity. Some such photocatalysts are the following: (1) metal sulfide semiconductors such as CdS and ZnS, (2) metal phosphide semiconductors such as GaP and InP, nd (3) others such as AgCl, AgBr, GaAs, and p-Si. Metal sulfides have relatively high CB states and are therefore more appropriate for better solar response than metal oxide semiconductors, which are facilitated by the higher VB states consisting of S 3p orbitals.

4.2.4 GRAPHENE-BASED SEMICONDUCTORS

Tan et al. (2013) used rGO–TiO_2 hybrid nanocrystals prepared by a simple solvothermal synthetic route. TiO_2 (anatase) particles (~12 nm) were uniformly dispersed on the GO sheet. As-prepared rGO–TiO_2 nanocomposites exhibited superior photocatalytic activity in the reduction of CO_2 as compared to GO and pure anatase also. The intimate contact between TiO_2 and rGO was considered responsible for the transfer of photogenerated electrons on TiO_2 to rGO, leading to an effective charge antirecombination.

Graphene–TiO_2 was obtained by the reduction of graphite oxide by the hydrothermal method (Zhang et al. 2014). Carbon dioxide was reduced to methanol and formic acid. It was observed that the graphene loading affects the absorption of light in the VIS light region. Its larger surface area also improved the catalytic activity. The largest yield of methanol and formic acid obtained was 160 and 150 μmol/g, respectively, with .5% graphene loading. Graphene loading enhanced photocatalytic performance up to a limit and after a certain limit, decrease in the reduction efficiency was observed.

A rapid one-pot microwave process was used by Shown et al. (2014) to prepare the GO decorated with Cu-NP, Cu/GO hybrids with various Cu contents and it has been used for photocatalytic CO_2 reduction under VIS light. Metallic Cu-NPs (~4–5 nm in size) in this GO hybrid significantly enhanced the photocatalytic activity of GO, basically through the suppression of electron–hole pair recombination, reduction of GO's band gap, and modification of its work function. It was also indicated that there was a charge transfer from GO to Cu. A strong interaction was observed between the metal content of the Cu/GO hybrids and the rates of formation and selectivity of the products. They observed about 60 times enhancement in CO_2 to fuel catalytic efficiency using Cu/GO (10 wt.% Cu) compared with that using pristine GO.

TiO_2–rGO nanocomposites were prepared by Liu et al. (2016) via a simple chemical method by using GO and TiO_2 NPs as starting materials. The morphologies and structural properties of the as-prepared composites were characterized by different techniques. TiO_2–rGO nanocomposites exhibited great photocatalytic activity toward the reduction of CO_2 into CH_4 (2.10 μmol/g/h) and CH_3OH (2.20 μmol/g/h), which was attributed to the synergistic effect between TiO_2 and graphene.

4.2.5 OTHERS

Ag-loaded Ga_2O_3 (Ag/Ga_2O_3) photocatalysts were prepared by Yamamoto et al. (2015) via an impregnation method and used for the photocatalytic reduction of CO_2 with water. CO, H_2, and O_2 were detected as products. It was revealed that around -nm-sized Ag clusters were formed predominantly in an active Ag/Ga_2O_3 sample. The presence of partially oxidized large Ag particles decreased its activity. This was due to the fact that small Ag clusters accepted more electrons in the d-orbitals as a result of the strong interaction with the Ga_2O_3 surface. The monodentate bicarbonate and/or the bidentate carbonate species changed to bidentate formate species unde UV light irradiation. It was not formed by the plasmonic excitation of the Ag NPs b by the photoexcitation of the Ga_2O_3 semiconductor. The process of formation wa promoted at the perimeter of the Ag clusters on the Ga_2O_3 surface by the effectiv separation of electron–hole pairs.

A ruthenium trinuclear polyazine complex was synthesized by Kumar et a (2014). They immobilized it through complexation to a GO support containing phen anthroline ligands (GO-phen). The photocatalyst reduced CO_2 to methanol in th presence of a 20 W white cold light-emitting diode (LED) flood light, in a dimeth formamide–water mixture containing triethylamine as a reductive quencher. Th yield of methanol was found to be 3977.57 ± 5.60 µmol/g catalyst after 48-hou illumination. This photocatalyst exhibited a higher photocatalytic activity than G(alone, which gave a yield of 2201.40 ± 8.76 µmol/g catalyst. The catalyst was easil recovered at the end of the reaction and was reused for four subsequent runs withou any significant loss of catalytic activity. In addition, no leaching of the metal/ligan was detected during the reaction.

A hybrid enzymatic/photocatalytic approach was proposed by Aresta et a (2014) for the reduction of CO_2 into methanol. They reported that production o 1 mol of CH_3OH from CO_2 required three enzymes and the consumption of 3 m(of nicotinamide adenine dinucleotide hydride (NADH). The cofactor NADH wa regenerated from nicotinamide adenine dinucleotide (NAD+) using VIS ligh active TiO_2-based photocatalysts. The regeneration efficiency of the process wa found to be enhanced by using a Rh(III)-complex for facilitating the electro and hydride transfer from some H-donor like water or water–glycerol solutio to NAD+. Production of 100–1000 mol of CH_3OH from 1 mol of NADH wa observed.

Reli et al. (2014) assessed the effect of a reaction media on CO_2 photocatalyti reduction yields in the presence of ZnS NPs deposited on montmorillonite (ZnS MMT). The four different reaction media used were NaOH, NaOH + Na_2SO_3 (1:1' NH_4OH, and NH_4OH + Na_2SO_3 (1:1). It was observed that the pure sodium hydroxid was better than ammonium hydroxide for the yields of product in both the phase: that is, gas phase (CH_4 and CO) and liquid phase (CH_3OH). The addition of Na_2SC was found to improve methanol yields due to the prevention of oxidation of methanc back to carbon dioxide, but gas phase yields were decreased by this addition.

Bonin et al. (2014) observed the photochemical catalytic reduction of CO_2 in a organic solvent using iron(0) porphyrins as homogeneous molecular catalysts unde VIS light irradiation. This photochemical process led mainly to the production o CO, while H_2 was obtained as a minor product. High catalytic selectivity for C(formation and turnover numbers (TONs) up to 30 were obtained. Addition of a wea acid led to the rapid deactivation of the catalyst. They observed lower performanc and higher proportion of H_2 with unmodified tetraphenylporphyrin as catalyst, whicl indicates that the reduction pathways are different.

Hong et al. (2014) constructed self-assembly of carbon nitride (C_3N_4) and lay ered double hydroxide (LDH) by electrostatic interaction. It was reported that pris tine nitrate-intercalated Mg–Al–LDH was turned to carbonate LDH through anio exchange during the photoreduction of CO_2 in aqueous solution. They reported tha carbonate anions enriched in the interlayer of LDH had a remarkably high reduction efficiency to CH_4 in the presence of a C_3N_4 photoabsorber and Pd cocatalyst.

GaP/TiO_2 composites exhibited a remarkable photocatalytic activity for CO_2 reduction in the presence of water vapor, where methane was obtained as the product (Marcì et al. 2014). The photocatalytic activity of the composite was observed up to 1:10 mass ratio, and it was found that activity increases on decreasing GaP:TiO_2 mass ratio. They attributed photocatalytic activity of the composite to the band structures of the solids as well as to the efficient charge transfer between the GaP and TiO_2 heterojunction.

Ono et al. (2014) synthesized four different rhenium complexes, Re(bpy-R)$(CO)_3$Cl (bpy = 2,2'-bipyridine and R = H, CH_3, COOH, or CN) as photocatalysts and used them for the reduction of CO_2 to CO. The effect of substituent groups on the absorption and photocatalytic properties for CO_2 reduction was observed under 365 nm light irradiation. The Re(bpy-R)$(CO)_3$Cl (R = H, CH_3 or COOH) reduced CO_2 to CO in CO_2-saturated dimethylformamide (DMF)–triethanolamine solution. The amount of CO produced was found to depend on the substituent R in the bipyridine moiety. They observed that with the introduction of the COOH group, the molar absorption coefficient becomes highest among the four rhenium complexes. It enhanced CO_2 to CO reduction capacity (6.59 mol·cat·mol in 2 hours) to five times that of Re(bpy-H)$(CO)_3$Cl with R as H.

Roldan et al. (2015) reported carbon dioxide reduction using Ru NPs supported on carbon on nitrogen-doped nanofibers. Here, the selectivity was decided by the concentration of ruthenium, which favored CO formation for lower Ru content and methane generation for higher Ru content.

Windle et al. (2015) reported the photocatalytic activity of phosphonated Re complexes [Re(2,2'-bipyridine-4,4'-bisphosphonic acid) $(CO)_3$(L)] (ReP; L = 3-picoline or bromide) immobilized on TiO_2 NPs. The heterogenized Re catalyst on the semiconductor (ReP–TiO_2 hybrid) displayed an improvement in CO_2 reduction. A high TON of 48 mol CO mol Re(−1) was observed in DMF with the electron donor triethanolamine at λ >420 nm. ReP–TiO_2 was compared with other homogeneous systems, and it was claimed that it had the highest TON reported for a CO_2-reducing Re photocatalyst under VIS light irradiation. Photocatalytic CO_2 reduction was observed with ReP–TiO_2 at wavelengths of λ >495 nm. It was observed that an intact ReP catalyst was present on the titania surface before and during catalysis. The high activity upon heterogenization was considered to be due to an increase in the lifetime of the immobilized anionic Re intermediate. The immobilization may also reduce the formation of inactive Re dimers.

The $ZnFe_2O_4$/TiO_2 heterostructure photocatalysts were synthesized with different mass percentages of $ZnFe_2O_4$ by Song et al. (2015) using a hydrothermal deposition method. It was observed that $ZnFe_2O_4$ NPs grow on the TiO_2 nanobelts, and the nanocomposites had high crystallinity. The photocatalytic activities of these nanocomposites were tested by photocatalytic reduction of CO_2 in cyclohexanol under UV light. The main products obtained were cyclohexanone and cyclohexyl formate. As-obtained nanocomposites showed much higher photocatalytic performance than with pure TiO_2 and $ZnFe_2O_4$ samples. The $ZnFe_2O_4$/TiO_2 heterostructure sample showed the highest activity for 9.78% loading of $ZnFe_2O_4$.

A robust and reliable method for improving the photocatalytic performance of InP, which is one of the best-known materials for solar cells, was reported by Zeng et al. (2015). A substantial improvement of up to 18 times was observed in the photocatalytic yields for CO_2 reduction to CO through the surface passivation of InP with TiO_2 deposited by atomic layer deposition (ALD). This enhancement was due to the introduction of catalytically active sites and the formation of a p–n junction. Photoelectrochemical reactions were studied in a nonaqueous solution consisting of ionic liquid, 1-ethyl-3-methylimidazolium tetrafluoroborate ([EMIM]BF$_4$), dissolved in acetonitrile. The photocatalytic yield was increased with the addition of a TiO_2 layer with a corresponding drop in the photoluminescence intensity, which indicates the presence of catalytically active sites causing an increase in the electron–hole pair recombination rate. [EMIM]$^+$ ions in solution form an intermediate complex with $CO_2{}^-$; thus, the energy barrier of this reaction was lowered.

Recently, three types of photocatalytic systems for CO_2 reduction were developed. Two-component systems containing different rhenium(I) complexes have different roles (Sahara and Ishitani 2015) as a redox photosensitizer and a catalyst in the reaction solution. The mixed system of a ring-shaped rhenium(I) trinuclear complex and fac-[Re(bpy)(CO)$_3$(MeCN)]$^+$ is currently the most efficient photocatalytic system for CO_2 reduction (Φ_{CO} = 0.82 at λ_x = 436 nm). The second one is a series of supramolecular photocatalysts having units with three different functions in one molecule, that is, redox photosensitizer, catalyst, and bridging ligand. The highest durability and speed of photocatalysis were achieved by using this system (Φ_{CO} = 0.45), while the third is a novel type of artificial Z-scheme photocatalyst, where photocatalysis was revealed by stepwise excitation of a semiconductor photocatalyst unit and the supramolecular photocatalyst unit. This system has both strong oxidation and reduction powers.

The photoreduction of CO_2 into CH_4 was reported by Kwak and Kang (2015) using Ca$_x$Ti$_y$O$_3$ perovskite NPs. Ca$_x$Ti$_y$O$_3$ NPs were successfully synthesized by them using a hydrothermal method. The photoreduction of CO_2 was performed with UV-lamp (6 W/cm^2) irradiation. The products were analyzed in a gas chromatograph. The rate of formation of methane from CO_2 and H_2O using 0.2 g of Ca$_{1.00}$Ti$_{1.00}$O$_3$ was 17 µmol/g on 7-hour irradiation.

Metal–organic frameworks (MOFs) with isolated metal-monocatecholato groups have been synthesized by Lee et al. (2015) via postsynthetic exchange (PSE) and used for CO_2 reduction under VIS light irradiation in the presence of 1-benzyl-1,4-dihydronicotinamide and triethanolamine. They reported that Cr-monocatecholato species were more efficient than the Ga-monocatecholato species.

Cheung et al. (2016) reported the use of Mn(CN)(bpy)(CO)$_3$ as a catalyst for CO_2 reduction. [Ru(dmb)$_3$]$^{2+}$ was used as a photosensitizer in mixtures of dry N,N- dimethylformamide–triethanolamine (N,N-DMF–TEOA) or acetonitrile–TEOA (MeCN–TEOA) with 1-benzyl-1,4-dihydronicotinamide as a sacrificial reductant. Yields of both CO and HCO$_2$H with maximum TONs as high as 21 and 127, respectively, were obtained after irradiation with 470 nm light for 15 hours. The product preference was found to depend on the solvent. The stability of the singly reduced [Mn(CN)(bpy) (CO)$_3$]$^{\cdot-}$ differ slightly in the N,N-DMF–TEOA solvent system as compared to the

Mecn–TEOA system and this was considered responsible for the observed selectivities for HCO_2H versus CO production.

A series of Re(I) pyridyl N-heterocyclic carbene (NHC) complexes were prepared and used in the photocatalytic reduction of CO_2 using a simulated solar spectrum (Huckaba et al. 2016). These complexes were compared for their activity with a known benchmark catalyst, Re(bpy)(CO)$_3$Br. It was found that an electron-deficient NHC substituent (PhCF$_3$) promoted catalytic activity as compared with electron-neutral and -rich substituents. Re(PyNHC-PhCF$_3$)(CO)$_3$Br was found to function without a photosensitizer with higher turnovers (32 TON) than the benchmark catalyst, which has lower activity (14 TON) under a solar-simulated spectrum.

Zhu et al. (2016) prepared monolayer SnNb$_2$O$_6$ 2D nanosheets with high crystallinity by a one-pot and eco-friendly hydrothermal method without using any organic additives. Then these SnNb$_2$O$_6$ nanosheets were applied for the photocatalytic reduction of CO_2 to CH_4 in the absence of cocatalysts and sacrificial agents under VIS light irradiation. The thickness of as-prepared SnNb$_2$O$_6$ samples with typical 2D nanosheets was about 1 nm. The surface area, photoelectrical properties, and the surface basicity of SnNb$_2$O$_6$ were greatly improved as compared with its counterpart prepared by solid-state reaction. The adsorption capacity of CO_2 on SnNb$_2$O$_6$ nanosheets was found to be much higher than that of layered SnNb$_2$O$_6$. As a result, the photocatalytic activity of SnNb$_2$O$_6$ nanosheets for the reduction of CO_2 was about 45 and 4 times higher than layered SnNb$_2$O$_6$ and N-doped TiO_2, respectively. The intermediates have also been detected by *in situ* Fourier transform infrared spectroscopy (FTIR) with and without VIS light irradiation to learn about the interactions between the CO_2 molecule and the surface of the photocatalyst, and the reactive species in the reduction process.

The production of renewable solar fuel through CO_2 photoreduction, namely artificial photosynthesis, has gained tremendous attention in recent times due to the limited availability of fossil fuel resources and global climate change caused by rising anthropogenic CO_2 in the atmosphere.

Overexploitation and use of fossil fuels have added to global warming and this is exacerbated by deforestation. The main culprit of global warming has been claimed to be carbon dioxide. Earlier, the amount of carbon dioxide was counterbalanced by plants using it in the process of photosynthesis, but this natural balance has been disturbed now and one has to face the consequences of global warming. Photocatalytic reduction of carbon dioxide can solve the problem of the energy crisis by providing synthetic fuels as well as combating against environmental pollution.

REFERENCES

Alves, D. C. B., R. Silva, D. Voiry, T. Asefa, and M. Chhowalla. 2015. Copper nanoparticles stabilized by reduced graphene oxide for CO_2 reduction reaction. *Mater. Renew. Sustain. Energy.* 4 (2): doi:10.1007/s40243-015-0042-0.
Ameta, R., S. Panchal, N. Ameta, and S. C. Ameta. 2013. Photocatalytic reduction of carbon dioxide. *Mater. Sci. Forum.* 764: 73–96.

Aresta, M. 2000. *Carbon Dioxide as Chemical Feedstock.* Weinheim, Germany: Wiley-VCH.

Aresta, M., A. Dibenedetto, T. Baran, A. Angelini, P. Łabuz, and W. Macyk. 2014. An integrated photocatalytic/enzymatic system for the reduction of CO_2 to methanol in bioglycerol–water. *Beil. J. Org. Chem.* 10: 2556–2565.

Bonin, J., M. Chaussemier, M. Robert, and M. Routier. 2014. Homogeneous photocatalytic reduction of CO_2 to CO using iron(0) porphyrin catalysts: Mechanism and intrinsic limitations. *ChemCatChem.* 6 (11): 3200–3207.

Cheung, P. L., C. W. Machan, A. Y. S. Malkhasian, J. Agarwal, and C. P. Kubiak. 2016. Photocatalytic reduction of carbon dioxide to CO and HCO_2H using fac-Mn(CN)(bpy) $(CO)_3$. *Inorg. Chem.* 55 (6): 3192–3198.

Cybula, A., M. Klein, A. Zielińska-Jurek, M. Janczarek, and A. Zaleska. 2012. Carbon dioxide photoconversion. The effect of titanium dioxide immobilization conditions and photocatalyst type. *Physicochem. Probl. Miner. Process.* 48(1): 159–167.

Das, S. and W. M. A. W. Daud. 2014. A review on advances in photocatalysts towards CO_2 conversion. *RSC Adv.* 4: 20856–20893.

De Brito, J. F., A. A. da Silva, A. J. Cavalheiro, and M. V. B. Zanoni. 2014. Evaluation of the parameters affecting the photoelectrocatalytic reduction of CO_2 to CH_3OH at Cu/Cu_2O electrode. *Int. J. Electrochem. Sci.* 9: 5961–5973.

Dey, G. R., A. D. Belapurkar, and K. Kishore. 2004. Photo-catalytic reduction of carbon dioxide to methane using TiO_2 as suspension in water. *J. Photochem. Photobiol. A Chem.* 163 (3): 503–508.

Habisreutinger, N., L. Schmidt-Mende, and J. K. Stolarczyk. 2013. Photocatalytic reduction of CO_2 on TiO_2 and other semiconductors. *Angew. Chemie. Int. Ed.* 52 (29): 7372–7408.

He, Z., Y. Yu, D. Wang, J. Tang, J. Chen, and S. Song. 2016. Photocatalytic reduction of carbon dioxide using iodine-doped titanium dioxide with high exposed {001} facets under visible light. *RSC Adv.* 6: 23134–23140.

Hong, J., W. Zhang, Y. Wang, T. Zhou, and R. Xu. 2014. Photocatalytic reduction of carbon dioxide over self-assembled carbon nitride and layered double hydroxide: The role of carbon dioxide enrichment. *ChemCatChem.* 6 (8): 2315–2321.

Huckaba, A. J., E. A. Sharpe, and J. H. Delcamp. 2016. Photocatalytic reduction of CO_2 with re-pyridyl-NHCs. *Inorg. Chem.* 55 (2): 682–690.

Indrakanti, V. P., J. D. Kubicki, and H. H. Schobert. 2009. Photoinduced activation of CO_2 on Ti-based heterogeneous catalysts: Current state, chemical physics-based insights and outlook. *Energy Environ. Sci.* 2: 745–758.

Inoue, T., A. Fujishima, S. Konishi, and K. Honda. 1979. Photoelectrocatalytic reduction of carbon dioxide in aqueous suspensions of semiconductor powders. *Nature.* 277: 637–638.

Ishitani, O., C. Inoue, Y. Suzuki, and T. Ibusuki. 1993. Photocatalytic reduction of carbon dioxide to methane and acetic acid by an aqueous suspension of metal-deposited TiO_2. *J. Photochem. Photobiol. A Chem.* 72 (3): 269–271.

Jacob, J., M. Mette, and F. C. Krebs. 2011. Flexible substrates as basis for photocatalytic reduction of carbon dioxide. *Sol. Energy Mater. Solar Cells.* 95: 2949–2958.

Jiang, Z., T. Xiao, V. Kuznetsov, and P. Edwards. 2010. Turning carbon dioxide into fuel. *Philos. T. Roy. Soc. A.* 368: 3343–3364.

Kang, Q., T. Wang, P. Li, L. Liu, K. Chang, M. Li, and J. Ye. 2015. Photocatalytic reduction of carbon dioxide by hydrous hydrazine over Au–Cu alloy nanoparticles supported on $SrTiO_3/TiO_2$ coaxial nanotube arrays. *Angew. Chem. Int. Ed.* 54 (3): 841–845.

Khodadadi-Moghaddam, M. 2014. Photocatalytic reduction of CO_2 to formaldehyde: Role of heterogeneous photocatalytic reactions in origin of life hypothesis. *Iran. J. Catal.* 4 (2): 77–83.

Kočí, K., L. Obalová, and Z. Lacný. 2008. Photocatalytic reduction of CO_2 over TiO_2 based catalysts. *Chem. Papers.* 62 (1): 1–9. doi:10.2478/s11696-007-0072-x.

očí, K., L. Obalová, L. Matéjová, D. Plachá, Z. Lacn, J. Jirkovsk, and O. Šolcová. 2009. Effect of TiO₂ particle size on the photocatalytic reduction of CO_2. *Appl. Catal. B Environ.* 89: 494–502.

u, Y., W. H. Lee, and W. Wang. 2003. Photocatalytic reduction of carbon dioxide in gas steams by UV/TiO₂ process. *J. Chin. Inst. Environ. Eng.* 13 (4): 243–249.

umar, P., B. Sain, and S. L. Jain. 2014. Photocatalytic reduction of carbon dioxide to methanol using a ruthenium trinuclearpolyazine complex immobilized on graphene oxide under visible light irradiation. *J. Mater. Chem. A.* 29 (2): 11246–11253.

wak, B. S. and M. Kang. 2015. Photocatalytic reduction of CO_2 with H_2O using perovskite $Ca_xTi_yO_3$. *Appl. Surface Sci.* 337: 138–144.

ee, Y., S. Kim, H. Fei, J. K. Kang, and S. M. Cohen. 2015. Photocatalytic CO_2 reduction using visible light by metal-monocatecholato species in a metal–organic framework. *Chem. Commun.* 51: 16549–16552.

i, H., C. Li, L. Han, C. Li, and S. Zhang. 2016. Photocatalytic reduction of CO_2 with H_2O on CuO/TiO_2 catalysts. *Energy Sources A.* 38 (3): 420–426.

i, H. Y. Lie, Y. Huang, and X. Li. 2011. Photocatalytic reduction of carbon dioxide to methanol by Cu_2O/SiC nanocrystallite under visible light irradiation. *J. Natural Gas Chem.* 20 (2): 145–150.

i, Q. Y., L. L. Zong, C. Li, Y. H. Cao, X. D. Wang, and J. J. Yang. 2014. Photocatalytic reduction of CO_2 to methane on Pt/TiO_2 nanosheet porous film. *Adv. Conden. Matter Phys.* 2014: 6 p. Article ID 316589. doi:org/10.1155/2014/316589.

iao, Y., Z. Hu, Q. Gu, and C. Xue. 2015. Amine-functionalized ZnO nanosheets for efficient CO_2 capture and photoreduction. *Molecules.* 20: 18847–18855.

iu, J., Y. Niu, X. He, J. Qi, and X. Li. 2016. Photocatalytic reduction of CO_2 using TiO_2-graphene nanocomposites. *J. Nanomater.* 2016: 5 p. Article ID 6012896. doi: org/10.1155/2016/6012896.

u, D., M. Zhang, Z. Zhang, Q. Li, X. Wang, and J. Yang. 2014. Self-organized vanadium and nitrogen co-doped titania nanotube arrays with enhanced photocatalytic reduction of CO_2 into CH_4. *Nanoscale Res. Lett.* 9: 272.

1arcì, G., E. I. García-López, and L. Palmisano. 2014. Photocatalytic CO_2 reduction in gas–solid regime in the presence of H_2O by using GaP/TiO_2 composite as photocatalyst under simulated solar light. *Catal. Commun.* 53: 38–41.

1ikkelsen, M., M. Jorgensen, and F. C. Krebs. 2010. The teraton challenge. A review of fixation and transformation of carbon dioxide. *Energy Environ. Sci.* 3: 43–81.

1urakami, N., D. Saruwatari, T. Tsubota, and T. Ohno. 2013. Photocatalytic reduction of carbon dioxide over shape-controlled titanium(IV) oxide nanoparticles with co-catalyst loading. *Curr. Org. Chem.* 17 (21): 2449–2453.

Jhat, H. N., H. Y. Wu, and H. Bai. 2015. Photocatalytic reduction of NO_2 and CO_2 using molybdenum-doped titania nanotubes. *Chem. Engg. J.* 269: 60–66.

)la, O. and M. Maroto-Valer. 2012. Solar fuel production using nickel based nanoparticles. *Prepr. Am. Chem. Soc. Div. Petrol. Chem.* 57: 238–240.

)la, O., and M. M. Maroto-Valer. 2015. Review of material design and reactor engineering on TiO_2 photocatalysis for CO_2 reduction. *J. Photochem. Photobiol. C Photochem. Revs.* 24: 16–42.

)la, O., M. M. Maroto-Valer, and S. Mackintosh. 2013. Turning CO_2 into valuable chemicals. *Energy Procedia.* 37: 6704–6709.

)lah, G. A., A. Goeppert, and G. K. S. Prakash. 2009. Chemical recycling of carbon dioxide to methanol and dimethyl ether: From greenhouse gas to renewable, environmentally carbon neutral fuels and synthetic hydrocarbons. *J. Org. Chem.* 74: 487–498.

)no, Y., J. Nakamura, M. Hayashi, and K. I. Takahashi. 2014. Effect of substituent groups in rhenium bipyridine complexes on photocatalytic CO_2 reduction. *Am. J. Appl. Chem.* 2 (5): 74–79.

Prasad, D. M. R., N. S. B. Rahmat, H. R. Ong, C. K. Cheng, M. R. Khan, and D. Sathiyamoorth 2016. Preparation and characterization of photocatalyst for the conversion of ca bon dioxide to methanol. *Int. J. Chem. Mol. Nucl. Metall. Eng.* 10 (5): 533–53 In: Proceedings of the 18th International Conference on Chemical and Environme Engineering, May 19–20, 2016, Berlin, Germany.

Qin, G., Y. Zhang, X. Ke, X. Tong, Z. Sun, M. Liang, and S. Xue. 2013. Photocatalytic redu tion of carbon dioxide to formic acid, formaldehyde, and methanol using dye-sensitize TiO_2 film. *Appl. Catal. B Environ.* 129: 599–605.

Reli, M., M. Šihor, K. Kočí, P. Praus, O. Kozák, and L. Obalová. 2014. Influence of reactic medium on CO_2 photocatalytic reduction yields over ZnS-MMT. *Geo. Sci. Eng.* 58 (1 34–42.

Richardson, R. D., E. J. Holland, and B. K. Carpenter. 2011. A renewable amine for phot chemical reduction of CO_2. *Nature Chem.* 3: 301–303.

Roldan, L., Y. Mareo, and E. Garcia-Bordeje. 2015. Function of the support and metal loa ing on catalytic CO_2 reduction using Ru nanoparticles supported on carbon nanofiber *ChemCatChem.* 7 (8): 1347–1356.

Sahara, G. and O. Ishitani. 2015. Efficient photocatalysts for CO_2 reduction. *Inorg. Chen* 54 (11): 5096–5104.

Sastre, F., A. V. Puga, L. Liu, A. Corma, and H. García. 2014. Complete photocatalytic redu tion of CO_2 to methane by H_2 under solar light irradiation. *J. Am. Chem. Soc.* 136 (19 6798–6801.

Shown, I., H. C. Hsu, Y. C. Chang, C. H. Lin, P. K. Roy, A. Ganguly et al. 2014. Highly efficie visible light photocatalytic reduction of CO_2 to hydrocarbon fuels by Cu-nanopartic decorated graphene oxide. *Nano Lett.* 14 (11): 6097–6803.

Slamet, H. W. Nasution, E. Purnama, S. Kosela, and J. Gunlazuardi. 2005. Photocatalyti reduction of CO_2 on copper-doped titania catalysts prepared by improved-impregnatic method. *Catal. Commun.* 6: 313–319.

Slamet, H. W. Nasution, E. Purnama, K. Riyani, and J. Gunlazuardi. 2009. Effect of cor per species in a photocatalytic synthesis of methanol from carbon dioxide over coppe doped titania catalysts. *World Appl. Sci. J.* 6 (1): 112–122.

Song, G., F. Xin, and X. Yin. 2015. Photocatalytic reduction of carbon dioxide over ZnFe2O TiO_2 nanobelts heterostructure in cyclohexanol. *J. Coll. Interface Sci.* 442: 60–66.

Tahir, B., M. Tahir, and N. A. S. Amin. 2015. Nitrogen-doped-TiO_2 nanocatalyst for selectiv photocatalytic CO_2 reduction to fuels in a monolith reactor. *Malays. J. Fund. Appl. Sc* 11 (3): 114–117.

Tan, L. L., W. J. Ong, S. P. Chai, and A. R. Mohamed. 2013. Reduced graphene oxide-TiC nanocomposite as a promising visible-light-active photocatalyst for the conversion c carbon dioxide. *Nanoscale Res. Lett.* 8: 465. doi:10.1186/1556-276X-8-465.

Tan, S. S., L. Zou, and E. Hu. 2006. Photocatalytic reduction of carbon dioxide into gaseou hydrocarbon using TiO_2 pellets. *Catal. Today.* 115 (1–4): 269–273.

Vaiano, V., D. Sannino, and P. Ciambelli. 2015. Steam reduction of CO_2 in a photocatalyti fluidized bed reactor. *Chem. Eng. Trans.* 43: 1003–1008.

Varghese, O. K., M. Paulose, T. J. L. Tempa, and C. A. Grimes. 2009. High-rate solar phc tocatalytic conversion of CO_2 and water vapor to hydrocarbon fuels. *Nano Lett.* 9 (2 731–737.

Wang, W. N., J. Soulis, Y. J. Yang, and P. Biswas. 2014. Comparison of CO_2 photoreductio systems: A review. *Aerosol Air Quality Res.* 14: 533–549.

Wang, Z., M. Jiang, J. Qin, H. Zhou, and Z. Ding. 2015. Reinforced photocatalytic reductio of CO_2 to CO by a ternary metal oxide $NiCo_2O_4$. *Phys. Chem. Chem. Phys.* 17 (24 16040–16046.

Windle, C. D., E. Pastor, A. Reynal, A. C. Whitwood, Y. Vaynzof, J. R. Durrant, R. N. Perutz, E. Reisner. 2015. Improving the photocatalytic reduction of CO_2 to CO through immobilisation of a molecular Re catalyst on TiO_2. *Chem. A Europ. J.* 21 (9): 3746–3754.

Woolerton, T. W., S. Sheard, E. Reisner, E. Pierce, S. W. Ragsdale, and F. A. Armstrong. 2010. Efficient and clean photo-reduction of CO_2 to CO by enzyme-modified TiO_2 nanoparticles using visible light. *J. Am. Chem. Soc.* 132 (7): 2132–2133.

Xie, S., Q. Zhang, G. Liu, and Y. Wang. 2016. Photocatalytic and photoelectrocatalytic reduction of CO_2 using heterogeneous catalysts with controlled nanostructures. *Chem. Commun.* 52: 35–59.

Xie, S., Y. Wang, Q. Zhang, W. Fan, W. Deng, and Y. Wang. 2013. Photocatalytic reduction of CO_2 with H_2O: Significant enhancement of the activity of Pt–TiO_2 in CH_4 formation by addition of MgO. *Chem. Commun.* 49: 2451–2453.

Yamamoto, M., T. Yoshida, N. Yamamoto, T. Nomoto, Y. Yamamoto, S. Yagi, and H. Yoshida. 2015. Photocatalytic reduction of CO_2 with water promoted by Ag clusters in Ag/Ga_2O_3 photocatalysts. *J. Mater. Chem. A.* 3: 16810–16816.

Yin, G., M. Nishikawa, Y. Nosaka, N. Srinivasan, D. Atarashi, E. Sakai, and M. Miyauchi. 2015. Nanosheets photocatalytic carbon dioxide reduction by copper oxide nanocluster-grafted niobatenanosheets. *ACS Nano.* 9 (2): 2111–2119.

Zeng, G., J. Qiu, B. Hou, H. Shi, Y. Lin, M. Hettick, A. Javey, and S. B. Cronin. 2015. Enhanced photocatalytic reduction of CO_2 to CO through TiO_2 passivation of InP in ionic liquids. *Chemistry.* 21 (39): 13502–13507.

Zhang, Q., C. F. Lin, Y. H. Jing, and C. T. Chang. 2014. Photocatalytic reduction of carbon dioxide to methanol and formic acid by graphene-TiO_2. *J. Air Waste Manag. Assoc.* 64 (5): 578–585.

Zhang, Z., Z. Huang, X. Cheng, Q. Wang, Y. Chen, P. Dong, and X. Zhang. 2015. Product selectivity of visible-light photocatalytic reduction of carbon dioxide using titanium dioxide doped by different nitrogen-sources. *Appl. Surface Sci.* 355: 45–51.

Zhu, S., S. Liang, J. Bi, M. Liu, L. Zhou, L. Wu, and X. Wang. 2016. Photocatalytic reduction of CO_2 with H_2O to CH_4 over ultrathin $SnNb_2O_6$ 2D nanosheets under visible light irradiation. *Green Chem.* 18: 1355–1363.

15 Artificial Photosynthesis

15.1 INTRODUCTION

Photosynthesis is a chemical process through which plants, algae, and some bacteria store energy from the Sun in the form of carbohydrates that can be used as conventional fuels such as wood, coal, oil, petrol, diesel, and so on. There are four main steps in the process of photosynthesis:

1. Light harvesting
2. Charge separation
3. Water oxidation
4. Fuel production

In the light harvesting step, antenna molecules absorb sunlight and transfer this energy among themselves, mostly chlorophyll but sometimes carotenes also. Then charge separation takes place through the reaction center. The energy from sunlight is used to separate positive and negative charges from each other in this step. The generated positive charges are used to oxidize water while electrons are transferred via cytochrome b_6f and mobile electron carriers to photosystem I, where these are again excited and thus used to produce fuel in the form of carbohydrate.

The chemical reactions involved for water splitting and production of fuel are the following:

- Water oxidation

$$2\ H_2O \xrightarrow{\ 4\ h\nu\ } O_2 + 4\ H^+ + 4\ e^- \qquad (15.1)$$

- Reduction of carbon dioxide to carbohydrate

$$CO_2 + 4\ H^+ + 4\ e^- \xrightarrow{\ 4\ h\nu\ } CH_2O \text{ or } (CH_2O)_n + H_2O \qquad (15.2)$$

These two chemical half-reactions on addition give the total chemical reaction for photosynthesis as

$$CO_2 + H_2O \xrightarrow{\ 8\ h\nu\ } H_2CO \text{ or } (CH_2O)_n + O_2 \qquad (15.3)$$

Four photons are required to drive each of these half-reactions. Thus, in total, eight photons are required for the complete chemical reaction. As four electrons are required for the reduction of CO_2 and eight photons are used, this process proceeds with two photons per electron. Nature uses two photosystems in tandem to drive these two chemical reactions, that is, water splitting and fuel production. Natural photosynthesis is not determined by solar insolation (the total amount of solar radiation collected per unit of time) but by the total light sum, that is, the number of photons from the overall visible spectrum—blue to red: 400–800 nm—collected per unit of time.

The individual parts of the natural photosynthetic process are quite efficient, even though the overall solar-to-carbohydrate efficiency is relatively low. Therefore, unmodified natural photosynthesis cannot serve our purpose for the production of fuel, but it can be used as a blueprint for developing an efficient process of artificial photosynthesis.

15.2 ARTIFICIAL PHOTOSYNTHESIS

Incident photon flux, transfer of energy and electrons, and catalysis all operate on different scales of time, energy, and length. Based on their limitations, one has to think how these components should be combined so that the most efficient solar-to-fuel conversion could be achieved, a conversion quite close to the theoretical limits of solar energy conversion.

The photosynthetic apparatus in plants absorbs light on the order of 700 nm, which means that plants are using half of the incident photons because they do not utilize infrared (IR) radiation. On the contrary, photovoltaic (PV) solar cells involving silicon absorb light around 1100 nm and, therefore, absorb more photons in comparison to plants. Nature uses two photosystems in tandem during the process of photosynthesis to drive the two chemical redox reactions: water oxidation and CO_2 reduction.

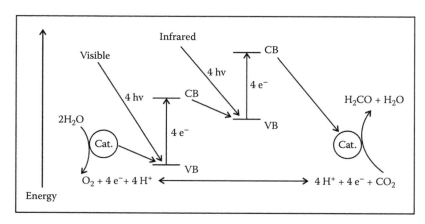

FIGURE 15.1 Tandem device for an artificial photosynthetic system.

The same can be done with an artificial device. One of the major weaknesses f the natural system is that in this process, both photosystems (PSI and PSII) bsorb light of approximately the same energy, and, therefore, these systems are ompeting for the same photons (~700 nm) while the IR photons (heat radiation) main almost unused. It is interesting to note that one photon absorber is present the visible region of the spectrum, and another is present in the IR region in n artificial system. As a consequence, the number of photons absorbed by such system is maximized. The cutoff wavelengths are also better matched. Optimal atching was obtained with cutoff wavelengths of 700 and 1100 nm. Therefore, fforts are being made with tandem devices having two absorbers, so that the best ossible use of the incident light may be made to drive reactions of water splitting s well as fuel production with two photons per electron. A scheme of an artificial hotosynthesis tandem device along with its light-absorbing properties is shown Figure 15.1. Artificial photosynthesis also occurs in four steps similar to natu- l photosynthesis, that is, light harvesting, charge separation, water oxidation, nd fuel production.

5.2.1 ARTIFICIAL PHOTOSYSTEM

n understanding of natural photosynthesis at the molecular level has been supported urther by the creation of artificial photosynthetic model systems such as donor– cceptor assemblies. The effort has been directed presently toward the development f an artificial photosynthesis system. The heterogeneous catalysts are generally ore robust toward oxidative degradation and it is also easier and more economi- al to fabricate such systems. Wen and Li (2013) demonstrated the feasibility of e hybrid photocatalyst, biomimetic molecular cocatalysts, and semiconductor light arvester for artificial photosynthesis and thus provided a promising approach for ational design and construction of highly efficient and stable artificial photosyn- etic systems. They give a strategy of hybrid photocatalysts using semiconductors s light harvesters with biomimetic complexes as molecular cocatalysts to construct fficient and stable artificial photosynthetic systems.

Semiconductor nanoparticles (NPs) were selected as light harvesters by them ecause of their broad spectrum absorption and relatively robust properties as com- ared with a natural photosynthetic system. Use of biomimetic complexes as cocata- ysts can significantly facilitate charge separation via fast charge transfer from the emiconductor to the molecular cocatalysts and also catalyze the chemical reactions f solar fuel production. The hybrid photocatalysts supply us with a platform to study e photocatalytic mechanisms of H_2/O_2 evolution and CO_2 reduction at the molecu- ar level. A comparison of mechanisms of natural and artificial photosynthesis has een given in Figure 15.2.

Alibabaei et al. (2013) also prepared a hybrid strategy for solar water splitting ased on a dye-sensitized photoelectrosynthesis cell. They used a derivative, core– hell nanostructure photoanode with the core high-surface area conductive metal xide film of indium tin oxide or antimony tin oxide coated with a thin outer shell f TiO_2 formed by atomic layer deposition. A chromophore–catalyst assembly

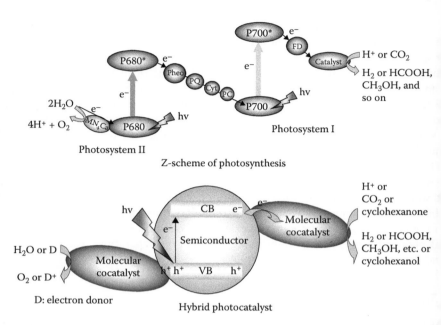

FIGURE 15.2 Comparison of mechanisms of natural photosystem and hybrid photosystem (Adapted from Wen, F., and Li, C., *Acc. Chem. Res.*, 46, 2355–2364, 2013. With permission.)

$[(PO_3H_2)_2bpy)_2Ru(4-Mebpy-4-bimpy) Rub(tpy)(OH_2)]^{+4}$, which combines both light absorber and water oxidation catalyst in a single molecule, was attached to the TiO shell. Visible photolysis of the resulting core–shell assembly structure with a Pt cath ode resulted in water splitting into hydrogen and oxygen with an absorbed photo conversion.

Nakamura and Frei (2006) reported that nanosized crystals of CoO_4 impregnate on mesoporous silica work efficiently as an oxygen-evolving catalyst. A wet-impreg nation procedure was used to grow a Co cluster within the mesoporous Si as tem plate. The yield for cluster of cobalt oxide–nanosized crystals was about 1600 time higher than that for micron-sized particles, and the turnover frequency (TOF) was about 1140 oxygen molecules per second per cluster.

Kalyanasundaram et al. (1981) observed that the conversion of light int chemical fuels in photochemical devices equipped with semiconductor electrode (e.g., n-CdS) is associated with a serious problem of photocorrosion, because hole produced in the valence band (VB) of a semiconductor upon irradiation migrat to the surface of the semiconductor, where photocorrosion occurs. It can be con trolled by a thin layer of RuO_2 in the microheterogenous CdS system. A CdS so prepared in the presence of maleic anhydride/styrene copolymer was loaded with RuO_2 and Pt. These CdS microelectrodes are surprisingly active catalysts for the cleavage of H_2O and H_2S.

15.2.2 ROLE OF PHOTOCATALYST

Irradiation of a photocatalyst (semiconductor) with light energy greater than its band gap generates photoexcited species such as electrons and holes in the CB and VB of semiconductor material, respectively. These migrate to the semiconductor surface. The photoexcited species such as electrons and holes may undergo the following events:

1. Recombination in bulk
2. Recombination at the surface
3. Reduction of suitable electron acceptor

Electron–hole recombination is promoted by defects in the semiconductor materials and as a consequence, they show little or negligible photocatalytic activity. No photocatalytic activity was observed when the recombination of an electron–hole pair takes place and in that case, it generates heat. If an electron donor molecule (D) is present at the surface, then the photogenerated hole can react with these molecules to generate an oxidizing product, D^+. Similarly, if there is an electron acceptor molecule (A) present at the surface, then the photogenerated electrons can react with them to generate a reduced product, A^- (Kamat and Meisel 2003).

$$A + D \xrightarrow{\text{Semiconductor}/h\nu} A^- + D^+ \qquad (15.4)$$

Moreover, heterogeneous catalysts can be integrated more easily into devices that are able to couple the water oxidation process to proton reduction, so as to achieve the splitting of water into molecular oxygen and hydrogen. The term "artificial photosynthesis" was formerly dedicated to molecular systems, but nowadays, it is also applied to water splitting at a semiconductor–electrolyte interface. Band gap excitation gives electron–hole pairs directly in semiconductors, which are the delocalized versions of reductive and oxidative equivalents. Once formed, electron–hole pairs are separated by internal electric fields and directed to spatially separate catalytic sites on the semiconductor surface.

TiO_2 is one of the most investigated photocatalysts in artificial photosynthesis, and it has been known from ancient times as white pigment. It is inexpensive, safe, and fairly stable. Stimulated by the recent breakthroughs in photocatalytic water splitting, research on CO_2 photoreduction is also developing fair quickly. Heterogeneous photocatalytic conversion of CO_2 was first demonstrated like photocatalytic water splitting using large band gap semiconductor materials (TiO_2, $SrTiO_3$, ZnO, and SiC) under strong ultraviolet (UV) light (Inoue et al. 1979). Even today, UV-sensitive materials like TiO_2 (Mori et al. 2012) and niobate perovskites (Shi et al. 2011) are popular starting points. Wide band gap photocatalysts are not ideal for this purpose, and efforts have to be made to utilize the wider solar spectrum. One of the popular routes toward visible-responsive materials is by top-down band gap engineering approach, that is, by introducing various dopants to the starting wide band gap material. A popular example is the nitridation of $ZnGe_2O_4$ (4.4 eV) (Liu et al. 2010). Various strategy have been discussed for photocatalysis in artificial photosynthesis (Figure 15.3).

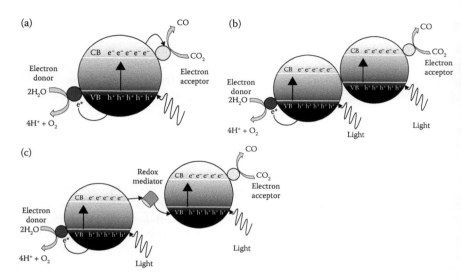

FIGURE 15.3 Various strategy of photocatalyst in artificial photosynthesis: (a) single exci-tation system comprising of a wide bandgap photocatalyst, (b) multiple excitation system comprising narrow bandgap photocatalyst pair connected with physical contact (junctions) and (c) multiple excitation system connected through electron (redox) mediator, imitating 'Z' scheme in natural photosynthesis system. (Adapted from Handoko, A.D. et al., *Curr. Opin. Chem. Eng.*, 2, 200–206, 2013. With permission.)

Manzanares et al. (2014) reported that magnesium-modified TiO_2 (Mg–TiO_2) pho-tocatalyst improved the CO_2 photoreduction reaction, with a high selectivity toward CH_4. Mg–TiO_2 has been synthesized with different compositions up to 2.0 wt.%. Unlike TiO_2, the surface reorganization originated by the presence of Mg enhances the formation of methane by a factor of 4.5 supporting the fact that complete reduction of CO_2 was achieved. It was found that the enhancement of the overall photocatalytic activity toward carbon dioxide reduction can be increased by a factor of 3, which reveals a straightforward correlation with the surface states induced by the presence of the doping element. The evolution of the selectivity versus methane formation against hydrogen was discussed for different magnesium loadings.

A novel inner-motile film for photocatalytic water splitting has been designed by Peng et al. (2015). The inner-motile photocatalyst film is a highly elaborate machin-ery and it mainly integrates three functional modules:

1. Magnetically actuated artificial cilia
2. ZnO nanowire arrays
3. CdS quantum dots

These can work synergistically to enhance the photocatalytic hydrogen evolution activity. The inner-motile film can mimic ciliary motion like nature-beating cilia under a rotational magnetic field through citing magnetically actuated artificial cilia. Hence, it exhibits a singular ability of microfluidic manipulation, which is helpful to solve the stubborn problem of desorption of hydrogen, and promotes the release of active sites. In contrast to the traditional external magnetic stirrer technologies, the photocatalytic activity can be greatly improved. Moreover, forest-like hierarchical-structured ZnO nanowire arrays have been constructed by grafting it on magnetically actuated artificial cilia, which increase the surface area and light absorption. Photocatalytic module–coupled ZnO/CdS heterostructures based on the Z-scheme mechanism have been devised to enhance electron–hole separation and interfacial charge transfer, where ZnO and CdS serve as PSII and PSI, respectively. Consequently, the H_2 evolution rates of ZnO nanowire arrays/CdS heterostructures were about 2.7 times and 2.0 times of CdS and ZnO NPs/CdS heterostructures, respectively. The design of the inner-motile system film is based on both the nature cilia and photosynthesis, which would broaden the horizon for designing the artificial photocatalyst system and also provide a new working prototype for photochemical hydrogen production.

Ehsan and He (2015) reported *in situ* synthesis of common cation heterostructure via the modification of zinc oxide by zinc telluride photocatalyst through a one-pot hydrothermal approach at a reaction temperature of 180°C. The heterostructure fabricated with different ZnO flower-like nanostructures exhibited different photocatalytic capability for the reduction of carbon dioxide into methane under visible light irradiation ($\lambda \geq 420$ nm). This was mainly due to the different exposed crystal planes of ZnO and different surface areas (15.0 and 5.6·m^2/g for sheet-like petals and rod-like petals, respectively). This photocatalytic system showed good visible light photoreduction capability even with ~3.35% ZnTe in terms of at%, and solar energy conversion efficiency can reach ~3.28% in the first 30 minutes of photoreduction. The formation of the heterojunction can also facilitate charge transfer and thus improve the photocatalytic activity.

Antoniadou et al. (2011) prepared a powdered composite CdS–ZnS with variable composition by a coprecipitation method and used it as a photocatalyst to produce hydrogen and as a photoelectrocatalyst to produce electricity. Band gap energy can be tuned between the band gaps of ZnS (3.5 eV) and CdS (2.3 eV) by varying the amount of cadmium and zinc. The composite materials can photocatalytically produce hydrogen using sulfide–sulfite ions as sacrificial electron donors. Photocatalytic performance was significantly improved on depositing small amounts of Pt crystallites on the surface of the photocatalyst. The rate of hydrogen production over the Pt-free CdS–ZnS powders was dependent on the content of Cd (or Zn) and it is generally much higher than with pure CdS or ZnS. Two specific photocatalyst compositions gave maximum hydrogen production rates, that is, 67% and 25% CdS.

Similar behavior was observed on using the same powders to make photoanode electrodes since the rate of hydrogen ion reduction and the current flow were proportional to the number of photogenerated electrons. Composite CdS–ZnS photocatalysts were also applied by successive ionic layer absorption and reaction on TiO_2 films deposited on fluorine-doped tin oxide (FTO) electrodes. As-obtained materials were used as photoanodes in a two-compartment photoelectrocatalysis cell, which was filled with a basic electrolyte and ethanol as the sacrificial electron donor (fuel). The (CdS–ZnS)/TiO_2 photoanodes demonstrated qualitatively similar behavior as CdS–ZnS photocatalysts, but 75%CdS–25%ZnS over TiO_2 was found to be a better electrocatalyst than 100%CdS over TiO_2. When CdS–ZnS photocatalysts were combined with titania, they mainly functioned as visible light photosensitizers of this large band gap semiconductor.

15.3 WATER SPLITTING

Solar-driven water splitting–generated hydrogen has the potential to be a clean, sustainable, and abundant energy source. Artificial solar water splitting devices are now being designed and tested as inspired by natural photosynthesis. In an attempt to mimic natural photosynthesis, recent developments based on molecular and/or nanostructure designs have led to some advances in our understanding of light-induced charge separation and subsequent catalytic water oxidation and reduction reactions. Some of these improvements toward developing artificial photosynthetic devices, together with their analogies to biological photosynthesis, include technologies that focus on the development of visible light active hetero-nanostructures and require an understanding of the underlying interfacial carrier dynamics. A vision for a future sustainable hydrogen fuel community based on artificial photosynthesis has also been proposed.

Overall decomposition of water into hydrogen and oxygen in the presence of a heterogeneous photocatalyst has received prodigious attention due to its potential for the production of clean and recyclable hydrogen energy. However, most of the efficient photocatalysts developed to date work primarily in the UV region of light. Efforts have already been made and the process is still continuing to develop photocatalysts that can decompose water under the more abundant visible range of light; these would have such a narrow band gap that they could utilize less energetic photons, that is, in the visible range. The catalysts that were prepared to exhibit high stability and to give good reaction rate and quantum efficiency are of extremely complex structure. A cumbersome synthesis route involving doping of different materials complicates core–shell nanostructure preparation.

Hydrogen has been advocated as the fuel of future and it is a natural choice of fuel, because water, which is abundantly available, is its raw material. Protons resulting from the splitting of water are reduced to molecular hydrogen.

$$4\,H^+ + 4\,e^- \xrightarrow{\ 4\,h\nu\ } 2\,H_2 \tag{15.5}$$

The simplest way of carrying out this reduction is on the surface of a noble metal like platinum. This approach will prove to be too expensive for commercialization on a large scale. Hydrogenase enzymes have iron and nickel centers that can efficiently catalyze the reversible reduction of protons. Although some success has been achieved in mimicking these catalytic centers, mostly they do not show very high catalytic rates like those of natural enzymes. Recent research has suggested that if the second coordination sphere is modified, it may improve their performance.

Hydrogen could play a key role in future renewable energy technology as it has been identified as an attractive zero-carbon energy carrier, but it is a gas and is highly explosive on mixing with oxygen (from air). The current fuel infrastructure is set up for liquid fuels and there are some problems in the way of home refueling due to safety and certain regulations. Artificial photosynthesis in the form of hydrogen generation from water splitting could play an important role in overcoming these problems before hydrogen is recommended as a fuel for any use worldwide.

Chiarello et al. (2014) demonstrated photocatalytic production of hydrogen from ethanol steam reforming over a series of fluorinated Pt/TiO_2 samples (F for O nomial molar substitution ranging from 5 to 15 at%) synthesized by flame spray pyrolysis a single step. X-ray photoelectron spectroscopy (XPS) analysis confirmed the presnce of both surface and bulk fluorine. It also revealed that Pt, mostly in oxidized rm in the prepared samples, was readily reduced to metallic form under gas-phase otocatalytic reaction conditions. Photocatalytic hydrogen production tests showed at 5 at% F for O substitution in TiO_2 led to an increase in the H_2 and CO_2 production tes, whereas the H_2 production rate decreased linearly with increasing the nominal F ading in the photocatalyst above this value.

Obregon et al. (2015) studied the influence of different TiO_2 supports on the u-active species. It was found that the photocatalytic H_2 evolution was much fected by the structural and electronic features of surface Cu species. Metal disperon and oxidation state appeared strongly conditioned by the structural and surface roperties of the TiO_2 support. They also examined three TiO_2 supports prepared by ifferent synthetic methods:

1. Sol–gel
2. Hydrothermal
3. Microemulsion

They have also induced structural and surface modifications by sulfate pretreatent over freshly prepared TiO_2 precursors followed by calcination. Different copr dispersion and oxidation states were obtained by using different TiO_2 supports. hey proposed that the occurrence of highly disperse Cu^{2+} species, surface area of e sample, and the crystallinity of the TiO_2 support are directly related to the phocatalytic activity for H_2 production reaction.

Vázquez-Cuchillo et al. (2013) prepared $Na_2Ti_6O_{13}$ and $Zr/Na_2Ti_6O_{13}$ by the sol–l method and impregnated these with different amounts of RuO_2 (0.1–10 wt.%). hese were used as photocatalysts for water splitting. The materials were calcined at 00°C. It was confirmed that both materials had the tunnel structure. However, the r/$Na_2Ti_6O_{13}$ sample revealed high Zr dispersion. The ideal amount of RuO_2 to load e samples was shown to be 2.0 wt.%, which produced hydrogen at a rate of 265 mol/h. The apparent quantum yield efficiency was about 21%. This improvement in e catalytic activity of the impregnated materials suggested that there is a synergisc effect between the Zr and RuO_2, which acted like an electron trap.

Sayama et al. (2002) reported water splitting into H_2 and O_2 using two differnt semiconductor photocatalysts and a redox mediator mimicked the Z-scheme echanism of the photosynthesis. It was found that the H_2 evolution took place on Pt–$SrTiO_3$ (Cr–Ta–doped) photocatalyst using I^- electron donor under visible light radiation. The Pt–WO_3 photocatalyst showed excellent activity of the O_2 evolution sing an IO_3^- electron acceptor under visible light. Both H_2 and O_2 gases were evolved the stoichiometric ratio ($H_2/O_2 = 2$) for more than 250 hours under visible light sing a mixture of the Pt–WO_3 and Pt–$SrTiO_3$ (Cr–Ta–doped) powders suspended Nal aqueous solution. They also proposed a two-step photoexcitation mechanism using a pair of I^-/IO_3^- redox mediators. The quantum efficiency of the stoichiometr water splitting was about 0.1% at 420.7 nm.

Obregón and Colón (2014) prepared $Pt-TiO_2$/graphitic carbon nitride (g-C_3N_4) MnO_x hybrid structures by means of a simple impregnation method of $Pt-TiO_2$ ar g-C_3N_4-MnO_x. They observed that TiO_2/g-C_3N_4 composites were formed by a effective covering of g-C_3N_4 by TiO_2. The modification of the composite by Pt an or MnO_x led to improved photoactivities for phenol degradation reaction. Enhance photoactivities have been obtained for the H_2 evolution reaction using these compo ite systems. The photocatalytic performance was related to the efficient separation charge pairs in this hybrid heterostructure.

A novel $TiO_2-In_2O_3$@g-C_3N_4 hybrid system was synthesized by Jiang et a (2015) through a facile solvothermal method. The photocatalytic activity of th $TiO_2-In_2O_3$@g-C_3N_4 hybrid material was evaluated via degradation of rhodamin B (RhB) and hydrogen production. It was found that $TiO_2-In_2O_3$@g-C_3N_4 ternar composites exhibit the highest RhB degradation rate, which was 6.6 times that pure g-C_3N_4. H_2 generation rate of the as-prepared ternary material was found increase by 48 times that observed with pure g-C_3N_4. The enhanced activities wer mainly due to the interfacial transfer of photogenerated electrons, and holes amon TiO_2, In_2O_3, and g-C_3N_4 led to the effective charge separation on these semicondu tors. The photocatalytic mechanism and photostability of the ternary hybrid mate rial were also proposed. This may provide a stepping stone toward the design an practical application of multifunctional hybrid photocatalysts in the photocatalyt degradation of pollutants and hydrogen generation.

Chen et al. (2014) prepared Cu_2O NPs on g-C_3N_4 via a one-pot in situ reductio method. The properties of these Cu_2O NPs–modified g-C_3N_4 photocatalysts wer observed to learn the effects of Cu_2O NPs on the photocatalytic activities of g-C_3N Close contact was formed between Cu_2O and g-C_3N_4, and the Cu_2O NPs were we dispersed on g-C_3N_4. The visible light photocatalytic hydrogen production activit over g-C_3N_4 was enhanced by more than 70% with Cu_2O NP modification. It wa revealed that efficient visible light absorption and type II band alignment–induce charge separation by Cu_2O NP modification should be the responsible key factors f improved photocatalytic performance.

Yu et al. (2016) developed a facile biomolecule-assisted one-pot strategy towar the fabrication of novel CdS/MoS_2/graphene hollow spheres. The molecular structur of cysteine was found to be crucial for controlling the morphology of composite Due to the unique hollow-shaped structure and improved charge separation abilit CdS/5 wt.% MoS_2/2 wt.% graphene hollow spheres exhibited superior high activit for visible light–driven water splitting without noble metals. The synergistic effect of graphene and MoS_2 on the photocatalytic hydrogen production were observe This method opens promising prospects for the rational design of high-efficienc and low-cost photocatalysts for hydrogen evolution based on graphene and MoS_2.

Kim et al. (2015) prepared a core–shell-structured Au@CdS on TiO_2 nanofiber (Au@CdS/TNF). It was used for photocatalytic H_2 production under the visibl ($\lambda > 420$ nm) irradiation. Photocatalysts, including Au-deposited CdS/TNF (Au CdS/TNF) and commercially available TiO_2 (P25) instead of TNF, were invest gated to know the effect of a core–shell-structured Au@CdS and the role of the TNI

The Au@CdS/TNF exhibited strongly enhanced H_2 production under visible irradiation as compared with CdS/TNF and Au/CdS/TNF. H_2 production on Au@CdS/TNF was significantly enhanced as compared to Au@CdS/P25. These results can be attributed to the synergistic effect of the core–shell-structured Au@CdS and the compact structure of TNF. The core–shell-structured Au@CdS may inhibit the formation of $CdSO_4$ from CdS and the Au core could transfer electrons from CdS to TiO_2 without any screening effect for light absorption by CdS. Moreover, interparticle electron transfer is more favorable for TNF than P25 due to the well-aligned TNF framework.

Banerjee and Mukherjee (2014) reported a facile and efficient approach to facilitate photocatalytic water splitting under visible light in a single step. They have modified Cu_2O, a well-known p-type semiconductor having a band gap ~2.1 eV, with RuO_2 NPs and used it as a photocatalyst. It was observed that there is a possibility of near-stoichiometric overall water decomposition under visible light with appreciable quantum efficiency by using this photocatalyst.

Tahir and Wijayantha (2010) deposited semiconducting nanocrystalline $ZnFe_2O_4$ thin films by aerosol-assisted chemical vapor deposition (AACVD) for photoelectrochemical (PEC) water splitting. The effect of various deposition parameters such as solvent type, temperature, and deposition time on PEC properties was investigated. The morphology of the films changes significantly with the change of solvent as evident from scanning electron microscope (SEM) analysis. The films deposited from the ethanolic precursor solution consisted of an interconnected cactus-like $ZnFe_2O_4$ structure growing vertically from the FTO substrate.

The current–voltage characteristics indicated that the nanocrystalline $ZnFe_2O_4$ electrode exhibits n-type semiconducting behavior and the photocurrent depends strongly on the deposition solvent, deposition temperature, and deposition time. The maximum photocurrent density of 350 $\mu A/cm^2$ at 0.23 V versus Ag/AgCl/3 M KCl (~1.23 V vs. RHE [reversible hydrogen electrode]) was obtained for the $ZnFe_2O_4$ electrode synthesized using the optimum deposition temperature of 450°C, a deposition time of 35 minutes, and 0.1 M solution in ethanol. The electrode gave an incident photon to electron conversion efficiency of 13.5% at an applied potential of 0.23 V versus Ag/AgCl/3 M KCl at 350 nm. The donor density of the $ZnFe_2O_4$ was 3.24×10^{24} per m^3, and the flat band potential was approximately −0.17 V, which remarkably agrees with the photocurrent onset potential of −0.18 V versus Ag/AgCl/3 M KCl.

Lixian et al. (2006) synthesized the mesoporous photocatalyst $InVO_4$ by the template-directing self-assembling method. It was observed that the crystal structure of $InVO_4$ could be controlled by changing the calcination temperature. The mesoporous $InVO_4$ was more responsive toward visible light compared with the anatase TiO_2 and conventional $InVO_4$. The evolution rate of H_2 from water over such mesoporous $InVO_4$ achieved was 1836 $\mu mol/g/h$ under UV light irradiation, which was much higher as compared to anatase TiO_2 and conventional $InVO_4$.

Kim et al. (2014) reported that various oxygen evolution electrocatalysts including cobalt phosphate, FeOOH, Ag^+, and metal oxides MO_x (M = Co, Mn, Ni, Cu, Rh, Ir, Ru, Pd) were loaded on $BiVO_4$ photoanode for PEC water splitting under simulated solar light. In all these cases, the electocatalysts increased photocurrent generation and brought cathodic shifts of current onset potential. The screening test led to the discovery of a novel PdO_x-loaded $BiVO_4$ electrode.

This showed the best performance with a five times increase in photocurrent, the largest onset potential shift, and improved stability dramatically, as compared to bare $BiVO_4$. Photooxidation of sulfite ion was investigated as a sacrificial agent to obtain charge separation yields in the bulk and on the surface of $BiVO_4$ to assess the role of the electrocatalyst in a quantitative way. They were able to demonstrate that electrocatalysts reduced surface charge recombination with no effect on bulk recombination.

15.4 CO_2 REDUCTION AND FUEL PRODUCTION

The whole world is facing a serious problem related to fossil resources, that is, a shortage of energy and carbon resources supported by global warming, as a consequence of their use. Development of practical systems for converting CO_2 to useful chemicals using solar light, that is, photocatalytic CO_2 reduction systems, seems to be one of the best solutions for these problems.

Nature is making carbon-based fuels through natural photosynthesis. The products are complex carbohydrates. However, the process of combining protons with carbon dioxide from air to produce these carbon-based fuels is much more different and challenging than producing hydrogen from splitting of water. This process of reduction of carbon dioxide by water to produce fuels involves difficult multielectron chemistry like methane, methanol, and syngas ($CO + H_2$). Efforts are being made to develop artificial molecular catalysts for the photosynthesis of carbon-based fuel and semisynthetic systems, which are based on the enzymes from microorganisms. H_2 and CO can be used as precursors for other fuels like methane, alcohols, and Fischer–Tropsch liquids. These fuels may be incorporated in the list of our current energy sources.

A renewable energy source as abundant and inexhaustible as the Sun has to be developed to drive a catalytic reaction like photocatalytic CO_2 reduction with the resultant fuel products. Many researchers have developed different approaches for the heterogeneous photocatalytic reduction of CO_2 on TiO_2, ZnO, and various metal oxides.

The photocatalytic reduction of CO_2 has been studied by Paulino et al. (2016) with an aim to find a useful application for such low-cost and abundant raw material. Apart from putting a check on increasing amount of CO_2 in the atmosphere, this process can contribute to the generation of high-energy products (CH_4 and CH_3OH). The reaction was performed in liquid phase at 25°C, with the photocatalyst (1 g/L) maintained in suspension. UVC lamp (18 W, 254 nm) was used as the radiation source. Photocatalysts were prepared using oxides of titanium, copper, and zinc. Commercial TiO_2 (P25, Degussa) was utilized as reference. Catalysts having specific area ranging from 36 to 52 m^2/g and band gap energies varying from 3.0 to 3.3 eV were obtained. Temperature-programmed desorption of CO_2 (TPD-CO_2) results showed different strengths of CO_2 adsorption for each photocatalyst. CH_4 production was achieved in the range of 126–184 μmol/g catalyst after 24-hour irradiation. It was observed that CH_4 formation increased in the following order:

$$TiO_2 \text{ (P25)} \sim TiO_2 < 2\%CuO/TiO_2 < 2\%CuO\text{-}19\%ZnO/TiO_2$$

Results indicated that the interaction between CO_2 and the photocatalyst influenced the photocatalytic activity.

Copper-doped titanium dioxide with anatase phase ($Cu–TiO_2$) with atomic Cu contents in the range of 0%–3% relative to the sum of Cu and Ti and particle sizes of 12–15 nm were synthesized by Gonell et al. (2016) through a solvothermal method. Ethanol was used as the solvent and small amounts of water was required to promote the hydrolysis–condensation processes. Photocatalytic CO_2 reduction was performed in aqueous $Cu–TiO_2$ suspensions under UV-rich light and in the presence of different solutes. Sulfide was found to promote the efficient production of H_2 from water and formic acid from CO_2. The effect of the Cu content on the photoactivity of $Cu–TiO_2$ was also studied and it was shown that copper plays an important role in the photocatalytic reduction of CO_2.

Phongamwong et al. (2015) synthesized visible light–reactive N-doped TiO_2 ($N–TiO_2$) catalysts using a simple sol–gel method, and chlorophyll in spirulina was consequently loaded onto the N-doped TiO_2 catalysts ($Sp/N–TiO_2$) in an attempt to enhance their photocatalytic efficiency. The effects of nitrogen and chlorophyll in *spirulina* and their loading amount on CO_2 photoreduction with water under visible light of $Sp/N–TiO_2$ catalysts were observed. The activities of catalysts were found in the following order:

$$\text{Undoped } TiO_2 < N–TiO_2 < Sp/N–TiO_2$$

Wang et al. (2015) found that Ag-modified $La_2Ti_2O_7$ with a layered perovskite structure exhibited activity for the photocatalytic conversion of CO_2 to CO in pure water. The evolution of O_2 supports the fact that H_2O works as an electron donor for the photocatalytic conversion of CO_2. CO was generated as the main product. H_2 and O_2 were also produced in stoichiometric amounts. The crystallite size of $La_2Ti_2O_7$ was found to increase and consequently the surface area decreased with increasing calcination temperature and time. Maximization of the photocatalytic activity of $La_2Ti_2O_7$ was obtained by making a trade-off between the crystallite size and surface area. It was observed that the loading amount and modification method of the Ag cocatalyst also influenced the amount of CO evolved.

Ehsan et al. (2014) demonstrated that ZnTe can be utilized as an efficient catalyst for the photoreduction of CO_2 into methane under visible light irradiation (≥ 420 nm). The results indicated that the combination of ZnTe with $SrTiO_3$ increases the formation of CH_4 by efficiently promoting electron transfer from the CB of ZnTe to that of $SrTiO_3$ under visible light irradiation, and it can be also a promising candidate for the photocatalytic conversion of CO_2 into hydrocarbon fuels.

The feasibility of applying a modified acidic photocatalyst (TiO_2/SO_4^{2-}) to reduce carbon dioxide was investigated by Lo et al. (2007). The photocatalytic reduction of CO_2 was conducted in a bench-scale batch photocatalytic reactor. Three near-UV black lamps with a maximal spectrum wavelength of 365 nm were assembled on the top of the reactor to provide an average irradiation intensity of 2.0 mW/cm^2. The TiO_2/SO_4^{2-} photocatalyst was prepared by a modified sol–gel process and coated on stainless steel substrates for the reduction of CO_2. The effect of experimental parameters such as reductants, the initial CO_2 concentration, and the reaction temperature were investigated.

The results indicated that the highest photoreduction rate of CO_2 was observed using H_2 as a reductant over TiO_2/SO_4^{2-}. The major gaseous products from CO_2 photoreduction were carbon monoxide and methane, while other products like ethene and ethane were also detected in minor amounts. The photoreduction rate of CO_2 increased with initial CO_2 concentration and reaction temperature. Fourier transform infrared spectroscopy (FTIR) spectra showed that formic acid, methanol, carbonate ions, formaldehyde, and methyl formate were formed on the surface of TiO_2/SO_4^{2-} photocatalyst. Two reaction pathways of CO_2 photoreduction over TiO_2/SO_4^{2-} were proposed. One reaction pathway described the formation of gaseous products CO, CH_4, C_2H_4, and C_2H_6, while another reaction pathway proposed the formation of CO_{3ads}^{2-}, CH_3OH_{ads}, $HCOO_{ads}^-$, $HCOOH_{ads}$, $HCOH_{ads}$, and $HCOOCH_{3ads}$ on the surface of the $TiO_2/TiO_2/SO_4^{2-}$ photocatalyst.

Yuan et al. (2013) reported a composite photocatalyst by coupling red phosphor (r-P) and g-C_3N_4. The introduction of g-C_3N_4 onto r-P surface led to considerable improvement on the photocatalytic activity for H_2 production and CO_2 conversion into valuable hydrocarbon fuel-like methane in the presence of water vapor. This enhancement may be due to the effective separation of photogenerated electrons and holes across the r-P/g-C_3N_4 heterojunction. This system had a number of advantages like nontoxicity, low cost, and abundance in nature, and, therefore, this active heterostructural r-P/g-C_3N_4 photocatalyst has great potential for efficient solar fuel production.

He et al. (2015) prepared ZnO/g-C_3N_4 composite photocatalyst and evaluated the conversion efficiency of CO_2 to fuel in its presence under simulated sunlight irradiation. The photocatalyst was synthesized by a simple impregnation method. The characterization indicated that ZnO and g-C_3N_4 were uniformly combined in this composite. The deposition of ZnO on g-C_3N_4 showed nearly no effect on its light absorption performance. However, the interactions between these two components promoted the formation of a heterojunction structure in the composite, which inhibited the recombination of electron–hole pairs and, finally, enhanced the photocatalytic performance of ZnO/g-C_3N_4. The optimal ZnO/g-C_3N_4 photocatalyst showed a CO_2 conversion rate of 45.6 μmol/h/g.cat, which was 4.9 and 6.4 times higher than those of individual components, g-C_3N_4 and P25, respectively. It is an important step toward artificial photocatalytic CO_2 conversion to fuel using cost-efficient materials.

Cao et al. (2014) reported *in situ* growth of In_2O_3 nanocrystals onto the sheet-like g-C_3N_4 surface. The resulting In_2O_3–g-C_3N_4 hybrid structures exhibited considerable improvement on the photocatalytic activities for H_2 generation and CO_2 reduction. The enhanced activities were attributed to the interfacial transfer of photogenerated electrons and holes between g-C_3N_4 and In_2O_3, thus leading to effective charge separation on both the parts. It was confirmed by transient photoluminescence (PL) spectroscopy that the In_2O_3–g-C_3N_4 heterojunctions remarkably promote charge transfer efficiency and as a result, increase the charge carrier lifetime for the photocatalytic reactions.

Shown et al. (2014) prepared graphene oxide (GO) decorated with copper NPs (Cu-NPs), that is, Cu/GO, and used it for enhancing photocatalytic CO_2 reduction under visible light. A rapid one-pot microwave process was used to prepare the Cu/GO hybrids with various Cu contents. The attributes of metallic Cu-NPs (about 4–5 m in size)in the GO hybrid were shown to significantly enhance the photocatalytic ctivity of GO, primarily through the suppression of electron–hole pair recombina- on, a reduction band gap of GO, and modification of its work function. A charge ansfer from GO to Cu was indicated by X-ray photoemission spectroscopy.

Wang et al. (2014) reported the synthesis of mesoporous Fe-doped CeO_2 catalysts ith different Fe doping concentrations through a nanocasting route using ordered esoporous SBA-15 as the template. The samples were prepared by filling meso- ores in silica template with an Fe–Ce complex precursor followed by calcination nd silica removal. These were tested for catalytic activity in photocatalytic reduc- on of CO_2 with H_2O under simulated solar irradiation. Fe species can effectively nhance photocatalytic performance in the reduction of CO_2 with H_2O compared ith nondoped mesoporous CeO_2 catalyst.

Various metal oxides were modified by different methods and several new efficient eterogeneous catalysts, that is, binary, ternary, and doped materials, were also used r the photocatalytic reduction of CO_2. Song et al. (2015) synthesized $ZnFe_2O_4/TiO_2$ eterostructure photocatalysts with different mass percentages of $ZnFe_2O_4$ through hydrothermal deposition method. The photocatalytic activities of the nanocompos- es were tested by photocatalytic reduction of CO_2 in cyclohexanol under UV light nain wavelength at 360 nm) irradiation. The experimental results showed that the nain products were cyclohexanone and cyclohexyl formate. As-obtained $ZnFe_2O_4/$ iO₂ nanocomposites showed much higher photocatalytic performance compared ith pure TiO_2 and $ZnFe_2O_4$ samples.

Yang and Jin (2014) synthesized Zn_2GeO_4 nanorods by a surfactant-assisted solu- on phase route. They were of the opinion that the cetyltrimethylammonium (CTA^+) ations preferentially adsorbed on the planes of Zn_2GeO_4 nanorods led to a prefer- ntial growth along the c-axis to form Zn_2GeO_4 rods with larger aspect ratio and igher surface area. This was the main reason for the improvement in photocatalytic ctivity for the photoreduction of CO_2. Jiang et al. (2014) prepared a series of novel iicrospheres of $CdIn_2S_4$ by a hydrothermal process. As-synthesized $CdIn_2S_4$ from -cysteine exhibited higher photocatalytic activity for CO_2 reduction and this system as potential application in using visible light. The mechanism of photocatalytic eduction of CO_2 in methanol over $CdIn_2S_4$ was also proposed. The narrow band gap f the as-prepared catalyst promoted the reduction of CO_2 to dimethoxymethane and nethyl formate in methanol.

Li et al. (2014) prepared a series of metal oxide complex by annealing and copre- ipitating Ni/Zn/Cr layered double hydroxides (LDHs) at different temperatures. heir activities in photocatalytic reduction of CO_2 with H_2O vapor were tested at mbient conditions. It was revealed that the sample calcined at 500°C possessed the ighest catalytic activity, giving CH_4 and CO as the major products. This was attrib- ted to the interaction of uniformly dispersed NiO, Cr_2O_3, $ZnCr_2O_4$, and $NiCr_2O_4$ /ith small grain size. Almeida et al. (2014) prepared pure and Cr(III)- and Mo(V)- oped $BiNbO_4$ and $BiTaO_4$ by the citrate method. Pure $BiNbO_4$ and $BiTaO_4$ were btained in triclinic phase at 600°C and 800°C, respectively. It was observed that netal doping influenced strongly the crystal structure as well as the photocatalytic ctivity of these ternary oxides. The results showed that Cr(III)-doped $BiTaO_4$ and iNbO₄ were more selective for hydrogen production, while Mo(V)-doped materials were more selective for CO_2 generation. Comparing the photocatalytic activity BiTaO$_4$ and BiNbO$_4$, BiTaO$_4$ showed higher activity for hydrogen production as we as for CO_2 generation. A negligible change of CB minimum (CBM) potential w found in the case of Mo(V)-doped materials, which indicates that there might l no improvement on the reduction power of the material following the substitution doping. There was a slight shift of the CBM potential increasing the reduction pow a little bit in the case of Cr(III)-doped BiNbO$_4$. However, the effect is much strong in the Cr(III)-doped BiTaO$_4$.

15.5 DIFFERENT APPROACHES TO ARTIFICIAL PHOTOSYNTHESIS

A number of different approaches to artificial photosynthesis have been tried by di ferent workers from time to time and many more will be attempted in years to con that can be employed on a global level. These all are based on four basic steps natural photosynthesis: light harvesting, charge separation, water splitting, and fu production. Similar scientific problems are encountered in one step or another. The different approaches are briefly based on the types of materials used. The materi may be organic, inorganic, hybrid, and semisynthetic in nature.

15.5.1 ORGANIC SYSTEMS

The components of molecular artificial synthesis are generally developed by biom metics. Energy conversion and storage in efficient natural enzymes has provided momentum to the development of the complex chemistry, which is helpful in min icking enzymatic catalytic functions. The major research problem is four-electro oxidation of water and it has remained a bottleneck in the development of any su cessful system of artificial photosynthesis.

It is very difficult to develop entire molecular systems, but there is always a po sibility of using a modular approach. In that case, individual components can be mac for different processes like light harvesting, charge separation, water oxidation, ar fuel production. Then they may be tried separately to get the maximum performanc of an individual component and integrated into the appropriate desired architectur Molecules having a well-defined structure can be prepared and can be modified to hav an improvement in the required property. Molecular systems are also readily amenab to study with different methods so as to provide structural and kinetic information.

All suitable processes can be followed, and structural and mechanistic detai of these molecular systems can be understood with different analytical technique This approach of molecular assembly looks elegant, but without any suitable strateg it is nearly impractical as it may involve a large amount of synthesis. It is also impo tant to note that most of the molecules have a tendency to degrade under extende exposure to sunlight.

15.5.2 INORGANIC SYSTEMS

In solar cells, semiconductor materials can also absorb sunlight, provided the have the desired band gap and are capable of separating the charges (electron–hol pairs). These materials are robust in nature under extended exposure to sunlight as compared to organic systems. Therefore, semiconductors (photocatalysts) seem to be prospective candidates for mimicking artificial photosynthesis. Those semiconductors with appropriate electronic properties can provide sufficient electrochemical potential and these can drive water oxidation or fuel production (by reduction of CO_2) on their surfaces. But, many such materials have the electronic properties for water splitting for absorption in the UV range, which is only a small portion of the incoming photons in sunlight. They also have another drawback in that catalysis on the semiconductor surface is not very efficient. This is quite appealing for its simplicity as one semiconductor performs all the tasks of absorption, charge separation, and catalysis, but many times, it is a tedious task to get all processes done using one material. No material with all these properties with good performance has been found to date, but the search is still on for a synthesizing semiconductor with some modifications. Nanotechnology is being applied to design certain new composite nanostructures, where individual component perform specialized functions to accept this significant challenge.

15.5.3 Organic–Inorganic Hybrids

Another appealing solution to this problem may be obtained from combining the best-performing organic and inorganic materials in a hybrid structure. It is proposed that the light absorption may be done either by a semiconductor or by a dye molecule on a semiconductor surface. It is then followed by separation of charges within the semiconductor and these are transported to optimized molecular catalysts tethered to the semiconductor surface. It seems to be a very good approach and quite promising also. A number of devices have been constructed using this hybrid concept, but many of these hybrids are still either not cost-effective or not efficient enough so that they can be tried for large-scale application to have commercial viability.

15.5.4 Semisynthetic Systems

A new approach has also been attempted involving a hybrid of biological and synthetic components. A biological photosynthetic component, which harvests solar energy and splits water in natural photosynthesis, can be purified and tethered to an appropriate scaffold. This photosynthetic enzyme is then linked to a hydrogen-producing enzyme (a hydrogenase) or a catalyst for the production of synthetic fuel. Chlorophylls may be separated from biological sources and then modified chemically and assembled to form semiartificial modules.

Nature has been successfully performing photosynthesis for about 3 billion years using these biological components. However, this method is still in its infancy, and it is very important to know whether such biological components can be made sufficiently robust by some chemical modifications outside their natural environment. What will be their fate, if they are extracted and modified on a large scale? If such modifications never yield commercially viable devices, even then the science involved will show the intricacies about the approach of nature in photosynthesis.

15.6 GOALS AND BENEFITS

Although natural photosynthesis is very important, it is inefficient for many reasons. Water is also wasted due to transpiration and the evaporation of water from plants, as the stomata must be open. It is also necessary to have carbon dioxide and to give out oxygen. The lower amount of CO_2 (0.036% in the air) also adds to this inefficiency as the stomata must be constantly open. Thus, it is necessary to develop artificial photosynthesis with better efficiency. The main goal of mimicking photosynthesis (artificial photosynthesis) is to develop photosynthetic processes on a large scale that have much greater efficiency in solar energy conversion and maximum use of water. Such an approach on successful completion would bring significant, life-changing benefits, which are important from an energy as well as environmental protection point of view.

As the human population is continually polluting our environment and conventional sources of energy are being rapidly depleted like fossil fuels, one has to look toward finding more eco-friendly and long-lasting solutions to overcome these problems. The water splitting process of photosynthesis generating hydrogen can solve this dilemma. Mass production of hydrogen is highly desirable as it is an eco-friendly energy source. Hydrogen has been viewed as the energy source of the future.

The most efficient solar cells have a conversion efficiency on the order of 30%. If the absorption of photon could be achieved to the level of perfection as in the case of photosynthesis, one could obtain efficient solar cells. These are highly convenient units as they are relatively low cost with minimal maintenance, smaller in size, and can be easily mounted anywhere to provide a regular power supply. If their efficiencies are increased to the level that batteries would no longer be required, then there are endless possibilities for their possible use in varied fields.

Burning fossil fuels has led to a large increase in the level of carbon dioxide in the atmosphere causing global warming. The use of photosynthesis (natural as well as artificial) on a massive scale will control the increasing amounts of carbon dioxide in the atmosphere so that the harmful effects of the greenhouse effect can be reversed to a desired extent.

Natural photosynthesis synthesizes carbohydrates as the major product in the form of biomass, which is the source of food for survival of life on Earth. If this process of photosynthesis is mastered, we will not be limited to carbohydrates only. Methods have been developed to manipulate genetics and some enzymes present during the Calvin cycle so as to produce sugars, protein, and even alcohol. The production of these compounds will be also quite beneficial.

15.7 LIMITATIONS

Materials used for artificial photosynthesis often undergo corrosion in water and these are less stable than PV materials over the passage of time. Most of the catalysts are very susceptible to attack by oxygen. These are either inactivated or degraded in the presence of oxygen. Photodamage may also occur with time. At present, the overall cost is high enough that this process cannot compete with fossil fuels as a viable source of energy. The reactions are either not that efficient or unsustainable to date. Major reactions need the heat generated from the solar insolation to power them and therefore they are not suitable for operation worldwide.

It is very important to design a catalyst in such a way that most of the incident light can be utilized by the system. The efficiency should be comparable with photosynthetic efficiency, that is, light-to-chemical energy conversion. Photosynthetic organisms are normally able to collect about 50% of incident solar radiation, but use of different materials in photochemical cells could make them more efficient in absorbing a relatively wider range of solar radiation.

It is slightly difficult to compare overall fuel production between natural and artificial systems. Plants have a theoretical threshold of 12% efficiency of glucose formation from natural photosynthesis, while a carbon-reducing catalyst may go beyond this limit. Plants are efficient in using CO_2 at atmospheric concentrations, which is not possible by artificial catalysts at present.

Different methods to make hydrogen and carbon-based fuels using water, carbon dioxide, and sunlight as raw materials are being searched for all over the globe by different research groups. These fuels provide the possibility of converting and storing solar energy in one part of the globe, which may be transported to other part of the world at any time.

The basic barrier to achieve this objective is to create a device, which is robust, low cost, and efficient. Large efforts are being made to solve these and other related problems so as to provide viable prototype devices for this purpose. After devising it on laboratory scale, the next step is to try an artificial photosynthesis process at pilot scale. A global group on artificial photosynthesis consisting of chemists, biologists, and engineers may give these efforts sufficient momentum and public awareness raising the visibility of this potentially game-changing technology in the coming era of energy crisis.

Such technology for the conversion and storage of sunlight in the form of fuels (hydrogen and carbon based) will contribute to the closing of loops on both scales: large as well as small scale. However, it is important to see what the social and economic consequences on application of these technologies will be. Science is marching ahead to make this scenario a possibility, but the time is far off at present to convert this dream into a reality. This should be duly supported by economic incentives and of course, the political will of all countries across the globe. The prospects of artificial photosynthesis are very exciting and one can look forward to a global project on artificial photosynthesis in the near future.

REFERENCES

Alibabaei, L., M. K. Brennaman, M. R. Norris, B. Kalanyan, W. Song, M. D. Losego, J. J. Concepcion, R. A. Binstead, G. N. Parsons, and T. J. Meyer. 2013. Solar water splitting in a molecular photoelectrochemical cell. *Proc. Natl. Acad. Sci. U S A.* 110: 20008–20013.

Almeida, C. G., R. B. Araujo, R. G. Yoshimura, and A. J. S. Mascarenhas. 2014. Photocatalytic hydrogen production with visible light over Mo and Cr-doped $BiNb(Ta)O_4$. *Int. J. Hydrogen Energy.* 39 (3): 1220–1227.

Antoniadou, M., V. M. Daskalaki, N. Balis, D. I. Kondarides, C. Kordulis, and P. Lianos. 2011. Photocatalysis and photoelectrocatalysis using (CdS-ZnS)/TiO$_2$ combined photocatalysts. *Appl. Catal. B Environ.* 107 (1–2): 188–188.

Banerjee, T. and A. Mukherjee. 2014. Overall water splitting under visible light irradiation using nanoparticulate RuO$_2$ loaded Cu$_2$O powder as photocatalyst. *Energy Proc.* 54: 221–227.

Cao, S. W., X. F. Liu, Y. P. Yuan, Z. Y. Zhang, Y. S. Liao, J. Fang, S. C. J. Loo, T. C. Sum, and C. Xue. 2014. Solar-to-fuels conversion over In$_2$O$_3$/g-C$_3$N$_4$ hybrid photocatalysts. *Appl. Catal. B Environ.* 147: 940–946.

Chen, J., S. Shen, P. Guo, M. Wang, P. Wu, X. Wang, and L. Guo. 2014. In-situ reduction synthesis of nano-sized Cu$_2$O particles modifying g-C$_3$N$_4$ for enhanced photocatalytic hydrogen production. *Appl. Catal. B Environ.* 152–153: 335–341.

Chiarello, G. L., M. V. Dozzi, M. Scavini, J. D. Grunwaldt, and E. Selli. 2014. One step flame-made fluorinated Pt/TiO$_2$ photocatalysts for hydrogen production. *Appl. Catal. B Environ.* 160–161: 144–151.

Ehsan, M. F. and T. He. 2015. In situ synthesis of ZnO/ZnTe common cation heterostructure and its visible-light photocatalytic reduction of CO$_2$ into CH$_4$. *Appl. Catal. B Environ.* 166–167: 345–352.

Ehsan, M. F., M. N. Ashiq, F. Bi, Y. Bi, S. Palanisamy, and T. He. 2014. Preparation and characterization of SrTiO$_3$-ZnTe nanocomposites for the visible-light photoconversion of carbon dioxide to methane. *RSC Adv.* 4: 48411–48418.

Gonell, F., A. V. Puga, B. Julián-López, H. García, and A. Corma. 2016. Copper-doped titania photocatalysts for simultaneous reduction of CO$_2$ and production of H$_2$ from aqueous sulphide. *Appl. Catal. B Environ.* 180: 263–270.

Handoko, A. D., K. Li, and J. Tang. 2013. Recent progress in artificial photosynthesis: CO$_2$ photoreduction to valuable chemicals in a heterogeneous system. *Curr. Opin. Chem. Eng.* 2 (2): 200–206.

He, Y., Y. Wang, L. Zhang, B. Teng, and M. Fan. 2015. High-efficiency conversion of CO$_2$ to fuel over ZnO/g-C$_3$N$_4$ photocatalyst. *Appl. Catal. B Environ.* 168–169: 1–8.

Inoue, T., A. Fujishima, S. Konishi, and K. Honda. 1979. Photoelectrocatalytic reduction of carbon dioxide in aqueous suspensions of semiconductor powders. *Nature.* 277: 637–638.

Jiang, W., X. Yin, F. Xin, Y. Bi, Y. Liu, and X. Li. 2014. Preparation of CdIn$_2$S$_4$ microspheres and application for photocatalytic reduction of carbon dioxide. *Appl. Surf. Sci.* 288: 138–142.

Jiang, Z., D. Jiang, Z. Yan, D. Liu, K. Qian, and J. Xie. 2015. A new visible light active multifunctional ternary composite based on TiO$_2$–In$_2$O$_3$ nanocrystals heterojunction decorated porous graphitic carbon nitride for photocatalytic treatment of hazardous pollutant and H$_2$ evolution. *Appl. Catal. B Environ.* 170–171: 195–205.

Kalyanasundaram, K., E. Borgarello, D. Duonghong, and M. Graetzel. 1981. Cleavage of water by visible-light irradiation of colloidal CdS solutions: Inhibition of photocorrosion by RuO$_2$. *Angew. Chem. Int. Ed.* 20: 987–988.

Kamat, P. V. and D. Meisel. 2003. Nanoscience opportunities in environmental remediation. *Compt. Rend. Chim.* 6: 999–1007.

Kim, F., Y. K. Kim, S. K. Lim, S. Kim, and S. L. In. 2015. Efficient visible light-induced H$_2$ production by Au@CdS/TiO$_2$ nanofibers: Synergistic effect of core-shell structured Au@CdS and densely packed TiO$_2$ nanoparticles. *Appl. Catal. B Environ.* 166–167: 423–431.

Kim, J. H., J. W. Jang, H. J. Kang, G. Magesh, J. Y. Kim, J. H. Kim, J. Leea, and J. S. Lee. 2014. Palladium oxide as a novel oxygen evolution catalyst on BiVO$_4$ photoanode for photoelectrochemical water splitting. *J. Catal.* 317: 126–134.

Li, B. J., Z. J. Wu, C. Chen, W. F. Shangguan, and J. Yuan. 2014. Preparation and photocatalytic CO$_2$ reduction activity of Ni/Zn/Cr composite metal oxides. *J. Mol. Catal. (China).* 3: 268–274.

iu, Q., Y. Zhou, J. Kou, X. Chen, Z. Tian, J. Gao, S. Yan, and Z. Zou. 2010. High-yield synthesis of ultralong and ultrathin Zn_2GeO_4 nanoribbons toward improved photocatalytic reduction of CO_2 into renewable hydrocarbon fuel. *J. Am. Chem. Soc.* 132: 14385–14387.

ixian, X. U., L. Sang, C. Ma, Y. Lu, F. Wang, Q. Li, H. Dai, H. He, and J. Sun. 2006. Preparation of mesoporous $InVO_4$ photocatalyst and its photocatalytic performance for water splitting. *Chin. J. Catal.* 27 (2): 100–102.

o, C. C., C. H. Hung, C. S. Yuan, and Y. L. Hung. 2007. Parameter effects and reaction pathways of photoreduction of CO_2 over TiO_2/SO_4^{2-} photocatalyst. *Chin. J. Catal.* 28 (6): 528–534.

lanzanares, M., C. Fàbrega, J. O. Ossó, L. F. Vega, T. Andreu, and J. R. Morante. 2014. Engineering the TiO_2 outermost layers using magnesium for carbon dioxide photoreduction. *Appl. Catal. B Environ.* 150–151: 57–62.

lori, K., H. Yamashita, and M. Anpo. 2012. Photocatalytic reduction of CO_2 with H_2O on various titanium oxide photocatalysts. *RSC Adv.* 2: 3165–3172.

akamura, R. and H. Frei. 2006. Visible light-driven water oxidation by Ir oxide clusters coupled to single Cr centers in mesoporous silica. *J. Am. Chem. Soc.* 128: 10668–10669.

bregón, S. and G. Colón. 2014. Improved H_2 production of Pt-TiO_2/g-C_3N_4-MnO_x composites by an efficient handling of photogenerated charge pairs. *Appl. Catal. B Environ.* 144: 775–782.

bregon, S., M. J. Muñoz-Batista, M. Fernández-García, A. Kubacka, and G. Colón. 2015. Cu–TiO_2 systems for the photocatalytic H_2 production: Influence of structural and surface support features. *Appl. Catal. B Environ.* 179: 468–478.

aulino, P. N., V. M. M. Salim, and N. S. Resende. 2016. Zn-Cu promoted TiO_2 photocatalyst for CO_2 reduction with H_2O under UV light. *Appl. Catal. B Environ.* 185: 362–370.

eng, F., Q. Zhou, D. Zhang, C. Lu, Y. Ni, J. Kou, J. Wang, and Z. Xu. 2015. Bio-inspired design: Inner-motile multifunctional ZnO/CdS heterostructures magnetically actuated artificial cilia film for photocatalytic hydrogen evolution. *Appl. Catal. B Environ.* 165: 419–427.

hongamwong, M. Chareonpanich, and J. Limtrakul. 2015. Role of chlorophyll in *Spirulina* on photocatalytic activity of CO_2 reduction under visible light over modified N-doped TiO_2 photocatalysts. *Appl. Catal B Environ.* 168–169: 114–124.

ayama, K., K. Mukasa, R. Abe, Y. Abe, and H. Arakawa. 2002. A new photocatalytic water splitting system under visible light irradiation mimicking a Z-scheme mechanism in photosynthesis. *J. Photochem. Photobiol. A Chem.* 148 (1–3): 71–77.

hi, H., T. Wang, J. Chen, C. Zhu, J. Ye, and Z. Zou. 2011. Photoreduction of carbon dioxide over $NaNbO_3$ nanostructured photocatalysts. *Catal. Lett.* 141: 525–530.

hown, I., H. C. Hsu, Y. C. Chang, C. H. Lin, P. K. Roy, and A. Ganguly. 2014. Highly efficient visible light photocatalytic reduction of CO_2 to hydrocarbon fuels by Cu-nanoparticle decorated graphene oxide. *Nano Lett.* 14: 6097–6103.

ong, G., F. Xin, and X. Yin. 2015. Photocatalytic reduction of carbon dioxide over $ZnFe_2O_4$/ TiO_2 nanobelts heterostructure in cyclohexanol. *J. Colloid Interf. Sci.* 442: 60–66.

ahir, A. and K. G. U. Wijayantha. 2010. Photoelectrochemical water splitting at nanostructured $ZnFe_2O_4$ electrodes. *J. Photochem. Photobiol. A Chem.* 216 (2–3): 119–119.

ázquez-Cuchillo, O. R. Gómez, A. Cruz-López, L. M. Torres-Martínez, R. Zanella, F. J. A. Sandoval, and K. D. Ángel-Sánchez. 2013. Improving water splitting using RuO_2-Zr/ $Na_2Ti_6O_{13}$ as a photocatalyst. *J. Photochem. Photobiol. A Chem.* 266: 6–11.

/ang, Y., F. Wang, Y. Chen, D. Zhang, B. Li, S. Kang, X. Li, and L. Cui. 2014. Enhanced photocatalytic performance of ordered mesoporous Fe-doped CeO_2 catalysts for the reduction of CO_2 with H_2O under simulated solar irradiation. *Appl. Catal. B Environ.* 147: 602–609.

/ang, Z., K. Teramura, S. Hosokawa, and T. Tanaka. 2015. Photocatalytic conversion of CO_2 in water over Ag-modified $La_2Ti_2O_7$. *Appl. Catal. B Environ.* 163: 241–247.

Wen, F. and C. Li. 2013. Hybrid artificial photosynthetic systems comprising semiconducto as light harvesters and biomimetic complexes as molecular cocatalysts. *Acc. Chem. Re* 46(11): 2355–2364.

Yang, M. and X. Q. Jin. 2014. Facile synthesis of Zn_2GeO_4 nanorods toward improved ph tocatalytic reduction of CO_2 into renewable hydrocarbon fuel. *J. Cent. South Univ.* 2 2837–2842.

Yu, X., R. Du, B. Li, Y. Zhang, H. Liu, J. Qu, and X. An. 2016. Biomolecule-assisted sel assembly of CdS/MoS_2/graphene hollow spheres as high-efficiency photocatalysts f hydrogen evolution without noble metals. *Appl. Catal. B Environ.* 182: 504–512.

Yuan, Y. P., S. W. Cao, Y. S. Liao, L. S. Yin, and C. Xue. 2013. Red phosphor/g-C_3N_4 heter junction with enhanced photocatalytic activities for solar fuels production. *Appl. Catc B Environ.* 140–141: 164–168.

16 Medical Applications

16.1 INTRODUCTION

Usually, antibacterial agents are used in the medication of bacterial infections. Antibacterial agents are originated from any substance that is of natural, semisynthetic, or synthetic origin. They can successively kill or inhibit the growth of microorganisms while causing little or no damage to the host cell. Generally, the word antimicrobial was obtained from the combination of the Greek words "anti" meaning against, "mikros" meaning little, and "bios" meaning life. Antimicrobial agents act against all types of microorganisms such as bacteria (antibacterial), viruses (antiviral), fungi (antifungal), and protozoa (antiprotozoal).

Nanoparticles have a unique quality of resistance toward an infection and this has been much utilized in the past few decades. Nanoparticles were employed to control the formation of biofilms in oral cavity as a function of their biocidal, antiadhesive, and delivery capabilities, and also were used in prosthetic devices as topical applied agents and in dental materials. Some particular nanoparticulate systems such as silver, copper, zinc, silicon, and their oxides have also been explored for their antibacterial effects.

Antimicrobial agents can be classified depending on their activity on the microorganism, whether they act against bacterial infection like antibacterials or as antifungals acting against the survival of fungi. The agents that kill microbes are termed in general as microbicidal, while those agents that only inhibit microbial growth are called biostatic. Antimicrobial chemotherapy is the use of antimicrobial medicines to treat infection, whereas antimicrobial prophylaxis is the use of antimicrobial medicines to prevent infection.

Wastewater is completely contaminated and it contains high levels of microorganisms along with some organic compounds; therefore, sterilization of water is being carried out in biological and biochemical industries by the use of some important and essential technologies. The chlorination of ground water containing high total organic carbon (TOC) produces invalid high levels of trihalomethanes (THMs) and also some carcinogenic disinfection by-products (DBPs). As a consequence, inactivation of organisms and decomposition of organic compounds is to be performed in such a way that the formation of DBP is minimized, if not completely stopped. Many approaches such as the use of chlorine dioxide, ozone, ultraviolet (UV) radiation, advanced filtration processes, and photocatalytic oxidation have been tried for this purpose. Among these approaches, photocatalytic oxidation is considered to be the most convenient, eco-friendly, and least expensive method for the inactivation of microorganisms.

Usually, microorganisms such as bacteria, virus, and fungi are not decomposed in contaminated materials (e.g., water and air) but it is interesting to note that in the majority of cases the whole microbe cell is decomposed itself in the presence of any photocatalyst. Titanium dioxide is a versatile photocatalyst, which shows relatively better performance. It is nonhazardous, less expensive, and shows good photocatalytic activity in UV light producing reactive oxygen species (ROS) including hydroxyl radicals ($^\bullet$OH). These ROS are responsible for the inactivation of microorganisms. Therefore, TiO_2 has been extensively used for the last two to three decades for the inactivation of microorganisms.

The use of photocatalysts as a antimicrobial agent has many advantages, that is, they can not only kill bacteria and fungi but also inhibit the growth of virus and moreover, they can decompose airborne toxic and volatile organic compounds (VOCs). In this catalytic degradation reaction, water and carbon dioxide are produced as by-products, which are then used again in the process of photosynthesis thus completing this process in an eco-friendly manner.

Photocatalytic techniques have been used for disinfection in recent years and they have enough potential for some widespread applications such as in indoor air and environmental health, biological, medical, laboratory, hospital, pharmaceutical and food industry, as well as use in plant protection, wastewater and effluent treatment, and drinking water disinfection to protect from microorganisms (Gamage and Zhang 2010).

Nanosized photocatalytic materials are now commonly used and these can be applied in many fields, one of which is in treatment of microbial infections. It was observed that the antimicrobial activity of a photocatalyst is enhanced when it is taken in nanometric range (1–100 nm). Numerous nanosized and modified photocatalysts have been used as antimicrobial and antitumor agents. These nanosized antimicrobial agents inactivate microorganisms more successfully as compared to the micro- or macrosized counterparts.

Presently, nanotechnology and nanomaterials are rapidly growing fields of medical research, chemotherapy treatment, antibiotics, and bactericidal materials. Allahverdiyev et al. (2011) suggested that metal oxide nanoparticles, especially TiO_2 and Ag_2O nanoparticles, show significant antibacterial activity, and therefore these can be used as antibacterial agents, eukaryotic infectious agents, and also in other antimicrobial applications. Metal oxide nanoparticles are toxic in nature and restricted for human use. However, metal oxide nanoparticles are the first choice for antibacterial and antiparasitic applications.

Titanium dioxide has resistant qualities toward UV rays, and it can also act as a UV absorbent. Nowadays, TiO_2 is extensively used in pharmaceutical industries, painting, food industries as a coloring agent, and sunscreens, as well as in cosmetics. The role of light on nonphototrophic microbial activity was studied by Lu et al. (2012). Photoelectrons were excited from metal oxide and metal sulfide on visible light irradiation. This process also stimulates chemoautotrophic and heterotrophic bacterial growth. It was suggested that the observed bacterial growth was dependent on intensity and wavelength of light, and also that the growth pattern matched with the light absorption spectra of the minerals.

16.1.1 Photocatalytic Sterilization of Bacteria/Microbial Cells

A novel idea of photochemical sterilization of microbial cells has been suggested by Matsunaga et al. (1985). Microbial cells were killed by using platinum-loaded titanium oxide (TiO$_2$/Pt) in this photoelectrochemical process. *Lactobacillus acidophilus, Saccharomyces cerevisiae*, and *Escherichia coli* bacteria incubated with TiO$_2$/Pt particles were completely sterilized under irradiation for 60–120 minutes. Bactericidal action of illuminated TiO$_2$ on *E. coli* K12 and bacterial consortium was studied by Rincón and Pulgarin (2004). They observed that *E. coli* concentration was increased after illumination without TiO$_2$, while it decreases in the presence of TiO$_2$ but it is sustained in the dark even with TiO$_2$. No regrowth was observed within the following 60 hours. The bactericidal effect on two different contaminated samples was observed in photolytic and photocatalytic experiments, and the effective disinfection time (EDT) was attained under both these conditions. It was also observed that the chemical oxygen demand (COD) decreases during photocatalytic treatment but not during the photolytic condition, while dissolved organic carbon (DOC) increases in both these cases and the overall COD/DOC ratio decreases during phototreatments.

Park et al. (2012) used TiO$_2$ nanoparticles to find out its toxicity for microorganisms. *E. coli, Bacillus subtilis*, and *S. cerevisiae* were undertaken to examine the antimicrobial activity of TiO$_2$. *E. coli* showed the lowest survival rate (36%), while *S. cerevisiae* showed the highest survival rate (71%). It was reported that antimicrobial effect of TiO$_2$ is dependent on UV wavelength, and therefore the survival ratio of *E. coli* was 40% and 80% at 254 nm and 365 nm wavelengths, respectively. *Aeromonas hydrophila* bacteria mediated by the biosynthesis of titanium dioxide nanoparticles (TiO$_2$ NPs) were reported by Jayaseelan et al. (2013). *A. hydrophila* bacteria consist of components like uric acid (2.95%), glycyl-L-glutamic acid (6.90%), glycyl-L-proline (74.41%), and L-leucyl-D-leucine (15.74%) and glycyl-L-proline as capping agent. The antibacterial activity of TiO$_2$ NPs was also examined toward *E. coli, Pseudomonas aeruginosa, Staphylococcus aureus, Streptococcus pyogenes*, and *Enterococcus faecalis*.

Tallósy et al. (2014) investigated the photocatalytic effect of nanosilver-modified TiO$_2$ and ZnO (Ag–TiO$_2$ and Ag–ZnO, respectively) photocatalysts/polymer (ethyl acrylate-co-methyl methacrylate nanohybrid films against methicillin-resistant *S. aureus* [MRSA]). It was observed that the photocatalyst/polymer nanohybrid thin films could inactivate 99.9% of MRSA after 2 hours (Figure 16.1).

16.2 ANTIMICROBIAL ACTIVITY

A photocatalyst initially damages the weak surface points of bacterial cells and then completely decomposes its membranes and the rest of the components leak out from the damaged sites of cell. In this way, debris cells are oxidized by a photocatalytic reaction. Thus, the photocatalysts are a promising candidate for the disinfection of pathogenic bacteria and also in the prevention of infecting diseases because they are more active in visible light than in UV light and do not require harmful UV light irradiation to function (Liou and Chang 2012).

The bactericidal efficiency and killing mechanism of Ag/AgBr/TiO$_2$ under visible light irradiation were discussed by Hu et al. (2006). The Ag/AgBr/TiO$_2$ was prepared

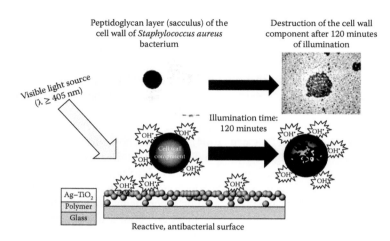

Peptidoglycan layer (sacculus) of the cell wall of *Staphylococcus aureus* bacterium

Destruction of the cell wall component after 120 minutes of illumination

Visible light source (λ ≥ 405 nm)

Illumination time: 120 minutes

Cell wall component

OH·

Ag–TiO₂
Polymer
Glass

Reactive, antibacterial surface

FIGURE 16.1 Antibacterial effects of photocatalysts embedded in polymer thin films. (Adapted from Tallósy, S.P. et al., *Environ. Sci. Poll. Res. Int.*, 21, 11155–11167, 2014. With permission.)

by the deposition–precipitation method and used in the degradation of nonbiodegradable azo dyes and in the killing of *E. coli* under visible light irradiation. It was found that the surface of Ag species exists as Ag^0 and it worked as scavenger for the hole in the valence band and then trapped electrons in the conduction band in the process of photocatalytic reaction to inhibit the decomposition of AgBr. Therefore, AgBr is the main photoactive species for the demolition of azo dyes and bacteria under visible light.

Evans and Sheel (2007) introduced a novel combination of titania thin films on stainless steel. In this process, a flame-assisted chemical vapor deposition (FACVD) method was used for silica deposition, and thermal atmospheric pressure chemical vapor deposition (APCVD) was used for titania deposition. Its photocatalytic and antibacterial activity was also investigated. Thin films were used for water purification, air cleaning, self-sterilizing, and self-cleaning surfaces such as those used in hospitals or food preparation.

Gondal et al. (2009) studied photocatalytic activity and disinfecting activity of nano-WO_3 against *E. coli* in water. The bacterial decay rate was estimated for different concentrations of catalytic and laser pulse energies. The decay rate increased as high as 0.94 per minute as compared to 0.65 per minute for the microstructured WO_3 under the same experimental conditions.

The photodegradation of *E. coli* bacteria in the presence of Ag–TiO_2/Ag/a-TiO_2 nanocomposite film was investigated by Akhavan (2009). Both Ag/a-TiO_2 and Ag–TiO_2/Ag/a-TiO_2 nanocomposite films were synthesized by sol–gel deposition of a 30 nm Ag–TiO_2 layer on approximately 200 nm anatase (a-)TiO_2 film. The antibacterial activity against *E. coli* bacteria of Ag–TiO_2/Ag/a-TiO_2 nanocomposite was 5.1 times more than a-TiO_2 in the dark. The photoantibacterial activity of the nanocomposite film was 1.35 and 6.90 times better than the activity of the Ag/a-TiO_2 and

TiO_2, respectively. It was observed that the stability of nanocomposite film was least 11 times higher than the Ag/a-TiO_2 film. Therefore, the Ag–TiO_2/Ag/a-TiO_2 ⸱otocatalyst has been suggested as one of the effective and long-lasting antibacte- al nanocomposite materials. Visible light-active anatase/rutile mixed phase carbon- ⸱ntaining TiO_2 was used in bacteria killing by Cheng et al. (2009). Its antimicrobial ⸱tivity was investigated against S. aureus, Shigella flexneri, and Acinetobacter bau- ⸱annii. Therefore, such materials, which have high bacterial interaction ability, are ⸱eful in improving the antimicrobial activity of TiO_2.

Yuan et al. (2010) reported the antibacterial properties of TiO_2, N–TiO_2, and 1% ⸱g–N–TiO_2 nanoparticles against E. coli and B. subtilis. It was observed that antibac- rial properties of TiO_2 nanoparticles were increased (codoping) under fluorescent ⸱ht irradiation after doping of Ag and N. Modified 1% Ag–N–TiO_2 showed highest ⸱tibacterial activity with a clear antibacterial circle of 33.0 and 22.8 mm toward E. coli ⸱d B. subtilis, respectively. Fe–N–doped TiO_2 nanocrystals have been synthesized by ⸱e et al. (2009). As-synthesized nanocrystals were in anatase phase with 10 nm size ⸱d the ratio of Fe/Ti and N/O in Fe–N–doped TiO_2 was 2.2 and 0.8 at%, respectively. ⸱he mixture of organic silicon and acrylic syrup with TiO_2 powders was prepared to ⸱st antibacterial performance by the colony counting method. It was observed that the ⸱erilization ratio of E. coli by heat-treated Fe–N–doped nanocrystalline TiO_2 powders ⸱ached up to 94.5%, while without any heat treatment, it was 91.1% with 8-hour irra- ⸱ation using 400 lux visible light and 55% relative humidity (RH).

Wu et al. (2010a) studied bactericidal activity of TiO_2 nanoparticles codoped ⸱ith nitrogen and silver (Ag_2O/TiON) against E. coli under visible light irradiation ⸱ > 400 nm). It was observed that production of ·OH under visible light irradiation was ⸱hanced by the addition of Ag. In photocatalytic reaction of Ag_2O/TiON, electrons in ⸱e conduction band were trapped by Ag_2O species, which inhibit the recombination of ⸱ectrons and holes, and as a result, it shows strong photocatalytic bactericidal activity. ⸱he killing mechanism of Ag_2O/TiON was related to oxidative damages in the forms ⸱f cell wall thinning and cell disconfiguration under visible light irradiation.

A montmorillonite (MMT)-supported Ag/TiO_2 composite (Ag/TiO_2/MMT) has ⸱een prepared through a one-step, low-temperature solvothermal technique by Wu ⸱ al. (2010b). As-prepared Ag/TiO_2/MMT composite had high photocatalytic activ- ⸱y and good recycling performance to degrade E. coli bacteria under visible light. ⸱his MMT layer prevents the loss of the catalyst during recycling test. Ag/TiO_2/ ⸱MT composite was responsible for the enhancement of surface active centers and ⸱e localized surface plasmon effect of the Ag nanoparticles and was found to be ⸱ighly active in visible light. Ag/TiO_2/MMT composite has excellent stability, recy- ⸱ability, and bactericidal activities, and therefore it may be considered as a promis- ⸱g photocatalyst for the application in decontamination and disinfection.

The antibacterial activities of neodymium and iodine doped TiO_2 (Nd:I:TiO_2) ⸱gainst E. coli and S. aureus have been studied by Jiang et al. (2010). The activity of ⸱. coli and S. aureus was completely inhibited by the destruction of its outer layer by ⸱d:I:TiO_2 photocatalyst under visible light irradiation. Other than this, Ag_2O/TiON ⸱anoparticles were also used for the inactivation of E. coli.

The TiO_2 photocatalyst is very capable of killing a wide range of Gram-negative ⸱nd Gram-positive bacteria, filamentous and unicellular fungi, algae, protozoa, mammalian viruses, and bacteriophages. Cell wall and cytoplasmic membrane we decomposed by ROS such as ˙OH and hydrogen peroxide in this photocatalytic pr cess. ROS species initially attack on the cellular contents of cell, then the whole c undergoes the lysis process, and finally mineralization of the organism takes place the end of reaction. A direct contact between organisms and TiO_2 catalyst as well the presence of other antimicrobial agents like Cu and Ag increases the decompos tion or killing of the cell (Foster et al. 2011).

Nano-NiO photocatalyst was synthesized by the sol–gel method and utilized disinfecting water infected with *E. coli* in conjunction with 355 nm laser radiatio The dependence of depletion rate of bacterial count in the infected water on t nano-NiO concentration and the irradiating laser pulse energy was carried out l Gondal et al. (2011). The bacterial decay rate constant for nano-NiO was 0.35 p minute, which is higher than 0.24 per minute for TiO_2 as a photocatalyst under t same catalytic concentration and laser pulse energy.

A novel series of layered niobate $K_4Nb_6O_{17}$ photocatalyst with Ag–Cu nanocon posite cocatalyst was designed by Lin and Lin (2012) for the inactivation of *E. c* under visible light irradiation. The effects of loading method of Cu species on t characteristics of Ag–Cu nanocomposites and photocatalytic antibacterial activi were studied. The loading of Ag, Cu, and Ag–Cu composite as cocatalyst leads increased antibacterial activity as compared to bare $K_4Nb_6O_{17}$. The presence of A$ Cu nanocomposite on $K_4Nb_6O_{17}$ surface significantly enhanced the electron–ho pair separation efficiency as well as the synergistic effects of coexisting Ag and C ions on antibacterial activity. The marked improvement of photocatalytic activi and photokilling ability of Ag–Cu/$K_4Nb_6O_{17}$ film was observed.

The antibacterial activity of V_2O_5-loaded TiO_2 nanoparticles against *E. coli* w reported by Kim et al. (2012). *Escherichia coli* was significantly reduced by van dium pentoxide–loaded TiO_2 at a rate better than pure TiO_2 under illumination wi fluorescent light. V_2O_5-loaded TiO_2 was found to be more active due to the chang in its surface conditions, which were created due to loading of V_2O_5. Thus, it w suggested that preferably V_2O_5–TiO_2 shows much better performance as compare to pure TiO_2 but under UVA (352 nm) irradiation, both V_2O_5-loaded TiO_2 and pu TiO_2 showed almost similar activity toward bacteria.

A core–shell-structured $In_2O_3@CaIn_2O_4$ substrate shows superior visible ligh induced bactericidal properties, as compared to other visible light–responsive ph tocatalysts, which are commercially available or synthesized (Chang et al. 2012 Photoexcited electrons were easily transferred between In_2O_3 and $CaIn_2O_4$ interfac in the photocatalysis process to minimize the electron–hole recombination as we as to enhance the performance of the photocatalyst. $In_2O_3@CaIn_2O_4$ did not induc significant cell death and tissue damage, implying a superior biocompatibility compared to TiO_2-based photocatalysts. Thus, $In_2O_3@CaIn_2O_4$ was suggested in th application for the development of a safer and highly bactericidal photocatalyst.

Cui et al. (2013) synthesized γ-$Fe_2O_3@SiO_2@TiO_2$–Ag nanocomposit with a core–shell structure using a hydrothermal method and a sol–gel metho As-synthesized γ-$Fe_2O_3@SiO_2@TiO_2$–Ag was also used to examine photocatalyt reaction toward methyl orange solution under UV irradiation. γ-$Fe_2O_3@SiO_2@TiO_2$ Ag nanocomposites showed remarkable antibacterial activity greater than that bare TiO_2 nanoparticles. The addition of silver nanoparticles into the TiO_2 matrix facilitates charge separation by trapping photogenerated electrons, thereby enhancing biological activity and photoactivity.

The antibacterial activities of Nd-doped and Ag-coated TiO_2 nanoparticles have been investigated against *S. aureus* and *E. coli* by Bokare et al. (2013). It was observed that the order of activity is as follows:

$$\text{Undoped } TiO_2 < \text{Nd-doped } TiO_2 < \text{Ag-coated } TiO_2$$

Pathakoti et al. (2013) reported visible light–activated sulfur-doped TiO_2 (S–TiO_2) and nitrogen–fluorine-codoped TiO_2 (N–F–TiO_2) and their performance in the photoinactivation of some bacteria. The photoinactivation performance of S–TiO_2 and N–F–TiO_2 was tested against *E. coli* under solar simulated light and visible light irradiation and it was compared with commercially available TiO_2. TiO_2 performed much better than others under solar light irradiation, while S–TiO_2 showed a moderate toxicity, whereas N–F–TiO_2 did not show any toxicity. Ag-modified carbon-doped TiO_2 (C–TiO_2) has also been used for the disinfection of microorganisms under visible light irradiation. The loaded Ag nanoparticles were uniformly distributed on the TiO_2 surface and they show enhanced absorption in visible light due to surface plasmon resonance yet inhibit the charge carrier recombination by conduction band electron trapping. Their performance in irradiation and disinfection of *E. coli* and *E. faecalis* was also observed in comparison with C–TiO_2. When loading of Ag nanoparticles was increased from 0.5 to 5.0 wt.%, absorption as well as charge carrier separation was also improved because of inherent bactericidal effect of metallic Ag (Zhang et al. 2013b).

Xu et al. (2014) proposed magnetic $Ag_3PO_4/TiO_2/Fe_3O_4$ heterostructured nanocomposite, which has excellent bactericidal activity and recyclability toward *E. coli* cells under visible light irradiation. The intrinsic cytotoxicity of Ag ions may elevate bactericidal efficiency of $Ag_3PO_4/TiO_2/Fe_3O_4$ with enhanced photocatalytic activity. The formation of $^{\bullet}OH$ and superoxide ions at interfaces was responsible for morphological changes in the cells of microorganism and led to the death of the bacteria.

The $Ag@CeO_2$ nanocomposites were synthesized by a biogenic and green approach using electrochemically active biofilms (EABs) as a reducing tool. These were characterized and utilized for antimicrobial and visible light photocatalytic activity and photoelectrode by Khan et al. (2014). As-synthesized nanocomposites have effective and efficient bactericidal activities and survival tests against *E. coli* O157:H7 and *P. aeruginosa*. However, $Ag@CeO_2$ nanocomposites also showed enhanced photocatalytic degradation of 4-nitrophenol and methylene blue compared to pure CeO_2 under visible light, which confirmed that the $Ag@CeO_2$ nanocomposites had excellent visible light photocatalytic activities as compared to pure CeO_2. Ag-NPs anchored at CeO_2 induced photoactivity in visible light by reducing recombination of photogenerated electrons and holes. It was suggested that as-synthesized $Ag@CeO_2$ nanocomposites are smart materials that can be used as an antimicrobial agent.

Yadav et al. (2014) investigated photocatalytic inactivation of pathogenic bacteria *E. coli* and *S. aureus* by using copper-doped titanium dioxide (Cu–TiO_2) nanoparticles under visible light irradiation. It was confirmed that bactericidal activity or bacterial survival was not affected even in contact with nanoparticles

TABLE 16.1
Some Nanosized Photocatalysts Used in Inactivating/Killing Microorganism$

Microorganism	Photocatalyst Nanoparticles	References
1 *Escherichia coli*	Cu/SiO$_2$ composite	Trapalis et al. (2003)
2 Gram-positive (*E. coli.* and *Pseudomonas aeruginosa*) and Gram-negative bacteria (*Staphylococcus aureus* and *Bacillus subtilis*)	ZnO, CuO, and Fe$_2$O$_3$	Azam et al. (2012)
3 Gram-positive/Gram-negative bacteria	CuO	Padil and Černík (2013
4 Gram-positive/Gram-negative bacteria	ZnO	Siddique et al. (2013)
5 *Prevotella intermedia, Porphyromonas gingivalis, Fusobacterium nucleatum,* and *Aggregatibacter actinomycetemcomitans*	Ag, Cu$_2$O, CuO, ZnO, TiO$_2$, WO$_3$, Ag + CuO composite, Ag + ZnO composite	Vargas-Reus et al. (2012)
6 Gram-negative bacteria *E. coli* and *P. aeruginosa*, Gram-positive bacterium *S. aureus*	ZnO	Premanathan et al. (2011)
7 Gram positive (*S. aureus* and *B. subtilis*) and Gram-negative (*E. coli* and *Aerobacter aerogenes*)	ZnO nanorods	Jain et al. (2013)
8 *Aeromonas hydrophila*	ZnO NPs	Jayaseelan et al. (2012)
9 Gram-negative (*P. aeruginosa*) and Gram-positive bacteria (*Staphylococcus epidermidis*) and pathogenic yeast (*Candida tropicalis*)	CaO NPs	Roy et al. (2013)
10 Gram-negative/Gram-positive bacteria	Fe$_2$O$_3$/Polyrhodanine NPs	Kong et al. (2010)
11 *E. coli* (Gram negative) and *S. aureus* (Gram positive)	ZnO nanocrystals	Perelshtein et al. (2009
12 *S. aureus*	ZnO NPs	Jones et al. (2008)
13 *Shigella flexneri, S. aureus, S. epidermidis, Salmonella typhimurium, B. subtilis, E. coli, Vibrio cholera, P. aeruginosa,* and *Aeromonas liquefaciens*	CuO nanoflakes	Pandiyarajan et al. (2013)
14 *E. coli, B. subtilis,* and *S. aureus*	CuO, NiO, ZnO, and Sb$_2$O$_3$	Baek and An (2011)
15 *E. coli* (Gram-negative) and *S. aureus* (Gram-positive) bacteria	Zn-doped CuO (Cu$_{0.88}$Zn$_{0.12}$O)	Malka et al. (2013)
16 *E. coli* and *S. aureus*	ZnO nanoparticles	Banoee et al. (2010)
17 *E. coli, S. aureus,* and *P. aeruginosa* bacteria	Sn doped ZnO	Jan et al. (2013)
18 *S. aureus*	Carbon-coated ZnO (ZnOCC)	Sawai et al. (2007)
19 *S. aureus*	Cadmium oxide	Salehi et al. (2014)
20 *Botrytis cinerea* and *Penicillium expansum* (postharvest pathogenic fungi)	ZnO NPs	He et al. (2011)
21 *Listeria monocytogenes, Salmonella enteritidis,* and *E. coli* O157:H7	ZnO QDs	Jin et al. (2009)

Notes: NPs, nanoparticles; QDs, quantum dots.

in the dark. Some more nanophotocatalysts have been used by various workers for inactivating or killing bacteria (Table 16.1).

16.3 ANTIFUNGAL ACTIVITY

The antifungal activity of TiO_2/TiO_2 coated on a plastic film against *Penicillium expansum* and its photocatalytic activity was investigated *in vitro* and in fruit tests by Maneerat and Hayata (2006). The mixture of *P. expansum* conidial suspension and TiO_2 powder was added to potato dextrose agar (PDA) plates for an *in vitro* test. The photocatalytic reaction of TiO_2 reduces conidial germination of fungal pathogen. It also suppresses the growth of *P. expansum*, and with increasing TiO_2 amount, a reduction of *P. expansum* colonies was observed. Both TiO_2 powder and TiO_2-coated film exhibit antifungal activity to control fruit rot in a fruit inoculation test. Therefore, it was suggested that photocatalytic reaction of TiO_2 and its antifungal activity against *P. expansum* may have potential for postharvest disease control.

Darbari et al. (2011) investigated and compared the antifungal effect of TiO_2/branched carbon nanotube (CNT) with thin films of TiO_2 on *Candida albicans* biofilms under visible light. The results showed that TiO_2/CNTs exhibited highly improved photocatalytic antifungal activity as compared to TiO_2 film and it was attributed to the generation of electron–hole pairs. The TiO_2/CNTs provide high surface area for the interaction between the cells and the nanostructures. They have low recombination rate under visible light excitation. The branched CNT arrays were synthesized by plasma-enhanced chemical vapor deposition on a silicon substrate. Ni was used as the catalyst in this process and played an important role in the realization of branches in vertically aligned nanotubes. The APCVD method followed by a 500°C annealing step was used to produce TiO_2 nanoparticles on the branched CNTs. The fungicidal activity of $CuO/Cu(OH)_2$ nanostructures was studied by Azimirad and Safa (2014). The nanostructures such as nanoflakes of $CuO/Cu(OH)_2$ grown in ammonia bath showed the strongest antifungal and photocatalytic activity.

Yaithongkum et al. (2011) synthesized $TiO_2/SnO_2/SiO_2$ nanocomposite powders and investigated the effect of 0.1–1 mol%Ag doping on crystallite size, morphology, and photocatalytic and fungal growth suppression activities. The photocatalytic activities of $TiO_2/SnO_2/SiO_2$ toward *P. expansum* growth suppression were correlated under UV radiation. It could completely kill *P. expansum* within 1 day of photocatalytic treatment under UV irradiation.

The antimicrobial activity of zinc sulfate (100 nm) was evaluated by a well diffusion method. The zinc sulfate was synthesized by a ball milling method and characterized. The zinc sulfate nanomaterial in 0.01%, 0.05%, 0.1%, and 1.0% was used to study its influence on growth and death kinetics of all phytopathogens. The microbicidal activity of zinc sulfate exhibited an inhibition zone of 18, 14, 12, and 10 mm against *Pseudomonas solanacearum*, *Pseudomonas syringae*, *Xanthomonas malvacearum*, and *Xanthomonas campestris*, respectively. The presence of 1% of $ZnSO_4$ nanomaterial was very effective to control and destroy all phytopathogens from 1 hour onwards. Thus, zinc nanomaterial is significantly promising as a photocatalyst to control phytopathogens (Indhumathy and Mala 2013).

Antifungal activity of palladium-modified nitrogen-doped titanium oxide photo-catalyst on agricultural pathogenic fungi *Fusarium graminearum* was investigated by Zhang et al. (2013a). Eco-friendly alternative TiON/PdO nanoparticulate photo-catalyst was used for the disinfection of *F. graminearum* macroconidia under visible light illumination. The photocatalytic disinfection of these macroconidia success-fully occurred because of opposite surface charges on both *F. graminearum* macro-conidium and TiON/PdO nanoparticles surface. The cell wall of macroconidia was damaged by the attack from ROS.

16.4   ANTIVIRAL ACTIVITY

Hajkova et al. (2007) reported the photocatalytic decomposition of organic matter and antibacterial and antiviral effects of TiO_2 thin films. A plasma-enhanced chemi-cal vapor deposition (PECVD) method was used to deposit TiO_2 films on glass sub-strates. Titanium isopropoxide (TTIP) was used as a precursor for oxygen plasma discharge in a vacuum reactor with radio frequency (RF) at low temperature. The effect of organic matter, acid orange 7, on photocatalytic decomposition of bacteria *E. coli* and viruses (herpes simplex virus [HSV-1]) was explained.

The photochemical sterilization ability of TiO_2 nanomaterial as an eco-friendly disin-fectant against avian influenza (AI) was studied by Cu et al. (2010). A neutral and viscous aqueous colloid of 1.6% TiO_2 was prepared from peroxotitanic acid solution using the Ichinose method. The TiO_2 particles were spindle-shaped with an average size of 50 nm. A photocatalytic film of nano-TiO_2 sol was used for inactivating H9N2 avian influenza virus (AIV). Such inactivation capabilities were observed with 365 nm UV light.

The antiviral activity of nanosized cuprous iodide (CuI) particles having an aver-age size of 160 nm was examined by Fujimori et al. (2012). CuI particles showed aqueous stability and generated ·OH, which was probably derived from monovalent copper (Cu^+). It was confirmed that CuI particles showed antiviral activity against an influenza A virus of swine origin (pandemic [H1N1] 2009) by plaque titration assay. The virus titer decreased in a dose-dependent manner upon incubation with CuI particles, with the 50% effective concentration being approximately 17 μg/mL after exposure for 60 minutes. It was confirmed that the inactivation of the virus was due to the degradation of viral proteins such as hemagglutinin and neuraminidase by CuI. CuI generates ·OH in aqueous solution, and radical production was found to be blocked by the radical scavenger *N*-acetylcysteine. These findings indicate that CuI particles exert antiviral activity by generating ·OH. Thus, CuI may be a useful mate-rial for protecting against viral attacks and may be suitable for applications such as filters, face masks, protective clothing, and kitchen cloths.

Ishiguro et al. (2013) investigated that photocatalytic active Cu^{2+}/TiO_2–coated cordierite foam inactivates bacteriophages and *Legionella pneumophila*. TiO_2–coated cordierite foam is generally used in air cleaners due to its antiviral activity. They also synthesized Cu^{2+}/TiO_2–coated cordierite foam and studied its utilization in the reduction in viral infection ratio. It was found that Cu^{2+}/TiO_2–coated cordierite foam reduced more efficiently the viral infection ratio as compared to TiO_2–coated cordierite foam. It was suggested that Cu^{2+}/TiO_2–coated cordierite foam can be used in air cleaners to reduce infection risk by polluted air.

5.5 ANTICANCER ACTIVITY

ancer has remained the top cause of death for a long time, but in the majority of ses, it remained undetected. Therefore, cancer treatment is an issue of high con-rn and now it has been integrated with photocatalysis. Certain treatments, such surgical, radiological, immunological, thermotherapeutic, and chemotherapeutic, ere developed for combating against this dreadful disease.

Intense research on the behavior of tumor cells on photocatalytic semiconductor rface has been carried out by Fujishima et al. (1986). They were quite interested killing the tumor cells via a very strong oxidizing agent generated in TiO_2 under umination. It was observed that the polarized, illuminated TiO_2 film electrode, well as TiO_2 colloidal suspension, was effective in killing HeLa cells. After this servation, a series of studies followed, where various tentative conditions were amined, including the effect of superoxide, which increases due to the production peroxide (Cai et al. 1991, 1992a, 1992b). Selective killing of a single cancerous 24 cell by using illuminated TiO_2 microelectrode was also reported by Sakai et al. 995). This anodically polarized TiO_2 microelectrode successfully inactivated the 24 cell under UV light irradiation. However, it was observed that when the micro-ectrode was located at 10 μm away from the cell surface, it was not able to kill the ll. Thus, it was concluded that the photogenerated holes and/or active oxygen spe-es with short diffusion length are responsible for the cell death process.

Rozhkova et al. (2009) reported high performance of nanobio-photocatalyst for rgeted brain cancer therapy (Figure 16.2). Here, TiO_2 particles have been used to ontrol the phototoxicity of antiglioblastoma cell. TiO_2 was covalently tethered to an itibody via a dihydroxybenzene bivalent linker. The phototoxicity is protected by OS, which is responsible for cancer cell death.

Liu et al. (2010) reported a noble metal Pt/TiO_2 nanocomposite for cancer cell eatment. Pt/TiO_2 nanocomposite is an extremely stable metal–semiconductor anomaterial and shows high photodynamic efficiency under mild UV irradiation.

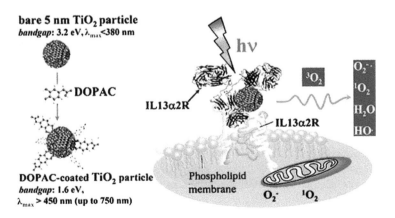

FIGURE 16.2 A nanobio-photocatalyst for targeted brain cancer therapy. (Adapted from ozhkova, E. A. et al., *Nano Lett.*, 9, 3337–3342, 2009. With permission.)

Pt/TiO$_2$ is a more promising nanocomposite than TiO$_2$ and Au/TiO$_2$ nanoparticl for cancer cell treatment. Fe-doped TiO$_2$ nanocomposite was synthesized by deposition–precipitation method and then used as a new "photosensitizer" f photodynamic therapy (PDT).

Rasmussen et al. (2010) reviewed the biomedical applications of metal oxi and ZnO nanomaterials at the experimental, preclinical, and clinical levels. Zn nanomaterials are considered a promising candidate for biomedical applications a therapeutic intervention. The inherent toxicity and selectivity of ZnO nanoparticl against cancer cells make them attractive as anticancer agents. The use of met oxide nanoparticles for cancer applications and drug delivery systems was discusse along with their advantages, approaches, and limitations. They proposed mech nisms of cytotoxic action of metal oxide and ZnO nanomaterials, as well as curre approaches to improve their targeting and cytotoxicity against cancer cells.

The photocatalytic inactivation of Fe-doped TiO$_2$ on human leukemic HL60 cel was investigated by Huang et al. (2012). This PDT reaction chamber was based on light-emitting diode (LED) light source, and the viability of HL60 cells was exar ined by cell counting kit-8 (CCK-8). It was found that the growth of HL60 cells w significantly inhibited by adding TiO$_2$ nanoparticles. The inactivation efficiency TiO$_2$ could be effectively increased by the surface modification of TiO$_2$ nanoparticl with Fe doping. The optimized conditions were obtained at 5 wt.% Fe/TiO$_2$, whe up to 82.5% PDT efficiency for the HL60 cells could be obtained under the irradi tion of 403 nm light (the power density is 5 mW/cm^2) within 60 minutes.

Sato et al. (2011) reported anticancer activity of TiO$_2$ that produces free radica on irradiation with near-UV light. They used the photocatalyst to reduce the contar ination due to anticancer agents in the biological safety cabinet (BSC). A stainle steel plate coated with TiO$_2$ was used with different concentrations of cyclophosph mide dropped or sprayed on it, and then it was irradiated with near-UV for 12 hou in the BSC.

For many years, cancer has continued to become a greater health concern ar there is an urgent need for effective medicines that can treat cancer and at tl same time are safe for human health also. When anticancer drugs are used in larg amounts, they become more harmful. Thus, a new technique was developed f cancer therapy. TiO$_2$ shows unique photocatalytic properties for some biomedic. applications, which motivated researchers to do intense research and experiments well as theoretical studies in this direction. TiO$_2$ is associated with such advantag like unique photocatalytic properties, excellent biocompatibility, and high chemic stability, accompanied by low toxicity. In the present scenario, TiO$_2$ is able to assi in solving some current problems of life sciences, which could be resolved or great improved by its application. Yin et al. (2013) reviewed recent advances in the bi medical applications of TiO$_2$, including PDT for cancer treatment, drug delivery sy tems, cell imaging, biosensors for biological assay, genetic engineering, and so on.

Chemotherapeutic drugs are frequently used in tumor cell treatment. Zhang et a (2015) suggested a novel injectable *in situ* photosensitive inorganic/organic hybr hydrogel as a localized drug delivery system. In this system, poly(ethylene glyco double acrylates (PEGDAs) were used as a polymeric matrix; doxorubicin (DO) as the model drug; and TiO$_2$-multiwalled CNT (TiO$_2$@MWCNT) nanocomposi

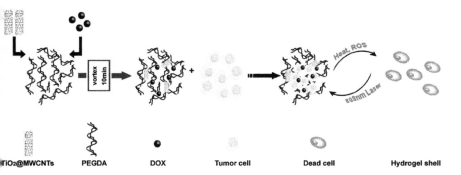

FIGURE 16.3 Use of poly(ethylene glycol) double acrylates, doxorubicin, and TiO_2-multiwalled carbon nanotube (TiO2@MWCNT) nanocomposite for tumor therapy. (Adapted from Zhang, H. et al., *J. Mater. Chem. B.*, 3, 6310–6326, 2015. With permission.)

as the photoinitiator, photosensitizer, and photothermal agent for tumor therapy possessing a multimechanism using a single near infrared (NIR) laser (Figure 16.3). When DOX is present with low accumulation in tumors, it has limited side effects with systemic administration; thus, limiting its therapeutic index. The PEGDA solution containing DOX and TiO_2@MWCNTs was injected into a tumor and this gel was rapidly converted *in vivo* via a photocrosslinking action, which is triggered by an NIR laser. DOX was released from DOX/TiO_2@MWCNTs/PEGDA hydrogel. It was sustained and long lasting, over 10 days, which indicates that the PEGDA gel acted as a drug depot. NIR laser light was absorbed in this process and converted into ROS or local hyperthermia by TiO_2@MWCNTs, leading to death of tumor cells. DOX/TiO_2@MWCNTs/PEGDA hydrogel showed remarkable antiproliferative activities against MCF-7 cancer cells *in vitro*. A single dose of this hydrogel with 808 nm laser irradiation was much effective than the other DOX systems. It was suggested that this novel photosensitive hybrid hydrogel system can afford high drug loading, sustained and stable drug release, as well as repeated phototherapy of the tumor with the administration of a single dose.

Intense research has been carried out with the combination of nanotechnology and material science on intrinsic antimicrobial and anticancer activity. Selvamani et al. (2016) introduced metal@metal tungstate ($Ag@Ag_8W_4O_{16}$) nanoroasted rice beads for antibacterial activity against *E. coli* and *S. aureus*. Müeller-Hinton broth and its anticancer activity against B16F10 cell line were also studied. The silver-decorated silver tungstate ($Ag@Ag_8W_4O_{16}$) was synthesized using a microwave irradiation method by cetyltrimethyl ammonium bromide (CTAB) and characterized.

A newer dimension has been added to photocatalysis by its use in killing some bacteria like *E. coli, B. subtilis, P. aeruginosa, A. hydrophilla* and fungi like *C. albicans, P. expansum, F. graminearum*, and so on, along with antiviral activity against viruses like H1N1, HSV-1, and H9N2. Cancer is a well-known dreadful disease and photocatalysts can also put a check on the growth of some of these cancer cells such as HeLa, T24, HL60, MCF-7, and so on. The time is not for off when photocatalysis will acquire an important position in these fields, particularly in cancer treatment.

REFERENCES

Akhavan, O. 2009. Lasting antibacterial activities of Ag-TiO$_2$/Ag/a-TiO$_2$ nanocomposite thin film photocatalysts under solar light irradiation. *J. Colloid. Interface. Sci.* 336: 117–124.

Allahverdiyev, A. M., E. S. Abamor, M. Bagirova, and M. Rafailovich. 2011. Antimicrobial effects of TiO$_2$ and Ag$_2$O nanoparticles against drug-resistant bacteria and *leishmania* parasites. *Future Microbiol.* 6: 933–940.

Azam, A., A. S. Ahmed, M. Oves, M. S. Khan, S. S. Habib, and A. Memic. 2012. Antimicrobial activity of metal oxide nanoparticles against Gram-positive and Gram-negative bacteria: A comparative study. *Int. J. Nanomedicine.* 7: 6003–6009.

Azimirad, R. and S. Safa. 2014. Photocatalytic and antifungal activity of flower-like copper oxide nanostructures. *Synth. React. Inorg. Metal-Org. Nano-Metal Chem.* 44 (6): 798–803.

Baek, Y. W. and Y. J. An. 2011. Microbial toxicity of metal oxide nanoparticles (CuO, NiO, ZnO, and Sb$_2$O$_3$) to *Escherichia coli, Bacillus subtilis,* and *Streptococcus aureus. Sci. Total Environ.* 409: 1603–1608.

Banoee, M., S. Seif, Z. E. Nazari, P. Jafari-Fesharaki, H. R. Shahverdi, A. Moballegh, K. M. Moghaddam, and A. R. Shahverdi. 2010. ZnO nanoparticles enhanced antibacterial activity of ciprofloxacin against *Staphylococcus aureus* and *Escherichia coli. J. Biomed. Mater. Res. B Appl. Biomater.* 93: 557–561.

Bokare, A., A. Sanap, M. Pai, S. Sabharwal, and A. A. Athawale. 2013. Antibacterial activities of Nd doped and Ag coated TiO$_2$ nanoparticles under solar light irradiation. *Colloids Surf. B Biointerfaces.* 102: 273–280.

Cai, R., K. Hashimoto, Y. Kubota, H. Sakai, and A. Fujishima. 1992a. Phagocytosis of titanium dioxide particles chemically modified by hematoporphyrin. *Denki Kagaku.* 60: 314–321.

Cai, R., K. Hashimoto, K. Itoh, Y. Kubota, and A. Fujishima. 1991. Photokilling of malignant cells with ultra-fine TiO$_2$ powder. *Bull. Chem. Soc. Jpn.* 64: 1268–1273.

Cai, R., K. Hashimoto, Y. Kubota, and A. Fujishima. 1992b. Increment of photocatalytic killing of cancer cells using TiO$_2$ with the aid of superoxide dismutase. *Chem. Lett.* 427–M430.

Chang, W. K., D. S. Sun, H. Chan, P. T. Huang, W. S. Wu, C. H. Lin, Y. H. Tseng, Y. H. Cheng, C. C. Tseng, and H. H. Chang. 2012. Visible light-responsive core-shell structured In$_2$O$_3$@CaIn$_2$O$_4$ photocatalyst with superior bactericidal properties and biocompatibility. *Nanomedicine.* 8: 609–617.

Cheng, C. L., D. S. Sun, W. C. Chu, Y. H. Tseng, H. C. Ho, J. B. Wang et al. 2009. The effects of the bacterial interaction with visible-light responsive titania photocatalyst on the bactericidal performance. *J. Biomed. Sci.* 16: 7. doi:10.1186/1423-0127-16-7.

Cu, H., J. Ji, W. Gu, C. Sun, D. Wu, and T. Yang. 2010. Photocatalytic inactivation efficiency of anatase nano-TiO$_2$ sol on the H9N2 avian influenza virus. *Photochem. Photobiol.* 86 (5): 1135–1139.

Cui, B., H. Peng, H. Xia, X. Guo, and H. Guo. 2013. Magnetically recoverable core–shell nanocomposites γ-Fe$_2$O$_3$@SiO$_2$@TiO$_2$–Ag with enhanced photocatalytic activity and antibacterial activity. *Separ. Purific. Technol.* 103: 251–257.

Darbari, S., Y. Abdi, F. Haghighi, S. Mohajerzadeh, and N. Haghighi. 2011. Investigating the antifungal activity of TiO$_2$ nanoparticles deposited on branched carbon nanotube arrays. *J. Phys. D Appl. Phys.* 44 (24): 245401. doi:10.1088/0022-3727/44/24/245401.

Evans, P. and D. W. Sheel. 2007. Photoactive and antibacterial TiO$_2$ thin films on stainless steel. *Surf. Coat. Technol.* 201 (22–23): 9319–9324.

Foster, H. A., I. B. Ditta, S. Varghese, and A. Steele. 2011. Photocatalytic disinfection using titanium dioxide: Spectrum and mechanism of antimicrobial activity. *Appl. Microbiol. Biotechnol.* 90: 1847–1868.

Fujimori, Y., T. Sato, T. Hayata, T. Nagao, M. Nakayama, T. Nakayama, R. Sugamata, and K. Suzuki. 2012. Novel antiviral characteristics of nanosized copper(I) iodide particles showing inactivation activity against 2009 pandemic H1N1 influenza virus. *Appl. Environ. Microbiol.* 78 (4): 951–955.

Fujishima, A., J. Ohtsuki, T. Yamashita, and S. Hayakawa. 1986. Behavior of tumor cells on photoexcited semiconductor surface. *Photomed. Photobiol.* 8: 45–46.

Gamage, J. and Z. Zhang. 2010. Applications of photocatalytic disinfection. *Int. J. Photoenergy.* 2010: 11 p. Article ID 764870. doi:10.1155/2010/764870.

Gondal, M. A., M. A. Dastageer, and A. Khalil. 2009. Synthesis of nano-WO_3 and its catalytic activity for enhanced antimicrobial process for water purification using laser induced photo-catalysis. *Catal. Commun.* 11 (3): 214–219.

Gondal, M. A., M. A. Dastageer, and A. Khalil. 2011. Nano-NiO as a photocatalyst in antimicrobial activity of infected water using laser induced photo-catalysis. In: Proceedings of the 2011 Saudi International on Electronics, Communications, and Photonics Conference (SIECPC), 24–26 April 2011, Riyadh, pp. 1–5.

Hajkova, P., P. Spatenka, J. Horsky, I. Horska, and A. Kolouch. 2007. Photocatalytic effect of TiO_2 films on viruses and bacteria. *Plasma Proc. Polym.* 4 (1): S397–S401.

He, L., Y. Liu, A. Mustapha, and M. Lin. 2011. Antifungal activity of zinc oxide nanoparticles against *Botrytis cinerea* and *Penicillium expansum*. *Microbiol. Res.* 166: 207–215.

He, R. L., Y. Wei, and W. B. Cao. 2009. Preparation of (Fe, N)-doped TiO_2 powders and their antibacterial activities under visible light irradiation. *J. Nanosci. Nanotechnol.* 9: 1094–1097.

Hu, C., Y. Lan, J. Qu, X. Hu, and A. Wang. 2006. Ag/AgBr/TiO_2 visible light photocatalyst for destruction of azo dyes and bacteria. *J. Phys. Chem. B.* 110 (9): 4066–4072.

Huang, K., L. Chen, M. Liao, and J. Xiong. 2012. The photocatalytic inactivation effect of Fe-doped TiO_2 nanocomposites on leukemic HL60 cells-based photodynamic therapy. *Int. J. Photoenergy.* 2012: 8 p. Article ID 367072. doi:10.1155/2012/367072.

Indhumathy, M. and R. Mala. 2013. Photocatalytic activity of zinc sulphate nano material on phytopathogens, *Int. J. Agri. Environ. Biotechnol.* 6: 737–743.

Ishiguro, H., Y. Yao, R. Nakano, M. Hara, K. Sunada, K. Hashimoto, J. Kajioka, A. Fujishima, and Y. Kubota. 2013. Photocatalytic activity of Cu^{2+}/TiO_2-coated cordierite foam inactivates bacteriophages and Legionella pneumophila. *Appl. Catal. B: Environ.* 129: 56–61.

Jain, A., R. Bhargava, and P. Poddar. 2013. Probing interaction of gram-positive and gram-negative bacterial cells with ZnO nanorods. *Mater. Sci. Eng. C Mater. Biol. Appl.* 33: 1247–1253.

Jan, T., J. Iqbal, M. Ismail, M. Zakaullah, S. H. Naqvi, and N. Badshah. 2013. Sn doping induced enhancement in the activity of ZnO nanostructures against antibiotic resistant *S. aureus* bacteria. *Int. J. Nanomedicine.* 8: 3679–3687.

Jayaseelan, C., A. A. Rahuman, A. V. Kirthi, S. Marimuthu, T. Santhoshkumar, A. Bagavan, K. Gaurav, L. Karthik, and K. V. Rao. 2012. Novel microbial route to synthesize ZnO nanoparticles using *Aeromonas hydrophila* and their activity against pathogenic bacteria and fungi. *Spectrochim. Acta A Mol. Biomol. Spectrosc.* 90: 78–84.

Jayaseelan, C., A. A. Rahuman, S. M. Roopan, A. V. Kirthi, J. Venkatesan, S. K. Kim, M. Iyappan, and C. Siva. 2013. Biological approach to synthesize TiO_2 nanoparticles using *Aeromonas hydrophila* and its antibacterial activity. *Spectrochim. Acta A.* 107: 82–89.

Jiang, X., L. Yang, P. Liu, X. Li, and J. Shen. 2010. The photocatalytic and antibacterial activities of neodymium and iodine doped TiO_2 nanoparticles. *Colloids Surf. B Biointerfaces.* 79: 69–74.

Jin, T., D. Sun, J. Y. Su, H. Zhang, and H. J. Sue. 2009. Antimicrobial efficacy of zinc oxide quantum dots against *Listeria monocytogenes*, *Salmonella enteritidis*, and *Escherichia coli* O157:H7. *J Food Sci.* 74: M46–M52.

Jones, N., B. Ray, K. T. Ranjit, and A. C. Manna. 2008. Antibacterial activity of ZnO nanoparticle suspensions on a broad spectrum of microorganisms. *FEMS Microbiol. Lett.* 279: 71–76.

Khan, M. M., S. A. Ansari, J. H. Lee, M. O. Ansari, J. Lee, and M. H. Cho. 2014. Electrochemically active biofilm assisted synthesis of Ag@CeO$_2$ nanocomposites for antimicrobial activity, photocatalysis and photoelectrodes. *J. Coll. Interface Sci.* 431: 255–263.

Kim, Y. S., M. Y. Song, E. S. Park, S. Chin, G. N. Bae, and J. Jurng. 2012. Visible-light-induced bactericidal activity of vanadium-pentoxide (V$_2$O$_5$)-loaded TiO$_2$ nanoparticles. *Appl. Biochem. Biotechnol.* 168: 1143–1152.

Kong, H., J. Song, and J. Jang. 2010. One-step fabrication of magnetic gamma-Fe$_2$O$_3$/polyrhodanine nanoparticles using in situ chemical oxidation polymerization and their antibacterial properties. *Chem. Commun.* 46: 6735–6737.

Lin, H. Y. and H. M. Lin. 2012. Visible-light photocatalytic inactivation of *Escherichia coli* by K$_4$Nb$_6$O$_{17}$ and Ag/Cu modified K$_4$Nb$_6$O$_{17}$. *J. Hazard. Mater.* 217–218: 231–237.

Liou, J. W. and H. H. Chang, 2012. Bactericidal effects and mechanisms of visible light-responsive titanium dioxide photocatalysts on pathogenic bacteria. *Arch. Immunol. Ther. Exp. (Warsz).* 60: 267–275.

Liu, L., P. Miao, Y. Xu, Z. Tian, Z. Zou, and G. Li. 2010. Study of Pt/TiO$_2$ nanocomposite for cancer-cell treatment. *J. Photochem. Photobiol. B.* 98: 207–210.

Lu, A., Y. Li, S. Jin, X. Wang, X. L. Wu, C. Zeng et al. 2012. Growth of non-phototrophic microorganisms using solar energy through mineral photocatalysis. *Nature Commun.* 3: 768. doi:10.1038/ncomms1768.

Malka, E., I. Perelshtein, A. Lipovsky, Y. Shalom, L. Naparstek, N. Perkas et al. 2013. Eradication of multi-drug resistant bacteria by a novel Zn-doped CuO nanocomposite. *Small.* 9 (23): 4069–4076.

Maneerat, C. and Y. Hayata. 2006. Antifungal activity of TiO$_2$ photocatalysis against *Penicillium expansum* in vitro and in fruit tests. *Int. J. Food Microbiol.* 107 (2): 99–103.

Matsunaga, T., R. Tomoda, T. Nakajima, and H. Wake. 1985. Photoelectrochemical sterilization of microbial cells by semiconductor powders. *FEMS Microbiol. Lett.* 29: 211–214.

Padil, V. V. T. and M. Černík. 2013. Green synthesis of copper oxide nanoparticles using gum karaya as a biotemplate and their antibacterial application. *Int. J. Nanomedicine.* 8: 889–898.

Pandiyarajan, T., R. Udayabhaskar, S. Vignesh, R. A. James, and B. Karthikeyan. 2013. Synthesis and concentration dependent antibacterial activities of CuO nanoflakes. *Mater. Sci. Eng. C Mater. Biol. Appl.* 33: 2020–2024.

Park, S., S. Lee, B. Kim, S. Lee, J. Lee, S. Sim et al. 2012. Toxic effects of titanium dioxide nanoparticles on microbial activity and metabolic flux. *Biotechnol. Bioprocess Eng.* 17: 276–282.

Pathakoti, K., S. Morrow, C. Han, M. Pelaez, X. He, D. D. Dionysiou, and H. M. Hwang. 2013. Photoinactivation of *Escherichia coli* by sulfur-doped and nitrogen-fluorine-codoped TiO$_2$ nanoparticles under solar simulated light and visible light irradiation. *Environ. Sci. Technol.* 47: 9988–9996.

Perelshtein, I., G. Applerot, N. Perkas, E. Wehrschetz-Sigl, A. Hasmann, G. M. Guebitz, and A. Gedanken. 2009. Antibacterial properties of an in situ generated and simultaneously deposited nanocrystalline ZnO on fabrics. *ACS Appl. Mater. Interfaces.* 1: 361–366.

Premanathan, M., K. Karthikeyan, K. Jeyasubramanian, and G. Manivannan. 2011. Selective toxicity of ZnO nanoparticles toward Gram-positive bacteria and cancer cells by apoptosis through lipid peroxidation. *Nanomedicine.* 7: 184–192.

Rasmussen, J. W., E. Martinez, P. Louka, and D. G. Wingett. 2010. Zinc oxide nanoparticles for selective destruction of tumor cells and potential for drug delivery applications. *Expert Opin. Drug Deliv.* 7 (9): 1063–1077.

ncón, A. G. and C. Pulgarin. 2004. Bactericidal action of illuminated TiO_2 on pure *Escherichia coli* and natural bacterial consortia: Post-irradiation events in the dark and assessment of the effective disinfection time. *Appl. Catal. B Environ.* 49: 99–112.

oy, A., S. S. Gauri, M. Bhattacharya, and J. Bhattacharya. 2013. Antimicrobial activity of CaO nanoparticles. *J. Biomed. Nanotechnol.* 9: 1570–1578.

ozhkova, E. A., I. Ulasov, B. Lai, N. M. Dimitrijevic, M. S. Lesniak, and T. Rajh. 2009. A high-performance nanobio photocatalyst for targeted brain cancer therapy. *Nano Lett.* 9: 3337–3342.

ıkai, H., R. Baba, K. Hashimoto, Y. Kubota, and A. Fujishima. 1995. Selective killing of a single cancerous T24 cell with TiO_2 semiconducting microelectrode under irradiation. *Chem. Lett.* 24: 185–186.

ılehi, B., S. Mehrabian, and M. Ahmadi. 2014. Investigation of antibacterial effect of cadmium oxide nanoparticles on *Staphylococcus aureus* bacteria. *J. Nanobiotechnol.* 25: 12–26.

ıto, J., K. Kudo, I. Takimoto, M. Sanbayashi, I. Kijihana, K. Takahashi, and I. Yakugaku. 2011. Study on use of photocatalyst for degradation of residual anticancer drugs in safety cabinet. *Japan. J. Pharmaceut. Health Care Sci.* 37: 57–61.

ıwai, J., O. Yamamoto, B. Ozkal, and Z. E. Nakagawa. 2007. Antibacterial activity of carbon-coated zinc oxide particles. *Biocontrol Sci.* 12: 15–20.

elvamani, M., G. Krishnamoorthy, M. Ramadoss, P. K. Sivakumar, M. Settu, S. Ranganathan, and N. Vengidusamy. 2016. $Ag@Ag_8W_4O_{16}$ nanoroasted rice beads with photocatalytic, antibacterial and anticancer activity. *Mater. Sci. Eng. C.* 60: 109–118.

ddique, S., Z. H. Shah, S. Shahid, and F. Yasmin. 2013. Preparation, characterization and antibacterial activity of ZnO nanoparticles on broad spectrum of microorganisms. *Acta Chim. Slov.* 60 (3): 660–665.

ıllósy, S. P., L. Janovák, J. Ménesi, E. Nagy, Á. Juhász, L. Balázs, I. Deme, N. Buzás, I. Dékány. 2014. Investigation of the antibacterial effects of silver-modified TiO_2 and ZnO plasmonic photocatalysts embedded in polymer thin films. *Environ. Sci. Poll. Res. Int.* 21: 11155–11167.

ıapalis, C. C., M. Kokkoris, G. Perdikakis, and G. Kordas. 2003. Study of antibacterial composite Cu/SiO_2 thin coatings, *J. Sol-Gel Sci. Technol.* 26: 1213–1218.

argas-Reus M. A., K. Memarzadeh, J. Huang, G. G. Ren, and R. P. Allaker. 2012. Antimicrobial activity of nanoparticulate metal oxides against peri-implantitis pathogens. *Int. J. Antimicrob Agents.* 40: 135–139.

'u, P., R. Xie, K. Imlay, and J. K. Shang. 2010a. Visible-light-induced bactericidal activity of titanium dioxide codoped with nitrogen and silver. *Environ. Sci. Technol.* 44 (18): 6992–6997.

'u, T. S., K. X. Wang, G. D. Li, S. Y. Sun, J. Sun, and J. S. Chen. 2010b. Montmorillonite-supported Ag/TiO_2 nanoparticles: An efficient visible-light bacteria photodegradation material. *ACS Appl. Mater. Interfaces.* 2: 544–550.

ı, J. W., Z. D. Gao, K. Han, Y. Liu, and Y. Y. Song. 2014. Synthesis of magnetically separable $Ag_3PO_4/TiO_2/Fe_3O_4$ heterostructure with enhanced photocatalytic performance under visible light for photoinactivation of bacteria. *ACS Appl. Mater. Interfaces.* 6 (17): 15122–15131.

adav, H. M., S. V. Otari, V. B. Koli, S. S. Mali, C. K. Hong, S. H. Pawar, and S. D. Delekara. 2014. Preparation and characterization of copper-doped anatase TiO_2 nanoparticles with visible light photocatalytic antibacterial activity. *J. Photochem. Photobiol. A Chem.* 280: 32–38.

aithongkum, J., K. Kooptarnond, L. Sikong, and D. Kantachote. 2011. Photocatalytic activity against penicillium expansum of Ag-doped $TiO_2/SnO_2/SiO_2$. *Adv. Mater. Res.* 214: 212–217.

in, Z. F., L. Wu, H. G. Yang, and Y. H. Su. 2013. Recent progress in biomedical applications of titanium dioxide. *Phys. Chem. Chem. Phys.* 15: 4844–4858.

Yuan, Y., J. Ding, J. Xu, J. Deng, and J. Guo. 2010. TiO₂ nanoparticles co-doped with silv
and nitrogen for antibacterial application. *J. Nanosci. Nanotechnol.* 10: 4868–4874.

Zhang, H., X. Zhu, Y. Ji, X. Jiao, Q. Chen, L. Hou et al. 2015. Near-infrared-triggered in si
hybrid hydrogel system for synergistic cancer therapy. *J. Mater. Chem. B.* 3: 6310–632

Zhang, J., Y. Liu, Q. Li, X. Zhang, and J. K. Shang. 2013a. Antifungal activity and mech
nism of palladium-modified nitrogen-doped titanium oxide photocatalyst on agric
tural pathogenic fungi *Fusarium graminearum. ACS Appl. Mater. Interfaces.* 5 (21
10953–10959.

Zhang, L., M. Han, O. K. Tan, M. S. Tse, Y. X. Wang, and C. C. Sze. 2013b. Facile fabricati
of Ag/C-TiO₂ nanoparticles with enhanced visible light photocatalytic activity for di
infection of *Escherichia coli* and *Enterococcus faecalis. J. Mater. Chem. B.* 1: 564–57

17 Other Applications

17.1 INTRODUCTION

With the increasing advancement in technology and commercialization, we are facing a problem of increasing uncleanliness all around us. The buildings, pavements, fabrics, glasses, mirrors, kitchenware, tiles, and so on, are hard to keep clean, dirt-free, and fog-free. The foul smell from industries and rotten garbage is also a major factor in these unpleasant feelings. Our hectic routine makes it a tedious task to keep our surroundings clean and healthy. The conventional methods to maintain a hygienic and pollution-free environment are time consuming, costly, and complicated. In this regard, photocatalysis comes to our rescue, as it has the potential for combating this problem of uncleanliness. Photocatalysts use the light energy and a semiconductor (photocatalyst) for antifogging, self-cleaning, self-sterilization, deodorization, and so on.

As the nonrenewable sources of energy are depleting day by day, a renewable energy source is urgently needed to fulfill our energy demands. Solar energy is a renewable source and it plays a key role in mass production of chemicals like synthetic fuels. Many environmental problems could be solved by the use of solar energy. One of its most important applications is solar photocatalysis, which can be used in fighting against water, air, and soil pollution. The commercial solar photocatalytic applications include technologies like antifogging and self-cleaning of glass, concrete, and ceramics.

The lotus effect was theoretically discussed by Marmur (2004) for superhydrophobicity of a model system resembling the lotus leaf. Superhydrophobicity is defined by two things: (1) a very high water contact angle and (2) a very low roll-off angle. Nature utilizes metastable states in the heterogeneous wetting regime for superhydrophobic properties of lotus leaves.

The lotus effect explains the self-cleaning properties of leaves of the lotus flower (Nelumbo) as a result of superhydrophobicity, that is, very high water repellence. A nearly spherical shape is achieved due to high surface tension of water droplets as a sphere has minimum surface area and therefore least surface energy (Figure 17.1).

Dirt particles are picked up by these water droplets due to the micro- and nanoscopic architecture on the surface, which reduce the droplet's adhesion to that surface and as a result, the leaves are self-cleaned (superhydrophobicity and self-cleaning). Such properties were also found in other plants, such as *Tropaeolum* (nasturtium), *Opuntia* (prickly pear), *Alchemilla*, and cane and also in the wings of certain insects like butterflies and others.

A water drop occupies a spherical shape on a surface if it is dropped on a hydrophobic surface, but it will be flattened with very low contact angle if the surface is hydrophilic in nature. It spreads further and forms a still thin sheet of water if the surface is superhydrophilic in nature.

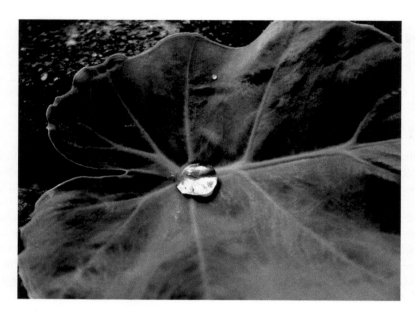

FIGURE 17.1   Water drop on a lotus leaf.

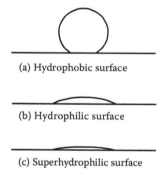

(a) Hydrophobic surface

(b) Hydrophilic surface

(c) Superhydrophilic surface

FIGURE 17.2   Different types of surfaces.

The surface of a material can be cleaned by hydrophobic as well as hydrophilic action. However, they differ in their mechanism of cleaning action. While a hydrophobic surface repels water molecules to form a spherical droplet on the surface, a hydrophilic surface spreads this water droplet in the form of a layer, which is turned into a very thin film on exposure to light. This ultimately results in self-cleaning of surface or antifogging as the water will take away the organic materials, dirt, and so on, along with it (Figure 17.2).

Superhydrophilic material has various advantages. It can defog glass and can also enable spots of oil or dirt to be swept away easily with water. Such materials have already become commercialized as rear view mirrors for cars, coatings for buildings, self-cleaning glass, tiles, and so on.

These properties are based on superhydrophilicity. The contact angle of water measures the degree by which water is repelled from a surface of a specific material. Glass and the other inorganic surfaces exhibit a contact angle ranging from 20° to 30°. Water dropped onto titanium dioxide forms no contact angle (almost 0°) under light irradiation. This effect was discovered in 1995 by the Research Institute of Toto Ltd., Japan, for titanium dioxide which was irradiated by sunlight, and it was termed superhydrophilicity. The mechanism of photoinduced superhydrophilicity can be explained by measuring the contact angle between photocatalyst surfaces and water molecules, which gradually decreases to almost zero.

When water is added to the surface of a photocatalyst, water droplets form a particular angle (very less) with the surface and remain there, but on adding water to the irradiated photocatalyst surface, spherical droplets of H_2O are not formed as water is not repelled, rather it makes a highly uniform thin film on the surface of the photocatalyst. This process of superhydrophilicity has many commercial applications like self-cleaning, antibacterial, and antifogging glass in windows, mirrors, and vehicle glasses. Dew drops do not stay on vehicle glass. Such self-cleaning and antibacterial fabrics may be used for tents, which remain clean for a very long time duration, thereby lowering their maintenance cost. Construction materials like concrete and cement may be used for buildings, hospital appliances, kitchenware, bathroom tiles, walls and roof materials, road pavement, sidewalks, and so on.

Several mechanisms of superhydrophilicity have been proposed by different groups of researchers. One proposal is the change of the surface structure to a metastable structure, while another is cleaning the surface by the photodecomposition of dirt or organic compounds adsorbed on the surface. After this, only water molecules can remain adsorbed to the surface. The mechanism of superhydrophilicity is still controversial, and it is not yet decided which particular suggestion is correct. Further studies are required to know the intricacies of it.

There were basically two ways to achieve self-cleaning material surfaces. The first one is to develop superhydrophobic materials and the second is to prepare superhydrophilic materials. As in the lotus effect where the leaf repels water, surfaces like tiles and paints may also do it. Superhydrophobic materials are well developed. Superhydrophilic materials can be generated by coating glass, ceramic tiles, or plastics with the photocatalytic semiconductor. Grease, dirt, and organic contaminants were decomposed on irradiating TiO_2 by lights and then these could be very easily removed by rainwater (Stamate and Lazar 2007).

TiO_2 is a kind of nontoxic, stable, and inexpensive building material, which has a high refractive index on the order of 2.7. Traditional TiO_2-based building materials (TBMs) were developed in the early nineties, when toxic lead oxides were slowly replaced by TiO_2. A number of attempts have been made to combine TiO_2 with building materials by mixing or coating to produce novel TBMs with photoactive functions, such as air cleaning, sterilization, self-cleaning, antifogging, decoration, and building cooling (Guo et al. 2009). These are widely used in exterior construction (paints, tiles, glass, plastic, and aluminum panels), interior furnishing (paints, tiles, wallpaper, and window blinds), and road construction (soundproof walls, tunnel walls, roadblocks, paints, traffic signs and reflectors, lamps, and coatings). Budde

(2010) discussed the self-cleaning, odor-reducing, and water-shedding properties of titanium dioxide photocatalytic oxidative coatings.

17.2 ANTIFOGGING

Fogging of glass surfaces is a result of condensation of water droplets, which scatters light. It causes poor visibility in mirrors, automobile windshields, windows and lenses for optical devices, and so on (Figure 17.3). Mirror surfaces fog up as the steam cools on these surfaces producing aqueous droplets. Cars are nowadays equipped with superhydrophilic antifogging side-view mirrors (Fujishima et al. 1999; Hata et al. 2000). Antifogging spectacles and goggles are also commercialized now.

Near-infrared (NIR) absorption spectroscopy showed that the water molecules adsorbed on the TiO_2 surfaces desorb on exposure to ultraviolet (UV) radiation. As

(a)

(b)

FIGURE 17.3   (a) Ordinary windscreen and (b) antifogging windscreen.

e amount of the adsorbed water molecules on the TiO_2 surfaces was decreased, the stribution of hydrogen bonds within the water molecules also decreased, thereby using a decrease in the surface tension of the H_2O clusters leading to the formation water thin layers. Another important factor leading to development of free spaces the surface, where the water clusters could spill over and spread out to form thin $_2O$ layers and results in partial elimination of hydrocarbon particles from TiO_2 rface by complete photocatalytic oxidation. The temperature of the TiO_2 samples ring UV light exposure had a relation with the changes in the contact angle of the ater droplets on the TiO_2 thin film surfaces. The rate of adsorption and desorption water molecules was found to be sensitive to the temperature changes of solid rfaces (Takeuchi et al. 2005).

WO$_3$-modified TiO_2 thin films were obtained by Hwang et al. (2005) on a glass bstrate by sol–gel and dip-coating processes using acetyl acetone as a chelating ent. Surface morphology was controlled with the change of precursor concentra-n. It was observed that 0.01 M of tungsten oxide–modified TiO_2 had the highest drophilicity after UV irradiation.

Multifunctional nanoporous thin films were fabricated by Cebeci et al. (2006) om layer-by-layer assembled silica nanoparticles (NPs) and a polycation. These ultilayer films were found to exhibit antifogging as well as antireflection proper-es. The antifogging property was a direct result of the development of superhy-rophilic wetting characteristics (water droplet contact angle of less than 5° within 5 seconds or even less). Nearly sheetlike wetting promoted by the superhydrophilic ultilayer allows light scattering water droplets to be formed on a surface, while the w refractive index of this multilayer film (as low as 1.22) due to the presence of anopores was considered to be responsible for excellent antireflection properties. lass slides coated on both sides with a nanoporous multilayer film exhibited 99.8% ansmission, which facilitates full visibility.

A spin-coating/sol–gel technique was used to prepare transparent TiO_2, WO$_3$–iO$_2$, and MoO$_3$–TiO_2 films by Chai et al. (2006). The coating of WO$_3$ on the sur-ce of TiO_2 enhanced the photocatalytic efficiency in the decomposition of gaseous -propanol, whereas on the addition of MoO$_3$, considerable retardation in the photo-atalytic reaction of TiO_2 was observed. The contact angle between the water drop nd the film surface was calculated as a function of UV exposure time to estimate e superhydrophilicity. It was observed that the contact angle of WO$_3$–TiO_2 film as less than that of TiO_2 before UV light irradiation, and it was reduced about four mes faster as compared to pure TiO_2 films. When MoO$_3$–TiO_2 films were studied, was observed that there was a low contact angle at the initial point, but there was a radual decrease in contact angle under the UV light. This observed trend indicated at the superhydrophilic behavior is closely related to the photocatalytic property.

Transparent mesoporous silica thin films (MSTFs) consisting of photocatalysts ke Ti, V, Cr, Mo, and W oxide (Me-MSTFs) were prepared by a sol–gel spin-oating method on quartz plates. All Me-MSTFs showed efficient hydrophilic prop-rties before and after UV light irradiation. It was observed that the W-MSTF had e highest hydrophilic behavior out of these samples (Horiuchi et al. 2008).

Tricoli et al. (2009) synthesized TiO_2 and SiO_2–TiO_2 nanofilms by using one-step ame spray pyrolysis (FSP). The pure TiO_2 films were composed of smooth lace-like nanostructures, whereas the SiO_2–TiO_2 films had nanostructures having spike-li termini. Antifogging performance of these films was also observed.

Two types of superhydrophilic surfaces were produced, which included polyes films treated by oxygen plasma and indium tin oxide-deposited glasses using a electrochemical method. These surfaces had almost zero water contact angles, whi confirmed their superhydrophilicity. The antifogging and antifouling properties these films were examined. The fluorescence microscopic study revealed a retard tion in adhesion of the fluorescein and fluorescent proteins on these surfaces indica ing their use as antifouling agents also (Patel et al. 2009).

Krylova et al. (2010) used a sol–gel method to synthesize Zn_2TiO_4, cubic sp nel (c)-$ZnTiO_3$, or hexagonal (h)-$ZnTiO_3$–ilmentite/rutile (r)-TiO_2 films. These film were superhydrophilic and cubic spinel-like in shape. Silica (5%) was added to fin tune the morphology and this results in superhydrophilicity. This doping results high dye absorption capacities and the water contact angle value was reduced to X-ray photoelectron spectroscopy (XPS) results indicated that a high level of Z and Si led to greater surface hydroxylation, which in turn improved water wettir capacity. h-$ZnTiO_3$–ilmenite/r-TiO_2 nanocomposite deposited on glass and Si-wafe exhibited high efficiency for the photomineralization of fatty acids.

Chekini et al. (2011) prepared pure TiO_2 and N-doped titanium dioxide (N-TiO thin films by the sol–gel method through spin coating on soda-lime glass substrate $TiCl_4$ and urea were used as precursors of Ti and N. X-ray diffraction (XRD) anal sis revealed that N doping decreased anatase to rutile phase conversion. The dopir also led to a retardation in roughness of the samples from 4 nm for pure TiO_2 to 1 n for doped N–TiO_2. A shift in optical band gap of thin films was observed from 3. eV for pure TiO_2 to 3.47 eV for N-TiO_2. It was concluded that N–TiO_2 thin film ha efficient hydrophilicity and photocatalytic properties under UV irradiation.

Kazemi and Mohammadizadeh (2012) synthesized anatase TiO_2 nanothin film on glass substrates using a sol–gel dip-coating method, where Tween-80 was use as a surfactant, $TiCl_4$ as the Ti precursor, and ethanol as a solvent. The changes structure and photocatalytic and superhydrophilicity behaviors of the films due the chemical aging time effect were also studied. They concluded that an optimur aging time of 2 hours was observed to show maximum values for both photocatalyt and superhydrophilic properties. This also finds use in the self-cleaning industry.

Kamegawa et al. (2012) designed superhydrophobic surfaces having self-cleanir properties. They coated nanocomposite TiO_2 and polytetrafluoroethylene (PTFE) a substrate using a codeposition method. It was observed that this coating showed photocatalytic property for self-cleaning and induced wettability.

Jesus et al. (2015) used TiO_2/SiO_2 composites having variable Ti content and con pared the activity of this composite with pure TiO_2 films. Both films were coate over low iron float glass surface by the sol–gel dip-coating method and various ca cination temperatures like 400°C, 500°C, and 600°C taking Si/Ti molar ratios Si86Ti14 and Si40Ti60. TiO_2/SiO_2 films inhibited a higher degree of transmittanc in the visible range as compared to pure TiO_2. TiO_2/SiO_2 films also had superhy drophilic character before and after UV irradiation having water contact angles almost 0°. It was also observed that the TiO_2/SiO_2 films retained the superhydrophil behavior even in dark environments, which was not observed in case of pure TiO_2 films. Both TiO_2 and TiO_2/SiO_2 films showed good adherence and it was observed that higher calcination temperatures and higher Ti amount increased the rate of adherence. These films achieved an abrasion-resistant property when touched with sponges and detergent. These properties show potential for their use in photovoltaic (PV) systems also.

The self-cleaning surfaces are used to prevent soiling accumulation on PV surfaces in the field of solar energy. TiO_2 has found extensive use due to its photocatalytic efficiency and photoinduced superhydrophilicity, but it has certain limitations also as it retards the glass transmittance and loses hydrophilicity by reestablishing the water contact angle in the dark. In order to overcome these limitations, composites like TiO_2/SiO_2 could be employed. Better transparency in the UV–visible region is required for use in solar cells. Besides this, efficient self-cleaning properties and long durability of the coating and sufficient adhesion to stand the outdoor conditions are also needed.

17.3 SELF-CLEANING

There are two ways by which a self-cleaning surface works: hydrophobic and hydrophilic coatings. Both these types of coatings clean themselves by the action of water. The hydrophobic surfaces do it by rolling droplets and the hydrophilic surfaces by spreading water that sweep away dirt. Titania-based hydrophilic coatings have an additional advantage in that they can decompose the absorbed dirt in sunlight by the process of photocatalysis.

A very high static water contact angle θ ($\theta > 160°$) is required for a hydrophobic self-cleaning surface along with a very low inclination angle (Marmur 2004). The hydrophobic self-cleaning surfaces also suffer from some drawbacks like

- Batch processing of a hydrophobic surface is costly and time consuming.
- The coatings developed were unclear, restricting their application for lenses, windows, and fragile materials.

The self-cleaning behavior in hydrophilic surfaces like glass shows two stages. The photocatalytic phase of the process breaks the organic dirt present on the glass chemically using UV light and makes the glass superhydrophilic, which was initially hydrophobic. In this stage, rainwater carries away the dirt, leaving the glass completely clean as water spreads evenly on superhydrophilic surfaces. Superhydrophilic material has the potential to be used in microfluidics, printing, PV, biomedical devices, antibacterial instruments, water remediation, and so on.

TiO_2 is the most widely used material for self-cleaning purposes due to its non-toxic nature, chemical inertness in the absence of light, and being inexpensive, handy, and easy to deposit into thin films. It is one of the most common household chemicals used as a paint, cosmetic pigment, and food additive.

Surface oxygen vacancies at bridging sites are developed on exposure to UV light leading to the conversion of Ti^{4+} to Ti^{3+}, which is favorable for the adsorption of dissociative (self-ionized or autodissociative) water. These result in forming hydro philic domains. Hydrophilic domains are areas where dissociative water is adsorbed and associated with oxygen vacancies (Wang et al. 1997). It is important to contro the surface wettability of solid substrates in different situations. They reported the photogeneration of a highly amphiphilic (both hydrophilic and oleophilic) surface o TiO_2. This unique character of titanium dioxide surface is due to the microstructure composition of hydrophilic and oleophilic phases, which are produced by the expo sure to UV radiation. This resulted in TiO_2-coated glass, which has antifogging and self-cleaning properties.

Glasses for architecture should have many other functions apart from thei transparency like self-cleaning, antibacterial, energy conversion, light control and UV reduction. Such glasses will find use in buildings in the near future. Zhac et al. (2008) discussed multifunctional photoactive glasses, which are based or multilayer coatings containing TiO_2 film and other functional coatings. The self cleaning of glasses can be realized by coating the photoinduced superhydrophilic nanoporous thin films based on TiO_2 photocatalysts via the sol–gel route. They developed a new method to enhance the photocatalytic activity of TiO_2 thin films. These films also have good photoinduced antibacterial properties, which were enhanced by doping with silver and without light. These TiO_2 thin films will ac as self-cleaning glasses and these are prepared by two layers of TiO_2–CeO_2 and TiO_2 thin films on soda-lime glasses, where these films can cut all the UV ligh with adjustment in the ratio of titania and ceria. It was also reported that TiO_2 TiN/TiO_2 type multilayer coated on glass substrate can act as low-E self-cleaning glass.

Banerjee et al. (2015) reviewed the self-cleaning applications of TiO_2. They con cluded that preparing a hybrid multifunctional photocatalytic substance and biologi cal structure having tunable wettability could prove to be a better way of treating the existing environmental problems.

Cu–Bi_2O_3 films were coated with SiO_2 by Shan et al. (2015) using a sol–gel, spin-coating technique. The photocatalytic efficiency and self-cleaning activity of the as-prepared films were studied by the decomposition of stearic acid. It was observed that these films exhibited excellent superhydrophilic properties even ir the dark and showed enhanced photocatalytic and self-cleaning properties in com parison to pure Bi_2O_3 films. It was concluded that the rate of photocatalytic deg radation and self-cleaning increased due to the interfacial charge transfer taking place between Bi_2O_3 and Cu and SiO_2. These films also worked as good antifog ging materials.

Hydrophobicity and self-cleaning are the important factors that influence the pre cision and environment resistance of quartz crystal microbalance (QCM) in detect ing various organic gas molecules. A ZnO nanorod array was prepared by Wei et al. (2015) via an *in situ* method on QCM coated with Au film by a hydrothermal pro cess. This ZnO nanorod array film on QCM was modified by β-cyclodexrin (β-CD in a hydrothermal process and then it was decorated by TiO_2 by impregnation in P25 suspension. As-prepared ZnO–TiO_2 nanocomposite exhibited excellent hydropho bicity for water molecules and superior self-cleaning property for organic molecules under UV exposure.

17.3.1 SELF-CLEANING GLASSES AND TILES

The most common applications of photocatalytic cement-based materials are the following:

- Concrete pavement
- Roofing tiles and panels
- Cement and concrete-based tiles
- Indoor and outdoor paints
- Finishing coatings, plasters, and other cement-based materials
- Sound-absorbent elements for buildings near roads

The first product based on solar photocatalysis with TiO_2 was a self-cleaning coating for window glasses (Stamate and Lazar 2007; Zhao et al. 2008). This glass consisted of a layer of nanocrystalline anatase TiO_2 deposited by a chemical vapor deposition technique on soda-lime silicate float glass. This glass had high visible transmission and reflectance properties (Mills et al. 2003). The cleaning of glass surface and tiles is ordinarily achieved by the use of chemical detergents, but this consumes more time and energy, leading to more cost as well. The inorganic and organic particles adsorbed on TiO_2-coated surfaces are easily degraded and then these could be washed away with water due to the high hydrophilicity of TiO_2 film as in self-cleaning tiles. The basic final product from degradation of organic molecules is carbon dioxide and water. This is referred to as cold combustion. Self-cleaning is operative mainly on the condition that the flux of the incident solar photons should be more than the rate of the adsorption of the organic pollutants on the surface (Parkin and Palgrave 2005).

Depolluting and self-cleaning coating compositions comprising an organic binder having dispersed photocatalytic titanium dioxide (anatase form) particles with an average crystallite size between 1 and 150 nm have been patented that exhibit photocatalytic activity in the presence of visible light (Bygott and Maltby 2007; Maltby and Bygott 2014). There is an added advantage that these coatings do not require any preactivation to achieve high initial photocatalytic activity against pollutants in the air, such as NOx compounds.

An example of self-cleaning spectacles is given in Figure 17.4.

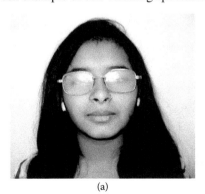
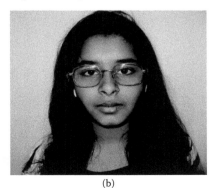

(a)　　　　　　　　　　　　　　　(b)

FIGURE 17.4　(a) Ordinary spectacles and (b) self-cleaning.

17.3.2 Self-cleaning Concrete, Cement, and Buildings

Cassar and Pepe (1998) described that the use of a combination of organic additives for the preparation of cementitious compositions is advantageous in having a high conservation of the degree of whiteness; this contains a photocatalyst in mass that is able to oxidize pollutants present in the environment in the presence of light, air, and ambient humidity. The photocatalyst titanium dioxide is present here, in particular, prevalently in the form of anatase.

The anatase form of TiO_2 has found wide use due to its strong oxidizing power under near-UV radiation, chemical stability on exposure to acidic and basic compounds, chemical inertness in the absence of UV light, and absence of toxicity. TiO_2 has proved to be very effective in the reduction of pollutants such as NO_X, aromatics, ammonia, and aldehyde in combination with cementitious materials. It has a synergistic effect in reduction of pollutants and, therefore, these new materials have found applications in self-cleaning building walls and in the reduction of pollutants (Figure 17.5).

Liu et al. (2000) prepared self-cleaning ceramic materials by coating photocatalytic membrane on ceramic matrix. They studied the photocatalytic behavior of these materials for the degradation of oleic acid and sterilization. The effect of preparation and reaction conditions like heating treatment and thickness of the membrane on photocatalytic activity of the self-cleaning ceramics was also observed. They concluded that the photodegradation of oleic acid and sterilization of the self-cleaning ceramics depend on crystal structure, particle size, and specific surface area of the supported photocatalyst membrane.

Walls are also self-cleaned by solar photocatalysis apart from window glass. The exterior building walls become dirty from many outward factors, the major one being automobile exhaust, containing mineral oils and unburnt pollutants. Dirt on walls could be washed away directly and easily by rain in the presence of sunlight by coating walls with superhydrophilic TiO_2. (Ramirez et al. 2010). If photocatalytic

FIGURE 17.5   Self-cleaning building. (By Silvia Ercoli [Own work] [CC BY-SA 3.0 {http://creativecommons.org/licenses/by-sa/3.0}], via Wikimedia Commons.)

ncrete is used in construction, then it keeps the building clean by decomposing emicals that led to soiling. The concrete also reflected much of the Sun's heat, and it is white, it also reduced heat intake (Puzenat 2009). Nowadays, white cement ntaining TiO_2 is used for the construction of buildings.

The photocatalytically active concrete has also been reported to promote the degdation of NOx (Bolte 2009). The materials made from such self-cleaning conete also proved to be more durable and required low maintenance. TiO_2-coated aterials have also been used as paving material by many companies. There is a ving stone called NOxerTS™, which is utilizing the catalytic properties of TiO_2 degrade NOx, mainly found in vehicle exhaust, converts these oxides into less rmful molecules that can be washed away by rainwater. Oxygen created on illuinating the surface of the NOxerTM oxidized NOx into HNO_3, which could then washed away by rainfall or could be neutralized by the alkaline property of the ncrete (Frazer 2001).

Automobiles have been causing serious air pollution problems in urban areas d roadsides due to exhaust gases containing nitrogen oxides and therefore NOx moval is urgently required. The pavement blocks on the roads may be used as r purifying material. Murata et al. (1999) developed an interlocking paving block hich has NOx removal capability by photocatalytic oxidation by titanium dioxide. performance were confirmed under UV intensity available at the outdoor level. hese photocatalytic concrete blocks worked in a humid atmosphere also and in Ox concentration ranges comparable to a roadside environment.

TiO_2-coated highways were also developed for converting pollutants like NOx and)x to their eco-friendly forms like nitrates and sulfates (Cassar 2004; Italcementi)05). Hamada et al. (2004) studied the use of TiO_2 materials applied in airports, hich reduced nitrogen oxides by 10%–30%.

7.3.3 SELF-CLEANING FABRICS AND POLYMERS

nt material containing polyvinyl chloride (PVC) is very difficult to clean. If phocatalytic PTFE and PVC membranes with TiO_2 were used, then the chemical sistance and self-cleaning properties were enhanced. Not only this, such strucres maintain their good appearance and light transmittance for a longer duration. lf-cleaning tent fabrics finds potential use in storage structures, bus stands and ilway stations, playgrounds, and canopies in parks and beaches (Kallio et al. 2006; uranova et al. 2007).

A Sol–gel process was used to prepare TiO_2–SiO_2@PDMS (polydimethylsiloxie) films showing remarkable superhydrophobic and photocatalytic behaviors)eng et al. 2014). These films had high thermal stability up to 400°C and showed perhydrophilicity upon calcination at 470°C. The TiO_2–SiO_2@PDMS hybrid)lution was used to coat polyester–cotton fabrics on a large scale, making them perhydrophobic in nature. This superhydrophobicity renders wash resistance d resistance to attack by strong acids to these fabrics. This fabric could be pplied as a filter cloth for both a oil–water separation purpose and colorful patrn printing. The TiO_2–SiO_2@PDMS hybrid solution was also used for degrading ye in wastewater on UV exposure. Small balls covered with the TiO_2–SiO_2@

PDMS hybrid solution moved faster as compared to the control sample due their smooth superhydrophobic surface. This multifunctional TiO_2-SiO_2@PDM hybrid material can therefore be used for fabric treatment and water-repellent sh coatings.

TiO_2 was coated on polycarbonate substrates by Yaghoubi et al. (2010). A chem cal surface treatment technique was employed to develop hydrophilic groups on th surface. TiO_2 coating was based on a wet coating method using a sol of anatase TiO_2 NPs of 30 nm size. The sol was prepared by the sol–gel method. A precoating peroxotitanium complex was made to increase adhesion and prevent substrate de radation. The photocatalytic activity and thickness of film had a linear relationshi It was also observed that the mechanical properties were improved after coating, evident from nanoindentation and nanoscratch tests. The rigidness was improved l 2.5, and ~6.4 times better scratch-proof ability was observed.

Fateh et al. (2014) prepared TiO_2-ZnO thin films on an SiO_2 interlayer and depo ited on the polycarbonate surface to prepare polymeric sheets having self-cleanin superhydrophilic, and photocatalytic properties. The polycarbonate sheets we exposed to UV light to increase adhesion of SiO_2 interlayers. It was analyzed th prepared films were transparent, having a thickness in the range 120–250 nm, al had superhydrophilic properties. They also showed good adhesion capability. It w observed that their mechanical strengths depend on the variation in the molar TiO ZnO ratio. The relation between superhydrophilicity and photocatalytic activity w studied by observing the change in angle of water contact. The prepared films we kept in the dark under ambient atmosphere and in the atmosphere of either aceto or isopropanol followed by UVA exposure. The best cleaning properties and efficie mechanical stability was obtained when the superhydrophilic coating with a mol TiO_2-ZnO ratio of 1:0.05 was made.

Lathe et al. (2014) coated SiO_2-TiO_2 on polycarbonate substrate, which had go potential for self-cleaning applications. It was found that this coating was optical transparent with efficient adherence capacity and was wettable toward water. Th influence of different vol% of SiO_2 in TiO_2 was also studied. The coatings havir 7 vol% of SiO_2 in TiO_2 resulted in smooth, crack-free surface morphology and lo surface roughness in comparison to the coatings having higher vol% of SiO_2 in TiO The contact angle achieved for 7% volume SiO_2 in TiO_2 was observed to be less th 10° after UV exposure for half an hour.

The capabilities of TiO_2 to purify/deodorize indoor air and industrial gaseo effluents were assessed by Pichat et al. (2000), using a laboratory photoreactor wi a lamp (~365 nm) and TiO_2-coated fiber glass mesh. The removal rate of three po lutants, for example, CO, n-octane, and pyridine, was determined as 5–10 μmol/W for 50–2000 ppm volume concentrations and 25–50 L/h flow rates (dry air or O It was inferred that this order of magnitude allows, by the use of a reasonable-si apparatus, the abatement of pollutants in constantly renewed indoor air, except C and CH_4, which are too concentrated. The average concentrations of benzene, tol ene, and xylenes were indeed reduced by a factor of 2–3 using a TiO_2 photocatalysi based individual air purifier prototype in an ordinary nonairtight room. It was show that addition of ozone in O_2 markedly increases the mineralization percentage n-octane.

TiO_2–SiO_2 composite film was prepared by the sol–gel method using $Si(OC_2H_5)$ and $Ti(OC_3H_7i)_4$ as starting materials. TiO_2–SiO_2 films were fabricated on the glass surface by the spin-coating method and heated at 500°C for an hour. Photocatalytic activity of TiO_2–SiO_2 films exhibited the degradation of ~97% of CH_3CHO in 2 hours and a water contact angle of approximately 10°. TiO_2–SiO_2 films showed more hydrophilic activity and less photocatalytic activity by increasing the content of SiO_2. It was also found that the amount of organic molecules adsorbed on the films was reduced with UV light exposure and SiO_2 addition, which was attributed to an increase in both the amount of OH group in films and decomposed organic contaminants on the surface of the films (Shin and Kim 2009).

17.4 DEODORIZATION

Peral et al. (1997) reviewed the decontamination and deodorization of air gas by solid heterogeneous photocatalysis. They have discussed oxygen and water vapor adsorption and also recent applications of photocatalysis used for pollutant removal in contaminated atmospheres.

An efficient deodorization system reduces or stops the emission of offensive odor from different sources. Nozawa et al. (2001) prepared a sheet material with TiO_2 photocatalyst fabricated on fiber-activated carbon (FAC) for a compact deodorization system. In this process, malodorants were absorbed and then decomposed using photocatalysis and UV light exposure. Methyl mercaptan, ammonia, and hydrogen sulfide were used as the malodorants. Two types of light sourcs were used. One was a blacklight bulb (BLB) having λ_{max} 365 nm and another a UV germicidal lamp (UV2) having λ_{max} 254 nm. A batch-type experiment was performed in this process. Photocatalytic degradation was studied from the rate of removal of methyl mercaptan, whereas the percent oxidation of NH_3 to NO_3^- and that of methyl mercaptan to sulfate was studied by analysis of the products, that is, NO_3^- and SO_4^{2-} ions. It was also concluded that the decomposition of malodorants was dependent on the wavelength of the light source used.

Song et al. (2001) reported that the stability of colloidal suspensions in aqueous solution can be greatly improved by monolayer coverage of highly acidic semiconductors, such as MoO_3 and WO_3 on the surface of TiO_2 NPs (Degussa P25). The average diameter of agglomerated MoO_3/TiO_2 and WO_3/TiO_2 particles in aqueous suspension was 85–110 nm, respectively, which is about 20%–25% lesser than that of pure TiO_2 suspension. Optically transparent photocatalytic films were prepared with the deposition of these colloidal suspensions. It was found that the activity of WO_3/TiO_2 film is 2.8–3 times that of pure TiO_2 film in photocatalytic decomposition of gas-phase 2-propanol, while MoO_3/TiO_2 film was relatively less effective.

Murakami et al. (2004) worked on the deodorization of kitchen exhaust of a Chinese restaurant using a silver-deposited photocatalyst. A photocatalytic filter was prepared from a porous ceramic body coated with TiO_2 photocatalyst. It had a three-dimensional (3D) network morphology. Nanosized Ag particles were superimposed on the TiO_2 surface. The main malodorant was H_2S. Keeping this in view, odor control apparatus was constructed with a nanosized Ag particle filter for H_2S.

Usuda et al. (2005) used light guard boards, which function as photocatalysts Two kinds of guide boards were constructed. One was developed of UV transparen acrylic board having V-shaped grooves on either sides, while the other was prepared of a similar acrylic board with a dot pattern silkscreen printed on one side. These boards were activated by light beams. It was observed that these boards were effec tive in photocatalytic deodorization.

Nishioka and Nishino (2014) used an air conditioner for deodorization using a photocatalyst. This air conditioner had an air conditioning case, heat exchanger, pho tocatalyst placed on the surface of the heat exchanger, and emitting section, which emits light for the activation of the photocatalyst. The photocatalyst degraded the odor components present in air.

A TiO_2–UV light-emitting diode (LED) system was also used for the purpose of deodorization (Jo et al. 2015). The effectiveness of the system was studied by instrumental determination as well as human sensory analysis. Two refrigerator sys tems were used, namely, control and treatment units. Ten types of food samples like raw beef, raw chicken, spam, onions, tomatoes, strawberries, boiled eggs, codfish mackerel, and French cheese were taken. The odorant samples were collected and analyzed for different periods of 0, 3, 6, 9, 24, 72, and 120 hours up to 5 days. It was observed that the average rates of removal for sulfur-containing compounds like H_2S, CH_3SH, dimethyl sulfide (DMS), dimethyl disulfide (DMDS), and other com pounds like styrene, isobutyl alcohol, and trimethyl aluminum (TMA) were in the range of 75.2%–94.2%, whereas carbon disulfide (CS_2), benzene, and methyl isobu tyl ketone (MIBK) were not efficiently removed by this photocatalytic system. In the human sensory test, the dilution-to-threshold ratio (D/T ratio) values were studied in reference to odor intensity (OI) and odor activity value (OAV). It was observed that 75%–94% odorants were removed in this study.

17.5 SELF-STERILIZATION

The silicone surface of silicone catheters and medical tubes were first treated with 5M solution of H_2SO_4 for about 3 hours. The TiO_2 photocatalyst was coated on these surfaces. The TiO_2 coated on the silicone substrate was stable against tensile and bending stresses. The photocatalytic bactericidal effect on *Escherichia coli* under UV light was also studied. It can be concluded that such a catheter could be sterilized and cleaned by exposure to low-intensity UV light. This can be considered a simple and efficient self-sterilization technique (Ohko et al. 2001).

López and Jacoby (2002) coated TiO_2 on metal fibrous mesh and used it as a self-sterilizing and disinfective filter for air. Mesh is produced by a roll-to-roll process in this process. The TiO_2 was coated from its aqueous suspension on the surface of the mesh by an airbrush. The coating increased the separation of *E. coli* from aqueous suspension, but it also resulted in an increase in the pressure drop in air stream flow through the mesh. A photocatalytic process was used by them involving the exposure of the mesh to UV light.

Interior paints contain aqueous acrylic dispersion, rutile TiO_2, extenders, and spe-cial additives, one of which is photocatalytic nano-ZnO. These paints contain a mix-ture of TiO_2 and ZnO, which have proved to be the best photocatalytic antimicrobial agent against *E. coli, Staphylococcus aureus, Pseudomonas aeruginosa,* fungi *Aspergillus niger,* and *Penicillium chrysogenum* (Hochmannova and Vytrasova 2010).

Sekiguchi et al. (2007) coated TiO_2 on catheters used for clean intermittent catherization (CIC). This catheter was studied for its antibacterial effect using only light energy. TiO_2-coated catheters were filled with bacterial cell suspension and exposed with a 15 W blacklight lamp for analyzing their antibacterial potency. Then tips of these catheters and zinc diethyldithiocarbamate, which was used as control toxic material, were soaked in M05 medium and the cell toxicity was determined from V79 colony count. It was observed that the survival rate of *E coli, S. aureus, P. aeruginosa,* and *Serratia marcescens* was negligible only within 1 hour of UVA exposure. V79 colonies had no toxic effect on these catheters. The positive bacterial culture rate on the TiO_2-coated catheters tips was only 20% as compared to 60% for conventional uncoated catheters after using them for about 4 weeks. It can be concluded that these sterilizing catheters have the potential to be used clinically for CIC after the required modifications are made.

Nakamura et al. (2007) developed a self-sterilizing lancet, which was coated with photocatalyst TiO_2 nanolayer. A lancet used for pricking the finger for self-monitoring of blood glucose was made self-sterilizing by coating it with photocatalytic TiO_2 in order to achieve an antibacterial property. The lancet was sealed in a capillary tube filled with a suspension of *E. coli* K-12 and was continuously rolled under UV exposure under blacklight irradiation. The antibacterial effect was studied in the TiO_2 layer coated on the lancet at 0.5 mW/cm^2 for 45 minutes. It was observed that the lancet with an unannealed TiO_2 layer exhibited a high lancing resistance in comparison to the other lancets. It was concluded that the lancet coated with a nanolayer of TiO_2 formed by annealing had efficient antibacterial properties than the lancing resistance of a bare lancet.

Removal of *E. coli, S. aureus,* SARS, and MS2 coliphage was quite effective via the oxidation of these species and these are almost completely reduced within an hour. Such an effect was observed using various light sources, natural sunlight, and indoor lighting sources also. Eco-friendly activities like self-cleaning, antifogging, deodorization, and self-sterilization have become easier and more convenient by the application of solar photocatalysis, which has made it possible to keep our environment clean and healthy.

REFERENCES

Banerjee, S., D. D. Dionysiou, and S. C. Pillai. 2015. Self-cleaning applications of TiO_2 by photo-induced hydrophilicity and photocatalysis. *Appl. Catal. B Environ.* 176–177: 396–428.

Bolte, G. 2009. Innovative building material-reduction of air pollution through TioCem®. *Nanotechnol. Constr.* 3: 55–61.

Budde, F. E. 2010. Self-cleaning, odor-reducing, water-shedding properties of titanium dioxide photocatalytic oxidizing coatings. *Metal Finishing.* 108 (1): 34–36.

Bygott, C. and J. E. Maltby. 2007. Photocatalytic coating. Patent EP2188125A1.

Cassar, L. 2004. Phocatalysis of cementitious materials: Clean buildings and clear air. *MRS Bull.* 29: 328–331.

Cassar, L. and C. Pepe. 1998. Use of organic additives for the preparation of cementitious compositions with improved properties of constancy of color. Patent US 6117229 A.

Cebeci, F. C., Z. Z. Wu, L. Zhai, R. E. Cohen, and M. F. Rubner. 2006. Nanoporosity-driven superhydrophilicity: A means to create multifunctional antifogging coatings. *Langmuir* 22: 2856–2862.

Chai, S. Y., J. K. Park, H. K. Kim, and W. I. Lee. 2006. Correlation of superhydrophilicity and photocatalytic activity in the TiO_2-based porous thin films. *Mater. Sci. Forum.* 510–511: 54–57.

Chekini, M., M. R. Mohammadizadeh, and S. M. V. Allaei. 2011. Photocatalytic and superhydrophilicity properties of N-doped TiO_2 nanothin films. *Appl. Surf. Sci.* 257 (16): 7179–7183.

Deng, Z.-Y., W. Wang, L.-H. Mao, C.-F. Wang, and S. Chen. 2014. Versatile superhydrophobic and photocatalytic films generated from TiO_2–SiO_2@PDMS and their applications on fabrics. *J. Mater. Chem. A.* 2: 4178–4184.

Fateh, R., R. Dillert, and D. Bahnemann. 2014. Self-cleaning properties, mechanical stability, and adhesion strength of transparent photocatalytic TiO_2-ZnO coatings on polycarbonate. *ACS Appl. Mater. Interfaces.* 6 (4): 2270–2278.

Frazer, L. 2001. Titanium dioxide: Environmental white knight? *Environ. Health Perspect.* 109 (4): A174–A177.

Fujishima, A., K. Hashimoto, and T. Watanabe. 1999. TiO_2 *Photocatalysis: Fundamentals and Applications.* Tokyo, Japan: BKC Inc.

Guo, S., Z. B. Wu, and W. R. Zhao. 2009. TiO_2-based building materials: Above and beyond traditional applications. *Chin. Sci. Bull.* 54 (7): 1137–1142.

Hamada, H., K. Komure, R. Takahashi, and T. Yamaji. 2004. NOx emission, local concentration and reduction by TiO_2-photocatalysis in airport area, in *RILEM International Symposium on Environment-Conscious Materials and Systems for Sustainable Development*, 6–7 September 2004, Koriyama, Japan, pp. 361–366.

Hata, S., Y. Kai, I. Yamanaka, H. Oosaki, K. Hirota, and S. Yamazaki. 2000. Development of hydrophilic outside mirror coated with titania photocatalyst. *JSAE Rev.* 21: 97–102.

Hochmannova, L., and J. Vytrasova. 2010. Photocatalytic and anti-microbial effects of interior paints, *Prog. Org. Coat.* 67: 1–5.

Horiuchi, Y., K. Mori, N. Nishiyama, and H. Yamashita. 2008. Preparation of superhydrophilic mesoporous silica thin films containing single-site photocatalyst (Ti, V, Cr, Mo, and W oxide moieties). *Chem. Lett.* 37 (7): 748–749.

Hwang, Y. K., K. R. Patil, H.-K. Kim, S. D. Sathaye, J.-S. Hwang, S.-E. Park, and J.-S. Chang. 2005. Photoinduced superhydrophilicity in TiO_2 thin films modified with WO_3. *Bull. Korean Chem. Soc.* 26 (10): 1515–1519.

Italcementi. 2005. *TX Millennium Photocatalytic binders (Technical Report).* Bergamo, Italy: Italcementi.

Jesus, M. A. M. L., J. T. S. Neto, G. Timò, P. R. P. Paiva, M. S. S. Dantas, and A. M. Ferreira. 2015. Superhydrophilic self-cleaning surfaces based on TiO_2 and TiO_2/SiO_2 composite films for photovoltaic module cover glass. *Appl. Adhesion Sci.* 3:5 1–9.

Jo, S.-H., K.-H. Kim, Y.-H. Kim, L. Min-He, B.-Won Kim, and J.-H. Ahn. 2015. Deodorization of food-related nuisances from a refrigerator: The feasibility test of photocatalytic system. *Chem. Engg. J.* 277: 260–268.

Kallio, T., S. Alajoki, V. Pore, M. Ritala, J. Laine, M. Leskelä, and P. Stenius. 2006. Antifouling properties of TiO_2: Photocatalytic decomposition and adhesion of fatty and rosin acids, sterols and lipophilic wood extractives. *Colloids Surf. A.* 291: 162–176.

Kamegawa, T., Y. Shimizu, and H. Yamashita. 2012. Superhydrophobic surfaces with photocatalytic self-cleaning properties by nanocomposite coating of TiO_2 and polytetrafluoroethylene. *Adv. Mater.* 24 (27): 3697–3700.

Kazemi, M., and M. R. Mohammadizadeh. 2012. Simultaneous improvement of photocatalytic and superhydrophilicity properties of nano TiO_2 thin films. *Chem. Engg. Res. Design.* 90 (10): 1473–1479.

ylova, G., A. Brioude, S. Ababou-Girard, J. Mrazek, and L. Spanhel. 2010. Natural super-hydrophilicity and photocatalytic properties of sol–gel derived $ZnTiO_3$-ilmenite/r-TiO_2 films. *Phys. Chem. Chem. Phys.* 12: 15101–15110.

the, S. S., S. Liu, C. Terashima, K. Nakata, and A. Fujishima. 2014. Transparent, adherent, and photocatalytic SiO_2-TiO_2 coatings on polycarbonate for self-cleaning applications. *Coatings.* 4: 497–507.

⅃, P., X.-C. Wang, and X.-Z. Fu. 2000. Processing and properties of photocatalytic self-cleaning ceramic. *J. Inorg. Mater.* 15 (1): 88–92.

pez, J. E. O. and W. A. Jacoby. 2002. Microfibrous mesh coated with titanium dioxide: A self-sterilizing, self-cleaning filter. *J. Air Waste Manage Assoc.* 52 (10): 1206–1213.

altby, J. E. and C. Bygott. 2014. Photocatalytic coating. Patent US 2014322116A1.

armur, A. 2004. The lotus effect: Superhydrophobicity and metastability. *Langmuir.* 20: 3517–3519.

lls, A., A. Lepre, N. Elliot, S. Bhopal, I. P. Parkin, and S. A. O'Neill. 2003. Characterisation of the photocatalyst Pilkington Activ™: A reference film photocatalyst? *J. Photochem. Photobiol. A Chem.* 160: 213–224.

ırakami, E., H. Kohno, S. Kato, S. Mizuno, M. Noguchi, and M. Hori. 2004. Deodorization of Chinese restaurant kitchen exhaust by an Ag-deposited photocatalyst. *Odor Environ. J.* 35 (3): 146–150.

ırata, Y., H. Tawara, H. Obata, and K. Takenchi. 1999. Air purifying pavement: Development of photocatalic concretebloces. *J. Adv. Oxid. Technol.* 4 (2): 227–230.

ıkamura, H., M. Tanaka, S. Shinohara, M. Gotoh, and I. Karube. 2007. Development of a self-sterilizing lancet coated with a titanium dioxide photocatalytic nano-layer for self-monitoring of blood glucose. *Biosens. Bioelectron.* 22 (9–10): 1920–1925.

shioka, Y. and T. Nishino. 2014. Air conditioner that enables deodorizing using a photocata-lyst. US Patent 8740420 B2.

ɔzawa, M., K. Tanigawa, M. Hosomi, T. Chikusa, E. Kawada. 2001. Removal and decom-position of malodorants by using titanium dioxide photocatalyst supported on fiber acti-vated carbon. *Water Sci. Technol.* 44 (9): 127–133.

ıko, Y., Y. Utsumi, C. Niwa, T. Tatsuma, K. Kobayakawa, Y. Satoh et al. 2001. Self-sterilizing and self-cleaning of silicone catheters coated with TiO_2 photocatalyst thin films: A pre-clinical work. *J. Biomed. Mater. Res.* 58 (1): 97–101.

ırkin, I. P. and R. G. Palgrave. 2005. Self-cleaning coatings. *J. Mater. Chem.* 15: 1689–1695.

ıtel, P., C. K. Choi, and D. D. Meng. 2009. Superhydrophilic surfaces for antifogging and antifouling microfluidic devices. *JALA.* 15: 114–119.

ɛral, J., X. Domènech, and D. F. Ollis. 1997. Heterogeneous photocatalysis for purifica-tion, decontamination and deodorization of air. *J. Chem. Technol. Biotechnol.* 70 (2): 117–140.

ιchat, P., J. Disdier, C. Hoang-Van, D. Mas, G. Goutailler, and C. Gaysse. 2000. Purification/deodorization of indoor air and gaseous effluents by TiO_2 photocatalysis. *Catal. Today.* 63 (2–4): 363–369.

ızenat, E. 2009. Photocatalytic self-cleaning materials: Principles and impact on atmosphere. *Eur. Phys. J. Conf.* 1: 69–74.

amirez, A. M., K. Demeestere, N. De Belie, T. Mantyla, and E. Levanen. 2010. Titanium dioxide coated cementitious materials for air purifying purposes: Preparation, character-ization and toluene removal potential. *Build. Environ.* 45: 832–838.

ɛkiguchi, Y., Y. Yao, Y. Ohko, K. Tanaka, T. Ishido, A. Fujishima, and Y. Kubota. 2007. Self-sterilizing catheters with titanium dioxide photocatalyst thin films for clean intermittent catheterization: Basis and study of clinical use. *Int. J. Urol.* 14 (5): 426–430.

ιan, W., Y. Hu, M. Zheng, and C. Wei. 2015. The enhanced photocatalytic activity and self-cleaning properties of mesoporous SiO_2 coated Cu-Bi_2O_3 thin films. *Dalton Trans.* 44: 7428–7436.

Shin, D. Y. and K. N. Kim. 2009. Effective of SiO_2 addition on the self-cleaning and photoca: lytic properties of TiO_2 films by sol-gel process. *Mater. Sci. Forum.* 620–622: 679–6:

Song, K. Y., M. K. Park, Y. T. Kwon, H. W. Lee, W. J. Chung, and W. I. Lee. 2001. Preparati of transparent particulate MoO_3/TiO_2 and WO_3/TiO_2 films and their photocatalytic pro erties. *Chem. Mater.* 13 (7): 2349–2355.

Stamate, M. and G. Lazar, 2007. Application of titanium dioxide photocatalysis to create se cleaning materials. *Rom. Technol. Sci. Acad.* 13 (3): 280–285.

Takeuchi, M., K. Sakamoto, G. Martra, S. Coluccia, and M. Anpo. 2005. Mechanism of ph toinduced superhydrophilicity on the TiO_2 photocatalyst surface. *J. Phys. Chem. B.* 1 (32): 15422–15428.

Tricoli, A., M. Righettoni, and S. E. Pratsinis. 2009. Anti-fogging coatings by flame synthe and deposition. *Langmuir.* 25 (21): 12578–12584.

Usuda, S., A. Chen, and M. Anpo. 2005. Modeling and simulation of light guide boards fo photocatalytic deodorizing system. *Res. Chem. Intermed.* 31 (4): 319–329.

Wang, R., K. Hashimoto, A. Fujishima, M. Chikuni, E. Kojima, A. Kitamura, M. Shimohigos and T. Watanabe. 1997. Light-induced amphiphilic surfaces. *Nature.* 388: 431–432.

Wei, Q., S. Wang, W. Li, X. Yuan, and Y. Bai. 2015. Hydrophobic ZnO-TiO_2 nanocompo ite with photocatalytic promoting self-cleaning surface. *Int. J. Photoenergy.* 201 6 p. Article ID 925638. doi.org/10.1155/2015/925638.

Yaghoubi, H., N. Taghavinia, and E. K. Alamdari. 2010. Self-cleaning TiO_2 coating on pol carbonate: Surface treatment, photocatalytic and nanomechanical properties. *Surf. Co Technol.* 204 (9–10): 1562–1568.

Yuranova, T., D. Laub, and J. Kiwi. 2007. Synthesis, activity and characterization of texti showing self-cleaning activity under daylight irradiation. *Catal. Today* 122: 109–117.

Zhao, X., Q. Zhao, J. Yu, and B. Liu. 2008. Development of multifunctional photoactive se cleaning glasses. *J. Non-Cryst. Solids.* 354: 1424–1430.

18 Photoreactors

18.1 INTRODUCTION

Photocatalytic degradation of organic and inorganic pollutants is a fast-growing technique, which is quite promising for wastewater purification. Selection of an efficient photooxidation reactor with a proper light source plays a significant role in the remediation of contaminated water. In this regard, designing a perfect model for a photoreactor has been a challenging problem for a long time. A variety of photochemical reactions are carried out by using different photoreactors. Some of the more common photoreactors are as follows:

1. Bed reactor
2. Batch reactor
3. Thin-film reactor
4. Annular reactor
5. Flow reactor
6. Immersion well reactor
7. Multilamp reactor
8. Slurry reactor
9. Merry-go-round reactor

18.2 BED REACTOR

Many types of bed photoreactors are used on laboratory scale, pilot scale, and large scale. The bed photoreactors find potential use in the purification of contaminated water, hydrogen production, and so on. Pure or modified photocatalysts are used in these photoreactors. A few bed photoreactors are discussed below.

18.2.1 FIXED BED PHOTOREACTOR

Alexiadis and Mazzarino (2005) worked at the design of pilot and industrial size fixed bed photocatalytic reactors for wastewater purification. The physical reactor model was employed to initiate the photocatalytic system. The cost of the water remediation was calculated by determining both the energy used during the process and the periodic replacement of commercial ultraviolet (UV) lamps. It was observed that optimum conditions for a photocatalytic wastewater treatment depend on the rate of degradation of the pollutants. Low-power UV sources were used for the reaction having fast rate of degradation using low absorption catalysts. Use of high-power UV sources and dense catalysts lowered the cost of the process when the degradation reaction was slow.

Esterkin et al. (2005) made a fixed bed photocatalytic reactor by coating TiO_2 o glass fiber meshes for the treatment of air pollution. The degradation of trichloroeth ylene (TCE) in an air stream was investigated. The reactor was so designed that could be effectively used for the purification of impure air. It allows uniform illum nation of the TiO_2-coated meshes and a good ratio of the irradiated surface area t the volume of the gaseous reacting mixture.

A fixed bed reactor (FBR) was used by Cloteaux et al. (2014) for the photodegra dation of formaldehyde, which has been considered a category 1 carcinogen. A fixe bed photocatalytic reactor was fabricated using TiO_2-coated Raschig rings an the light source was a UVA lamp. The hydraulic behavior of the reactor depend on experimental residence time distribution (RTD). This model accounted for th hydraulics, distribution of light, chemical kinetics, and mass transfer taking place i the reactor. The model in combination with a Langmuir–Hinshelwood (L–H) kineti model could determine variations in concentration at the reactor output. It was con cluded that the fixed bed photocatalytic reactor effectively degraded formaldehyd in the aqueous phase. This study suggested a novel route to design a photoreactor b integrating mass transfer limitations and light distribution.

Nitrogen-doped TiO_2 photocatalyst mounted on glass spheres was used fo photocatalytic removal of pollutants from wastewater using an FBR. A flat plate structured bed photoreactor was designed by Vaiano et al. (2015). Ananpattaracha and Kajitvichyanukul (2014) coated nitrogen-doped TiO_2 on stainless steel plate an used it for the purification of tannery wastewater using a fixed bed photocatalyti reactor. Visible light radiations were used for this study. The change in the concen tration of chromium and total organic carbon was determined to observe the photo catalytic efficiency. Kinetic studies were made and mass transfer was analyzed b studying the change in rate of flow and the hydraulic retention time of contaminate water sample in the photoreactor. The efficiency of N-doped TiO_2 decreased in th fixed bed photoreactor after undergoing six cycles of tannery wastewater remedia tion. This photocatalyst could, therefore, be reused in wastewater purification.

18.2.2 Annular Fluidized Bed (AFB) Photoreactor

Fluidization is the process by which solid particles are made to behave like a flui under certain conditions. A unit facilitating this process of fluidization is terme a fluidized bed. A fluidized bed finds applications for carrying out chemical reac tions, heat transfer, mixing, drying, and so on. A novel development of the fluid ized bed is the AFB with a large central nozzle containing a stationary fluidize bed around it.

The main components of an AFB are the central nozzel, riser, cyclone, and mix ing chamber. The riser has a narrow bottom part to prevent accumulation of solid at the bottom. The riser walls are made up of membrane waterwall surfaces, whicl affect the solid flow patterns, thereby ensuring efficient gas–solid mixing. The exit of the riser are divided into two types: (1) once-through exits and (2) internal reflu exits.

The first one has curved exits, while the second one has an abrupt exit. The cyclone separates the particles on the basis of their sizes by changing the velocity o the feed gas. The particles are then returned to the bed, depending on the particle size. These particles are trapped and sent back to the bottom of the riser through a vertical standpipe (Son et al. 2009). The distinguishing component of the AFB reactor is its central nozzle, which is not present in other fluidized bed reactors. The fluidization of gas takes place in the annular, which causes the solid to overflow on the upper edge of the central nozzle. This solid is then sent to the mixing chamber by the flow of the gas stream.

An AFB allows gas to enter the reactor at high speed from the bottom of the central nozzle. In addition to this, a fluidized gas is permitted through an annular nozzle ring. Consequently, gas and solids are thoroughly mixed at the bottom of the mixing chamber and allowed to flow upward in the riser. The gas and solids after leaving the riser get separated in a cyclone depending on the fixed velocities. This separated gas then flows through a bag filter and the solids go downward in the downer, which is introduced into the bottom of the plant and the same process is repeated again.

Faramarzpour et al. (2009) immobilized TiO_2 nanoparticles on a support called "perlite." The as-prepared TiO_2-coated perlite particles were used in a floating bed photoreactor for the photocatalytic treatment of wastewater contaminated by furfural. The concentration of furfural was reduced by more than 95% within 120 minutes.

Liu et al. (2011) prepared bismuth titanate photocatalyst hydrothermally and developed a three-phase internal circulating fluidized bed photoreactor (TPICFBP). The as-prepared photocatalyst was used for the photodecomposition of acid red G in this TPICFBP. It was observed that 92% of dye was removed, when the wastewater flux was 10 L/h and operating time was 3 hours.

Chiara et al. (2014) worked on the inhibition of ethylene activity during postharvest handling of fresh products by using TiO_2 under UV light. A fluidized bed photoreactor (FBP) was designed for this purpose to remove ethylene in the cold storage room atmosphere. SiO_2/TiO_2-coated alumina microspheres were used as the photocatalyst. Ethylene concentration was reduced by 72% in 40 ppm ethylene gas mixture, after 4.5 hours under 36 W UV light illumination.

Activated carbon fibers (ACFs) were modified with acidic, alkaline, and neutral solutions. These were then employed in the anaerobic fluidized bed photoreactor (AFBPR) for immobilization of bacteria and to enhance hydrogen production. It was concluded that an AFBPR could be used with nitric acid–modified ACFs for the large-scale generation of biohydrogen (Ren et al. 2014).

Some of the advantages of an AFB are as follows:

- An intense mixing zone is available on the bed.
- Combined advantage of long solid residence time and efficient transfer of heat and mass giving it the potential for being used in heat transfer processes like cooling, heating, and so on.

An AFB cannot be used for reactions that require shorter residence times and less intense mixing. The introduction of the nozzle also increases the cost of an AFB. It also needs repeated maintenance of its complicated components such as the central nozzle, which easily clogged by unwanted particles entering the nozzle.

18.2.3 PACKED BED PHOTOREACTOR (PBR)

Arabatzis et al. (2005) designed a novel gas-phase photocatalytic PBR. The PBR was used for the degradation of volatile organic compounds (VOCs) using porous foaming TiO_2 as a photocatalyst.

A continuous PBR was fabricated by Borges et al. (2015) for wastewater purification. Paracetamol was chosen as a model contaminant. Two photocatalytic reactor configurations were studied. These are a (1) stirred photoreactor using TiO_2 solid suspension and a (2) PBR using TiO_2 immobilized on glass surfaces (TGS). The surface morphology and texture of the TGS were monitored by scanning electron microscope. The effect of the amount of titania and pH of wastewater was determined in the stirred photoreactor and the influence of wastewater was investigated in the PBR in order to obtain the optimal operation conditions. It was concluded that the PBR was more effective for the decomposition of pollutants.

18.3 BATCH REACTOR

Batch reactors are commonly used in the process industries and in particular pharmaceutical industries. They also possess many laboratory applications like small-scale production. They find extensive use in wastewater remediation. A batch reactor works as a batch of reactants is introduced in the reactor initially. The reactants are permitted to react for a particular duration of time, and at the end of the reaction, the products are removed from the reactor. Nothing is put in or taken out from the batch reactor during the progress of the reaction. In an ideal batch reactor, the properties of the reacting mixture are almost stable throughout the reaction.

Synthetic organic photochemical reactions are usually performed in the solution phase employing an immersion well batch reactor. There are fixed-volume types of batch reactors irradiated by mercury vapor discharge lamps. Mercury vapor lamps are commonly used sources of light. These lamps can be categorized into three types depending on the pressure used and the output spectrum obtained. As compared to the low-pressure and high-pressure lamps, the medium-pressure (1–10 atm) lamps are more useful for synthetic photochemistry. Such lamps possess a working temperature of 600°C–800°C and thus can be utilized in cold water–cooled immersion jackets to avoid the thermal effects of the photochemical substrate. Spectrum wavelengths of around 200 nm are produced by medium-pressure lamps, so different glassware filters like quartz, vycor, or pyrex can be used to stop high-energy radiations from approaching the photochemical substrate and permit transmission of only specific regions of the mercury emission spectrum. This helps in monitoring radiations by which chromophores of the substrate are excited.

Batch reactors have certain advantages over others such as:

• They are easy and quick to set up, and reactions are monitored by a flip of the switch.
• Reactions on a scale up to 1 mmol/100 mL can be conducted.

- Volumes range from 100 mL to 1 L for an immersion well.
- The power supply can be varied from available lamps of different powers like 6, 125, 400, and 600 W.

But a photochemical batch reactor also has some disadvantages such as:

- Lamps acquire a very high temperature and therefore a continuous cold water flow is required to keep them cool.
- Medium-pressure lamps are broadband emitters and they can excite more than just the desired chromophore, resulting in some undesired photoproducts, polymerization, relatively smaller yields, and lowering in the purification rate.
- Large-scale photochemical synthesis is not possible with these batch reactors.

Divya et al. (2009) photodegraded acidic orange G dye using a batch photoreactor der UV radiation with a combination of UV exposure and H_2O_2. Various parameters such as pH, concentration, amount of H_2O_2, TiO_2, and light intensity/light source ith reflecting or nonreflecting surface of the photoreactor were analyzed to study e photodegradation of dye.

Behnajady et al. (2012) gave a design equation for a batch-recirculated photore-tor for the removal of acid red 17 dye. They presented a reactor composed of an nular photoreactor and a continuous stirred tank reactor (CSTR) with a batch-circulated current. The photooxidation of acidic dye acid red 17 was done by eans of a UV/H_2O_2 system. The kinetics of this reaction was also studied and a eudo-first-order kinetics was observed.

TiO_2 was used as the catalyst for the photodegradation of reactive yellow (RY) a batch and continuous photoreactor under UV light exposure. TiO_2 was immobi-zed on the ceramic plate using cement as a binder. Various parameters were varied. comparative study of photocatalytic efficiency between the batch and continuous stem was performed and it was concluded that the batch mode showed better deg-dation capacity. The dye was degraded to about 60% in 360 minutes for 200 ppm Y solution in batch mode (Alam et al. 2012).

Sheidaei and Behnajady (2015) used TiO_2-P25 nanoparticles for the removal of odel contaminant acid orange 7 (AO7) using a batch-recirculated photoreactor. ifferent experimental parameters like initial AO7 concentration, volume of solu-on, volumetric flow rate, reaction time, and power of light source were investi-ated and optimized using the Taguchi method. Sixteen experiments were carried ut to study the effects of these parameters on the removal of AO7. Results showed at the power of the light source was the most significant factor as compared to e others.

Chanathaworn et al. (2012) reported photocatalytic decolorization of basic dyes ke rhodamine B (RhB) and malachite green (MG) in aqueous solution using TiO_2 a photocatalyst on UV blacklight exposure. A 0.5 L batch photoreactor contain-g the dye solution was placed in a chamber made of stainless steel with air cool-g under light exposure. The influence of different parameters was investigated.

The rate of dye photobleaching depends on the morphology of the pollutant, and M dye was removed at a faster pace than the RhB dye.

Diclofenac (DCF), a pollutant in surface waters and drinking water, was treat by subjecting it to photolysis and TiO_2-catalyzed degradation in a circulating bat photoreactor by Devagi et al. (2014). Under optimum conditions, complete D(removal was observed within 15 minutes in the immersion well reactor. When ph tochemical degradation was carried out in presence of sunlight, an exposure peri of up to 360 minutes was needed for complete DCF removal. It was shown that t photocatalytic degradation kinetics of DCF depends on both the geometry of t photoreactor and the nature of the water matrices.

A batch photoreactor was used for photobleaching AO7 dyes and basic violet dye (BV14) by the UV/H_2O_2 process. The light source used was UV lamps emitti light of 254 nm. The highest decolorization rates observed for AO7 and BV14 we 98.2% and 97.5%, respectively. These rates were obtained at a peroxide concentr tion of 10–30 mmol/L. It was observed that the decolorization was higher in neut pH medium than for acidic or basic media.

18.4   THIN-FILM REACTOR

A thin-film reactor is designed so that it works best for small volumes of reacta (50 – 500 mL) for concentrated solutions, where radiation only penetrates a fracti of a millimeter. The reactant solution is introduced from reservoir by a glass jet a a thin film of liquid falls under gravity over a quartz or borosilicate closed tut A low-pressure mercury lamp or phosphor-coated lamp is placed inside the tut which guarantees uniform irradiation of this falling film of liquid. On reversing t flow of direction, the reactor is used to irradiate a 5 mm path length of solution.

In a thin-film photoreactor, five standard emission lamps are present. All the lamps are of the same size and electrical characteristics in this reactor. A thi film photoreactor has 350 nm, 410 nm, and white lamps in it. Phosphor coating not done in the 254 nm lamp and it shows over 90% emission of its radiation. Tl 350 nm lamp generates 4×10^{18} photons/second in the thin-film photoreactor determined by benzophenone/isopropanol actinometry. All these lamps work fro the same power supply, which is constructed into the control unit for the thin-fil photoreactor. This control unit runs the pump as well as the lamp.

Double-walled quartz or borosilicate glass irradiation tubes are used in this rea tor. Tubes have threaded glass entrance and exit ports and the liquid jet tube and ex tubing are fixed. However, the jet tube is removed for the reverse flow arrangeme Flow velocity through the jet varies from 0.1 to 1 mL/s depending on the size of tl jet and setting of the bypass control. Film thickness ranging from 0.1 to 0.3 mm observed depending on flow rate and the solvent used. A self-priming liquid pump used in this photoreactor to handle mixtures of liquid and air. The pumping solutic remains uncontaminated by the lubricants present in the pump as a diaphragm pun is used and stays only in contact with the stainless steel or polytetrafluoroethylei (PTFE) tubing.

A round-bottomed flask of capacity 250 mL works as the standard flask in tl thin-film photoreactor. The flask is composed of one central port and two side port ockets having entrance tubes and sinters are available for connection to the pump, radiation tube, and an external gas cylinder. Loss of solvent is prevented by using double-surface condenser. This condenser is fitted into one of the side ports. The minimum volume used in this reactor is about 50 mL of solution, but flasks of larger capacity are also available.

The thin-film reactor works best for monitoring the changes produced in a photochemical reaction during reactant depletion or product formation. The reactor is laced near the spectrometer and the flow cell remains connected to the reactor with PTFE tubing. There are two cuvettes, one having a path length of 10 mm and composed of far-UV quartz rectangular cell with three sides polished for absorption/fluorescence and the other with a 2 mm path length far-UV quartz rectangular cell with two sides polished for absorption.

All these components of the reactor are positioned inside metal housing having heavy metal base. The irradiation tube is held by a central metal support. This metal support has a reservoir on one side and pump assembly on the other side. Connections are supplied for (1) water cooling for the reflux condenser, (2) gas inlet, and (3) flow-through connections for spectrometric monitoring. Detachable covers are present on either side of the support plate. The reactor assembly and control unit are electrically connected.

Damodar et al. (2007) carried out the degradation of four dyes using TiO_2 and solar irradiation in a thin-film-immobilized surface photoreactor. Batch experiments were performed at initial concentrations varying in the range from 25 to 100 mg/L and at a catalyst loading of 0.5–1 g/L. Almost 30%–70% color removal was observed depending on the initial concentration of dye, dye morphology, and the amount of catalyst. The thin-film-immobilized surface photoreactor provided 90%–98% bleaching effect depending on the initial concentration and time of exposure. Flow rate also influences the extent of color removal at higher concentrations.

Platinum- and silver-incorporated TiO_2 (Pt-TiO_2 and Ag-TiO_2) were coated on sapphire tubes present in a thin-film photoreactor by Kuo et al. (2011) using a photoreduction process. It was observed that Ag and Pt particles enhanced the optical absorption and excitation of coatings in the visible region. The results showed that the photodegradation rate of o-cresol by Pt-TiO_2 coating under visible light exposure was more than the photodegradation rates of o-cresol by Ag-TiO_2 and pure TiO_2 coatings.

Adams et al. (2013) developed a TiO_2-immobilized thin-film multitubular photoreactor for use in liquid and gas phase media. TiO_2 was doped with a rare earth element for the photocatalytic degradation of methyl orange, which was taken as a model pollutant. It was observed that there was a linear relationship between increasing reactor volume and degradation, which could not be observed in a suspended reactor system.

18.5 ANNULAR REACTOR

In this type of photoreactor, the irradiation lamps (6 W or higher up to 400 W) are placed in large quartz and borosilicate glass immersion wells. The immersion well is positioned in the center of a rotating carousel assembly, which is composed of up to 24 quartz or borosilicate glass tubes. A motor rotates this carousel assembly.

This motor is present at the base of the reactor. Low-pressure mercury lamps of 6 an 12 W or medium-pressure mercury lamps of 125 and 400 W along with their respec tive power supplies are used in an annular reactor. Either a quartz well or a borosil cate glass immersion well is used in an annular reactor.

Glass and quartz tubes are held vertically by an assembly containing two meta discs having a diameter of 170 mm placed on three vertical rods. The separation c discs may be changed from 10 to 300 mm. Each metal disc comprises two concentri rings having equally spaced holes of diameter 16 mm. The inner and outer rings hav 12 holes each and these are placed so that their centers fall in a line aligning ther to the center of the lamp. This alignment makes it possible to obscure one tube wit the other when it is illuminated in the center. Twenty-four open-ended quartz tube 300 mm long with 12 mm outer diameter are available in this type of reactor.

A 20 rpm motor is present at the base of the reactor, which facilitates the rotatio of the carousel assembly around the immersion well. The carousel assembly is als surrounded by a cylindrical steel tube attached to the base of the housing with swi release catches. A central flange on the housing carries the immersion well.

Krishnan and Swaminathan (2010) used an annular tube reactor to conduct exper iments using TiO_2 as the photocatalyst. The effect of various parameters like cataly: load (5–20 g/m²), concentration of benzene (0.2–6 g/m³), and flow rate (0.2–1 L/mir on the removal of benzene was studied. Rate of removal of benzene ranged fror 7% to 96%, which depends on the limit levels of these parameters. An L–H kineti model was proposed for the removal of benzene. It was observed that a plug-flow type L–H kinetic model produced an equation which formed the basis for the pho toreactor scale-up and allow estimation of the mass transfer and reaction resistance in the photoreactor. It was shown that the ratio of reaction rate resistance to overal resistance played a crucial role in establishing the predominant resistances betwee mass transfer and rate of reaction taking place in the photoreactor.

Peres et al. (2015) designed an annular photoreactor having tangential inlet an outlet tubes. The fluid flow was determined by residence time distribution (RTD experiments, which were repeated by computational fluid dynamics taking int consideration four relevant turbulence models like the k-ϵ, the k-ω, the shear stres transport, and the Reynolds stress. Inlet effects resulted into helical flow withi the reactor, leading to plug flow, which depends on the rate of flow and the turbu lence model. The k-ω model efficiently dealt with viscous effects and generate experimental RTD curves with correlation coefficients higher than 0.9566, agains 0.8705 from the k-ϵ model.

18.6 FLOW REACTOR

When production is to be done on a large scale, chemical reactions are to be carrie out in a continuous flow stream. Chemical reactions taking place in fluid state con stitute the flow chemistry and the photoreactors designed to achieve such reaction: are called flow reactors. Many types of flow reactors are known such as spinnin; tube reactors, oscillatory flow reactors, microreactors, multicell flow reactors, he: reactors, aspirator reactors, continuous reactors, and so on.

Continuous reactors have a tube-shaped structure and are made up of less-reactive materials like steel, glass, and so on. Mixing is accomplished by two methods:

• If the diameter of the reactor is less than 1 mm, like in microreactors, mixing is done by diffusion.
• In the other method, mixing is done by the use of static mixers.

The reaction conditions like heat transfer, residence time, and mixing are efficiently controlled in continuous flow reactors. Residence time is defined as the time required for the reaction to heat up or cool down. The residence time of the reagents in the reactor is the ratio of reactor volume to the flow rate through it. So, reagents should be introduced slowly or the volume of the reactor should be increased to obtain a longer residence time. These factors will affect the production rate to a greater extent. Though these flow reactors are used to perform flow processes at ton scale, on a small scale microreactors can also be used for process development experiments.

Behnajady et al. (2007) photodegraded an azo dye acid red 27 using a tubular continuous-flow photoreactor with TiO_2 supported on glass plates. The length of photoreactor determines the percentage of removal. The decomposition of the dye also depends on volumetric flow rate and light intensity. The removal efficiency increases linearly with the light intensity but it decreases with increase in the flow rate. Ammonium, nitrate, nitrite, and sulfate ions were detected in the final outlet stream from the reactor as mineralization products of N and S, respectively. Pseudo-first-order kinetics was followed in the photocatalytic degradation of acid red 27. They also designed and constructed a new tubular continuous-flow photoreactor and used it for advanced oxidation processes (AOPs) (Behnajady et al. 2009). This photoreactor consisted of six quartz tubes and a UV light source lamp, mounted in the center of the quartz tubes. TiO_2 P25 was immobilized on glass plates by a heat attachment method. These glass tubes were placed in each quartz tube. 4-Nitrophenol was selected as a model pollutant in aqueous solutions and its photodegradation was observed. The equation proposed for the designed photoreactor and the observations had a linear relationship between a pseudo-first-order reaction rate constant and reciprocal of liquid volumetric flow rate.

Toosi et al. (2013) degraded pollutants like mercaptan including 80% *t*-butyl mercaptan and 20% methyl ethyl sulfide using a continuous photoreactor with TiO_2 and UV light exposure. These odorant sulfur compounds were efficiently degraded in continuous photoreactor in the presence of UV irradiation.

Chanathaworn et al. (2014) studied the photobleaching of wastewater contaminated by dye MG using TiO_2-coated glass tube media in a continuous photoreactor. TiO_2 was prepared by three different procedures: (1) titanium tetraisopropoxide (TTIP) sol–gel, (2) TiO_2 powder–modified sol, and (3) TiO_2 powder suspension coating on Raschig ring glass tube media. A continuous photoreactor packed with titania coating was developed. This photoreactor can be employed on an industrial scale due to its high recyclability.

Rasoulifard et al. (2015) investigated the performance of UV-light emitting diodes (UV-LEDs) for the decomposition of the dye direct red 23 (DR23) in a continuous photoreactor. The effect of initial concentration of dye, concentration of peroxydisulfate, temperature, and photonic efficiency were studied for the bleaching of DR23. The current intensity and irradiation efficiency of the UV/LEDs were also determined.

Anthony et al. (2015) constructed a continuous flow photoreactor to carry out cobalt-mediated radical polymerization (CMRP). It was shown that the use of flow photoreactors helped in the speeding up of the polymerization process of vinyl acetate without losing control on polymerization. UV light irradiation was used in this study.

18.7 IMMERSION WELL REACTOR

Synthetic photochemistry carried out in classic batch reactors has, for over half a century, proved to be a powerful but underutilized technique in general organic synthesis. Recent developments in flow photochemistry have the potential to allow this technique to be applied in a more mainstream setting. This review highlights the use of flow reactors in organic photochemistry, allowing a comparison of the various reactor types to be made.

An immersion well photoreactor with mercury vapor discharge lamps has been the most efficient apparatus for laboratory scale photochemical reactions for the last five decades or so. An immersion well functions as a compact batch reactor particularly for synthetic photochemistry halogenations, oxidation, and so on, in a range of milligrams up to a few grams. The name was coined as the lamp in such photoreactors was placed in a double-jacketed water-cooled immersion well. This double-walled immersion well allows cooling of lamps by water. The cooling water is connected to a flow sensor, which helps to shut down the lamp if the pressure of the water supply drops. The whole apparatus is protected against UV light by shielding it in a cabinet. Aluminum foil can also be wrapped around the glassware for effective shielding.

Mercury discharge lamps are the most common UV light source in these reactors. These are composed of empty glass tubes with mercury vapor. An electrical discharge is passed through these tubes, which causes excitation of the Hg atoms, resulting in the emission of UV radiation. The input powers of low-pressure lamps ranges from 6 to 300 W and above. The low range (6–16 W) converts 30% of input power into UV radiations. Uncoated lamps emitting 90% of the spectral output at a wavelength of 254 nm are used for carbonyl and arene photochemistry and also in halogenations chemistry. Phosphor coating is also done on these lamps. These lamps have a longer lifetime. The input power of medium-pressure lamps ranges from 125 to 60 kW. These are used for industrial purposes. Lamps of 125 and 400 W are commonly used on laboratory scale. UV output is obtained in the wavelength range of 300–370 nm. Strong emissions in the infrared region result in high operating temperatures, making a water-cooled immersion well a necessary requirement. The choice of lamp and lamp power is determined by the nature of photoreaction and the volume of reaction mixture.

The UV lamp is placed inside an immersion well. The immersion well contains a uble wall and is made up of quartz or borosilicate glass. Inlet and outlet tubes are esent for air or water cooling. A thin inlet tube reaches to the bottom of the annu-′ space to permit coolant flow upward from the bottom of the well. A 2 to 3 mm acing is provided between the walls, which helps in filtering certain wavelengths minimize secondary photochemical reactions of the products. It should be noted at when borosilicate glass is used, 254 nm radiations are not transmitted. There ₂ two types of outer reaction flasks: (1) type A: standard flask and (2) type B: gas et flask.

Both are composed of borosilicate glass and possess one central ground socket, lere an immersion well is placed and other smaller sockets are mounted by means laboratory stands and clamps. A standard flask is cylindrical in shape and has a t bottom for magnetic stirrer bar. It contains one angle socket and one vertical cket. Such a flask is best suited for low or constant temperature irradiation. An ditional sintered glass disc is fitted at the bottom of the flask to allow gas agitation d keep the reaction mixture under an inert or reactive atmosphere. A glass PTFE pcock is fitted.

A double-surface reflux condenser is present in the photoreactor, which reduces ₂ loss of vapor when using low boiling point liquids. The solvent chosen must be le to dissolve a variety of substrates but it should not strongly absorb UV radia-n itself. Acetonitrile acts as a versatile solvent as it is low cost, dissolves polar bstrates, does not show absorption above 200 nm, and can be easily removed on rotatory evaporator. If unshielded apparatus is operated, UV goggles should be rn especially while sampling of the reaction mixture is done. After all the initial rangements, the lamp is switched on and the progress of the reaction is monitored ′ thin layer chromatography (TLC), gas chromatography (GC), and so on. As often ₂ reagents are employed, workup is by evaporation of solvent, and ultimately, the oduct is purified by traditional methods.

3.8 MULTILAMP REACTOR

ung et al. (2005) determined the simulation of a radiation field in a multilamp pho-reactor. This is a difficult process as the problem of blockage of light arises due to ₂ neighboring lamps in the reactor. Shadow zones were created in the photoreactor ′ the lamp blockage and these zones vary depending on different lamp arrange-ents. Along with this, the shape of the reactor also influences the light intensity as actor walls also absorb radiation, if it reaches the wall. Reactor configuration and mp arrangement pattern were studied so as to obtain the maximum average light tensity. A line source with the diffused emission (LSDE) model, having sufficient curacy, was chosen to simulate the light field in the reactor. Simulation demon-rated that the square and cylindrical reactors having two and four lamps placed mmetrically had better working efficiency as compared to elliptical and rectan-ular reactors. They concluded that a multilamp reactor with a triangular pitch of mp arrangement with 33° should be preferred for maximum average light intensity.

Imoberdorf et al. (2008) studied the distribution of radiation inside the reactor the kinetics of photocatalytic reactions depend on the local incident radiation.

A Monte Carlo multilamp radiation model was proposed. The reflection on the s face of lamps, the refraction in the quartz lamp envelope, the adsorption, and t isotropic reemission of radiation in the mercury vapor of the lamp were considere

The mixing time in an agitated UV multilamp cylindrical photoreactor was det mined using electrical resistance tomography (ERT) by Zhao et al. (2008). It w observed that the location of the UV tubes affected the mixing time, when the imp ler speeds were less, that is, 45 and 150 rpm. The time required for mixing w maximum when UV tubes were situated at $r = 13$ cm ($r/R = 0.68$) and $\theta = 0°$, a it was minimum for location $r = 16$ cm ($r/R = 0.83$) and $\theta = 45°$. Thus, it could concluded that the mixing time had an inverse relation with the rotational speed, a this effect was dominant at lower speeds.

Johnson and Mehrvar (2008) developed a kinetic model for the removal of m ronidazole using UV/H_2O_2 in a single and multilamp tubular photoreactor. The r constant for the reaction between metronidazole and $^{\cdot}OH$ was found to be 1.98×1 per M/s. The rates of removal for metronidazole were 4.9%–13% for a single lan and 14%–41% for a multilamp photoreactor. The radius of the photoreactor w varied with H_2O_2 concentration used for maximum metronidazole removal. It w observed that the use of a stronger UV lamp (output 36 W) and low influent H_2 concentration (25 mg/L) were cost-effective factors.

18.9 SLURRY REACTOR

A slurry reactor is one of the most commonly used apparatuses in water purificatic These reactors require separation of the TiO_2 particles from the treated wat which obscures the technique. Different methods were performed for posttreatme separation like using settling tanks for overnight particle settling or an extern cross-flow filtration process. The use of such filtration techniques results in high cost of the treatment process.

Tokumura et al. (2006) designed a slurry photoreactor with external light irradi tion considering the average light intensity in the photoreactor. UV light or sunlig photo-Fenton bleaching of azo dye orange II in aqueous medium was perform using an external light irradiation cylindrical column photoreactor and with ir ion eluted from tourmaline powder (a natural mineral) containing 4.49 wt.% Fe_2C It was observed that the efficiency of discoloration was enhanced by decreasing t initial dye concentration, and increasing the amount of tourmaline and UV lig intensity. This model can be commonly used for both UV light and sunlight a employed for the treatment of textile effluents.

1,3-Dinitrobenzene (m-DNB) is released in water as a contaminant during t manufacture of explosives. It poses a threat to wildlife and biological systems by presence in water. Therefore, there is an urgent need to remove m-DNB from aqu ous solution. TiO_2 irradiated with solar radiation and artificial UV light was used a slurry photoreactor for this study (Kamble et al. 2006).

Chong et al. (2009) used an annular slurry photoreactor (ASP) system for t photodegradation of the model dye Congo red. Titania-doped kaolinite photocataly (TiO_2-K) was prepared. First-order kinetics was observed from the L–H model. T concentratio of Congo red played the most important role in the decompositic of the dye. It was concluded that such a model has the potential for scaling up a photocatalytic process for water remediation.

Sivagami et al. (2015) studied the photodegradation of two pesticides, commonly used in Indian agriculture, namely, endosulfan (ES) and chlorpyrifos (CPS), using an LSP under UV radiation at 254 nm. It was found that rate of degradation of catalyst was affected by the initial concentration of pesticide, pH of the solution, and concentration of catalyst. A batch degradation method was used for ES and CPS degradation in the concentration range of 5–25 mg/L at a pH ranging from 3.5 to 10.5 and catalyst loading of 0.5–2 g/L. The removal efficiency of ES was about 80%–99%, while CPS removal rate was only in the range of 84%–94%.

8.10 MERRY-GO-ROUND REACTOR

A merry-go-round reactor or turntable reactor is an apparatus which enables the rotation of several samples around a radiation source. Sample tubes are placed at an angle of $360/n$ (where n = number of tubes) and the radiation source is placed at the center. This provides an equal amount of radiation exposure to all the samples. In this case, about 6–10 samples depending on tubes containing reactants can be run at a time. Merry-go-round photoreactors help to carry out more than one photoreaction at a small scale simultaneously. They are employed in the determination of quantum yields of reaction, where identical irradiation conditions are a necessary factor.

Choy and Chu (2001) studied the photodegradation and photosensitization of TCE in the presence of hydrogen source of surfactant Brij 35 and photosensitizer (acetone). Photolysis was carried out in a merry-go-round photoreactor at 253.7 nm. A reaction mixture containing a fixed amount of TCE and surfactant Brij 35 was irradiated with UV light with varying doses of acetone. It was observed that the quantum yield in solution having surfactant Brij 35 and optimum concentration of acetone was 25 times higher than the solution having Brij 35 only. An excess of acetone concentration acted as a light barrier, which satisfies the light intensity for TCE decomposition.

Tsui and Chu (2001) also determined the quantum yield for the photodegradation of hydrophobic disperse dyes in the presence of photosensitizer acetone in a merry-go-round photoreactor having monochromatic UV lamps (253.7 nm). The selected model dyes were disperse yellow 7 (DY7) and an anthraquinone disperse dye disperse orange (DO11). The results showed that the addition of acetone enhanced the solubility of these dyes and also increased the photosensitization process. The quantum efficiency increased 10 times in the presence of acetone compared to water alone. The photobleaching of DY7 and DO11 was mainly due to photoreduction, which followed pseudo-first-order kinetics, and the reaction rates depend on the acetone/water ratio and the initial pH.

Kahveci et al. (2010) prepared polyacrylamide cryogels by irradiating acrylamide monomer and N,N'-methylene(bis)acrylamide (crosslinker) in the presence of 1-[4-(2-hydroxyethoxy)phenyl]-2-hydroxy-2-methyl-1-propane-1-one (Irgacure 2959) (photoinitiator) by employing freezing–thawing methods in a merry-go-round photoreactor. It was observed that gelation was initiated by light. The obtained cryogels demonstrated a rapid swelling property.

Palantöken et al. (2015) prepared hydrogels and investigated their antibacteria properties. These hydrogels were prepared by TiO_2 nanoparticles in a merry-go-roun photoreactor. TiO_2 nanoparticles were prepared using the sonochemical methoc The antibacterial activities of these hydrogels were studied against *Escherichi coli* bacteria using airborne testing and a modified Kirby–Bauer disc diffusio technique. In this merry-go-round photoreactor, the samples were encircled by si lamps radiating light at 300 nm.

Although a number of photoreactors are used for the study of photocatalyti reactions, only some of the major ones have been discussed; however, the list is nc complete as sometimes other photoreactors are also used apart from or in combina tion with other reactors.

REFERENCES

Adams, M., N. Skillen, C. McCullagh, and P. K. J. Robertso. 2013. Development of a dope titania immobilised thin film multi tubular photoreactor. *Appl. Catal. B Environ.* 130 131: 99–105.

Alam, M. M., M. Z. Bin Mukhlish, S. Uddin, S. Das, K. Ferdous, M. R. Khan, and M. A Islam. 2012. Photocatalytic degradation of reactive yellow in batch and continuous pho toreactor using titanium dioxide. *J. Sci. Res.* 4 (3): 665–674.

Alexiadis, A. and I. Mazzarino. 2005. Design guidelines for fixed-bed photocatalytic reactors *Chem. Engg. Process. Process Inten.* 44 (4): 453–459.

Ananpattarachai, J. and P. Kajitvichyanukul. 2014. Kinetics and mass transfer of fixed bec photoreactor using n-doped TiO_2 thin film for tannery wastewater under visible light *Chem. Engg. Trans.* 42: 163–168.

Anthony, K., W. Benjamin, D. Antoine, J. Christine, J. Thomas, and D. Christophe. 2015 Improved photo-induced cobalt-mediated radical polymerization in continuous flov photoreactors. *Polym. Chem.* 6: 3847–3857.

Arabatzis, I. M., N. Spyrellis, Z. Loizos, and P. Falaras. 2005. Design and theoretical study o a packed bed photoreactor. *J. Mater. Process. Technol.* 161 (1–2): 224–228.

Behnajady, M. A., S. Amirmohammadi-Sorkhabi, N. Modirshahl, and M. Shokri. 2009. Desig and construction of a tubular continuous-flow photoreactor with immobilized TiO nanoparticles on glass support and kinetic analysis of the removal of 4-nitrophenol i this photoreactor. In: Proceedings of the Ninth International Multidisciplinary Scientifi GeoConference, Vol. 2, pp. 479–486.

Behnajady, M. A., N. Modirshahla, N. Daneshvar, and M. Rabbani. 2007. Photocatalytic deg radation of an azo dye in a tubular continuous-flow photoreactor with immobilized TiO on glass plates. *Chem. Engg. J.* 127 (1–3): 167–176.

Behnajady, M. A., E. Siliani-Behrouz, and N. Modirshahla. 2012. Combination of desig equation and kinetic modeling for a batch-recirculated photoreactor at photooxidativ removal of C. I. acid red 17. *Int. J. Chem. React. Engg.* 10 (1): 1542–6580.

Borges, M. E., D. M. García, T. Hernández, J. C. Ruiz-Morales, and P. Esparza. 2015 Supported photocatalyst for removal of emerging contaminants from wastewater in a continuous packed-bed photoreactor configuration. *Catalysts.* 5 (1): 77–87.

Chanathaworn, J., C. Bunyakan, W. Wiyaratn, and J. Chungsiriporn. 2012. Photocatalyti decolorization of basic dye by TiO_2 nanoparticle in photoreactor. *Songklanakarin J. Sci Technol.* 34 (2): 203–210.

Chanathaworn, J., J. Pornpunyapat, and J. Chungsiriporn. 2014. Decolorization of dye ing wastewater in continuous photoreactors using TiO_2 coated glass tube media *Songklanakarin J. Sci. Technol.* 36 (1): 97–105.

Chiara, M. L. V. de, M. L. Amodio, F. Scura, L. Spremulli, and G. Colelli. 2014. Design and preliminary test of a fluidised bed photoreactor for ethylene oxidation on mesoporous mixed SiO_2/TiO_2 nanocomposites under UV-A illumination. *J. Agricult. Engg.* 45 (4): 435. doi: 10.4081/jae.

Chong, M. N., B. Jin, C. W. K. Chow, and C. P. Saint. 2009. A new approach to optimise an annular slurry photoreactor system for the degradation of Congo red: Statistical analysis and modeling. *Chem. Engg. J.* 152 (1): 158–166.

Choy, W. K. and W. Chu. 2001. The study of rate improvement of trichloroethene (TCE) decay in UV system with hydrogen source. *Water Sci. Technol.* 44 (6): 27–33.

Cloteaux, A., F. Gérardin, D. Thomas, N. Midoux, and J.-C. André. 2014. Fixed bed photocatalytic reactor for formaldehyde degradation: Experimental and modeling study. *Chem. Engg. J.* 249: 121–129.

Damodar, R. A., K. Jagannathan, and T. Swaminathan. 2007. Decolourization of reactive dyes by thin film immobilized surface photoreactor using solar irradiation. *Solar Energy.* 81 (1): 1–7.

Devagi, K., C. A. Motti, B. D. Glass, and M. Oelgemoeller. 2014. Photolysis and TiO_2-catalysed degradation of diclofenac in surface and drinking water using circulating batch photoreactors. *Environ. Chem.* 11 (1): 51–62.

Divya, N., A. Bansal, and A. K. Jana. 2009. Degradation of acidic orange G dye using $UV-H_2O_2$ in batch photoreactor. *Int. J. Biol. Chem. Sci.* 3 (1): 54–62.

Esterkin, C. R., A. C. Negro, O. M. Alfano, and A. E. Cassano. 2005. Air pollution remediation in a fixed bed photocatalytic reactor coated with TiO_2. *AIChE J.* 51 (8): 2298–2310. doi: 10.1002/aic.10472.

Faramarzpour, M., M. Vossoughi, and M. Borghei. 2009. Photocatalytic degradation of furfural by titania nanoparticles in a floating-bed photoreactor. *Chem. Engg. J.* 146 (1): 79–85.

Imoberdorf, G. E., F. Taghipour, and M. Mohseni. 2008. Radiation field modeling of multi-lamp, homogeneous photoreactors. *J. Photochem. Photobiol. A Chem.* 198 (2–3): 169–178.

Johnson, M. B. and M. Mehrvar. 2008. Aqueous metronidazole degradation by UV/H_2O_2 process in single-and multi-lamp tubular photoreactors: Kinetics and reactor design. *Ind. Eng. Chem. Res.* 47 (17): 6525–6537.

Kahveci, M. U., Z. Beyazkilic, and Y. Yagci. 2010. Polyacrylamide cryogels by photoinitiated free radical polymerization. *J. Polym. Sci. A.* 15 (22): 4989–4994.

Kamble, S. P., S. B. Sawant, and V. G. Pangarkar. 2006. Photocatalytic degradation of m-dinitrobenzene by illuminated TiO_2 in a slurry photoreactor. *J. Chem. Technol. Biotechnol.* 81 (3): 365–373.

Krishnan, J. and T. Swaminathan. 2010. Kinetic modeling of a photocatalytic reactor designed for removal of gas-phase benzene: A study on limiting resistances using design of experiments. *Latin Am. Appl. Res.* 40 (4): 359–364.

Kuo, Y.-L., T.-L. Su, K.-J. Chuang, H.-W. Chen, and F.-C. Kung. 2011. Preparation of platinum- and silver-incorporated TiO_2 coatings in thin-film photoreactor for the photocatalytic decomposition of o-cresol. *Environ. Technol.* 33 (15–16): 1799–1806.

Liu, H., H. Shon, Y. Okour, W. Song, and S. Vigneswaran. 2011. Photocatalytic degradation of acid red G by bismuth titanate in three-phase fluidized bed photoreactor. *J. Adv. Oxid. Technol.* 14 (1): 116–121.

Palantöken, A., M. S. Yilmaz, M. A. Yapaöz, and S. Pişkin. 2015. Investigation of antibacterial properties of hydrogel containing synthesized TiO_2 nanoparticles. *Sigma J. Engg. Nat. Sci.* 33: 1–7.

Peres, J. C. G., U. Silvio, A. C. S. C. Teixeira, R. Guardani, and A. S. Vianna Jr. 2015. Study of an annular photoreactor with tangential inlet and outlet: I. Fluid dynamics. *Chem. Engg. Technol.* 38 (2): 311–318.

Rasoulifard, M. H., M. Fazlid, and M. Eskandarian 2015. Performance of the light-emitting diodes in a continuous photoreactor for degradation of direct red 23 using UV-LED/$S_2O_8^2$ process. *J. Indust. Engg. Chem.* 24: 121–126.

Ren, H.-Y., B.-F. Liu, G.-J. Xie, L. Zhao, and N.-Q. Ren. 2014. Carrier modification and its application in continuous photo-hydrogen production using anaerobic fluidized bed photo-reactor. *GCB Bioenergy.* 6 (5): 599–605.

Sheidaei, B. and M. A. Behnajady. 2015. Determination of optimum conditions for removal of acid orange 7 in batch-recirculated photoreactor with immobilized TiO_2-P25 nanoparticles by Taguchi method. *Desalin. Water Treat.* 56 (9): 2417–2424.

Sivagami, K., B. Vikraman, R. R. Krishna, and T. Swaminathan. 2015. Chlorpyrifos and endosulfan degradation studies in an annular slurry photoreactor. *Ecotoxicol. Environ. Saf* doi:10.1016/j.ecoenv.2015.08.015.

Son, S. M., V. Y. Kim, I. S. Shim, Y. Kang, S. H. Kang, B. T. Yoon, and M. J. Choi. 2009. Analysis of gas flow behaviour in an annular fluidized-bed reactor for polystyrene waste treatment. *J. Mater. Cycles. Waste Manage.* 11: 138–143.

Tokumura, M., H. T. Znad, and Y. Kawase. 2006. Modeling of an external light irradiation slurry photoreactor: UV light or sunlight-photoassisted Fenton discoloration of azo-dye orange II with natural mineral tourmaline powder. *Chem. Engg. Sci.* 61 (19): 6361–6371

Toosi, M. R., M. H. Peyravi, J. Sajadi, M. J. Bayani, and H. Manghabati. 2013. Photocatalytic purification of wastewater polluted by odorant sulfur compounds using titanium oxide in a continuous photoreactor. *Int. J. Chem. Reactor Engg.* 11 (1): 561–567.

Tsui, S. M. and W. Chu. 2001. Quantum yield study of the photodegradation of hydrophobic dyes in the presence of acetone sensitizer. *Chemosphere.* 44 (1): 17–22.

Vaiano, V., O. Sacco, D. Pisano, D. Sannino, and P. Ciambelli. 2015. From the design to the development of a continuous fixed bed photoreactor for photocatalytic degradation of organic pollutants in wastewater. *Chem. Engg. Sci.* 137: 152–160.

Yang, Q., S. O. Pehkonen, and M. B. Ray. 2005. Simulation of radiation field in multilamp photoreactor using the LSDE model. *Canad. J. Chem. Engg.* 83 (4): 705–711.

Zhao, Z. F., M. Mehrvar, and F. Ein-Mozaffari. 2008. Mixing time in an agitated multi-lamp cylindrical photoreactor using electrical resistance tomography. *J. Chem. Technol. Biotechnol.* 83 (12): 1676–1688.

19 Future Trends

search on photocatalytic materials had been a field of continuous expansion in
last few decades. There are three different strategies to increase the efficiency of
otocatalysts: (1) combination with other functional materials, (2) combination of
ferent semiconductors, and (3) morphological modifications.

Humanity is facing two major challenges and these are renewable energy harvest-
; methods and maintaining a sustainable environment. One of the most promising
utions to meet the challenge is to convert solar energy to some fuels or electricity.
sically, there are three key processes for utilizing solar energy efficiently: (1) max-
um photon absorption, (2) efficient charge separation, and (3) effectively utilizing
:se separated charges. Photocatalysis will have a bright future in coming years,
t more focus should be on the development of solar light–activated photocatalysts
ving higher surface area, so that they can be useful for converting solar energy into
:ctrical or chemical energy.

Photovoltaic cells have great potential for supplying eco-friendly energy, but they
ve a disadvantage of lacking an efficient and cost-effective energy storage process,
ich can supply energy for transportation, particularly at night. If solar energy
n be converted into chemical fuels, then it may solve the problem of energy stor-
e. The photon-driven electrolysis of water to produce hydrogen and oxygen can be
hieved either with self-supported catalysts or photoelectrochemical (PEC) cells.
w metal oxide visible light–absorbing semiconductors may be designed to explore
: potential of using nanomaterials for this purpose.

PEC cells do not only generate electricity but also decompose water effectively
o hydrogen and oxygen. Such water splitting is the requirement of the day to pro-
ce hydrogen, because it has been advocated as the fuel of the future. A number of
:C cells have been fabricated with different materials either in the form of elec-
des or electrolytes. Efforts have also been made to achieve efficient and stable
:C cells with modification of the photoelectrode and/or electrolyte, photoetching
layered semiconductors, variation in configuration of PEC solar cells, dye sensiti-
tion, and so on, but there are still many possibilities to develop newer materials in
ars to come so as to achieve the PEC cells of the next generation.

Dye-sensitized solar cells (DSSCs) have proved to be an attractive potential
urce of renewable energy because of their eco-friendliness, easy fabrication, and
st-effectiveness. However, in DSSCs, the rarity and high cost of some electrode
aterials (transparent conducting oxide and platinum), slow electron transport, poor
ght harvesting efficiency, and significant charge recombination are the main rea-
ns for their inefficient performance. Carbon nanotubes (CNTs) are quite promising
inforcements to overcome such issues as they have some unique electrical, optical,
emical, physical, and catalytic properties.

Water splitting an area which is being explored by many workers around the globe
t is still in a premature stage. The generation of hydrogen from photocatalytic

water splitting is a clean and sustainable technique to produce renewable fu⬤
like hydrogen. Solar water splitting has an added advantage that it does not suf⬤
from an electricity storage problem as is the case with photovoltaics. However, t⬤
energy conversion efficiency of water splitting is much lower at present as compar⬤
to that of photovoltaics. Hydrogen and oxygen evolution are two steps in the so⬤
water splitting process. Thus, new materials need to be designed or fabricated ⬤
efficient solar energy harvesting as well as for active cocatalysts.

Two-dimensional (2D) nanomaterials have been tried for solar water splitting
they have potential as catalysts because of their high surface to volume ratio. Su⬤
materials can harvest solar energy and generate electrons and holes and also provi⬤
separation and diffusion of photoexcited carriers. Functionalized graphene (G⬤
oxides can perform overall water splitting without using any cocatalysts or sacri⬤
cial reagent.

Photocatalytic water splitting is a clean and sustainable technique for rene⬤
able fuels. Also, it can provide an important feedstock—hydrogen for the chemic
industry. Plasmonic nanostructures have emerged as a totally new type of visib⬤
light energy–harvesting materials. Surface plasmon resonances are widely used
enhance local electromagnetic fields to guide light and funnel energy to the acti⬤
regions of devices.

Photocatalysis has also been used for the reduction of carbon dioxide to use⬤
synthetic fuels such as methane, methanol, and ethanol along with clean fuel hydr⬤
gen. The present research is attempting to mimic the work of nature, that is, ph⬤
tosynthesis. However, so far the best known and widely used catalyst is TiO_2. Ev⬤
though various methods have been tried to enhance the activity of this or other ca⬤
lysts, the amount of CO_2 reduced is still quite less than the amount of generated C⬤

GR has emerged as a sparkling rising star in the field of material sciences, becau⬤
its planar structure, excellent transparency, superior electron conductivity and mob⬤
ity, high-specific surface area, and high chemical stability make it unique in natu⬤
GR has also been regarded as an ideal candidate to prepare GR-based nanocor⬤
posites for better performance in solar energy storage and conversion. Present⬤
GR-based photocatalysts have been attracting increasing attention as an emergi⬤
and prospective candidate as GR-based nanocomposites have many applications li⬤
nonselective processes for degradation of various pollutants, selective transform⬤
tions for organic synthesis, water splitting for clean hydrogen energy, and so on. T⬤
existing challenges for future exploitation and development of GR-based nanocor⬤
posites are to design smarter and even more efficient GR-based photocatalysts.

Following the discovery of CoPi as an efficient oxygen evolution catalyst, cob⬤
oxides (CoO and Co_3O_4) have emerged as new promising oxygen evolution catalys⬤
They greatly the increased the lifetime of photoexcited electrons, leading to enhanc⬤
oxygen evolution efficiency, particularly when these are decorated on the surface
other photocatalysts. Nanoparticles exhibit surprisingly higher activity than their bu⬤
counterparts. They can be synthesized using laser ablation without any precursors
surfactants. Co_3O_4 nanoparticles have exhibited a significant enhancement in oxyg⬤
evolution activity compared to micropowders. Nanoparticles of CoO have a high ef⬤
ciency toward overall water splitting without any cocatalysts and sacrificial reagen⬤

hile CoO is not active in bulk form. The efficiency has to reach the benchmark of ore than 10% in order to be competitive with other methods.

Photocatalysis had been instrumental in the development of air and water urification technologies along with cleaning, antifogging, and antistaining applica-ons. The cleansing effect of photocatalysis was observed due to the light-sensitive tanium dioxide embedded in the surface layer in the form of tiny particles. If titania articles are exposed to ultraviolet (UV) light, oxygen present in surrounding air ; chemically activated and it destroys organic matter along with other pollutants vithout attacking the surface, or in other words, it may be said that the dirt literally issolves in the air. Thus, TiO_2 may be called the ultimate self-cleaner, and is eco-riendly and sustainable also. This cleansing action of titania will be ideal for areas 1at require outstanding hygiene, such as bathrooms and other sanitary facilities.

Toilet hygiene plays an important role in sanitary facilities. Oxygen activated in 1e process of photocatalysis naturally breaks down all organic contaminants on the urface and such an effect is never depleted, and works effectively over a long period f time. Zirconium coatings on the toilet bowl further support the elimination of vaste and bacteria due to their superhydrophilic properties.

The process of self-cleaning using photocatalysis helps us in developing a new eneration of coatings and glazes. Such photocatalytic materials enable progressive unctions that will be beneficial to people as well as the environment. Such glazes nd coatings will become indispensable to the bathroom, toilet, or shower. Their use ramatically reduces the required amount of chemical cleansers.

One of the important applications of photocatalysis is in the construction of elf-cleaning buildings as tiles and other surfaces coated with TiO_2 remain clean. unlight causes oxidation that breaks down grime, while the superhydrophilicity of iO$_2$ causes the water to spread, washing surfaces clean. Other popular applications re air purifiers and antifog mirrors. TiO_2 can deodorize the smell of tobacco or pets lso.

Roads, footpaths, and car parking areas all act as catchments for runoff water. "he collected water contains heavy metals as well as hydrocarbons, which can either each into soils and groundwater and/or be flushed out to the sea. Concrete pave-nent containing titanium dioxide have the ability to remove certain contaminants rom water and air, resulting in reduction in the level of pollution in that area. The ncorporation of TiO_2 into construction materials is quite helpful in degrading vari-us water and air pollutants. The photoinduced redox reaction and superhydrophilic ature of TiO_2 assists in the degradation of organic pollutants.

The efficiency of the photocatalytic self-cleaning property can be enhanced by imploying various techniques like doping with metals and nonmetals, formation of leterojunctions between TiO_2 desiging, low band gap semiconductors, fabrication of GR-based semiconductor nanocomposites, and so on. Band gap narrowing by the iormation of localized energy levels within the band gap and intrinsic defects such is oxygen vacancies have been considered responsible for the improved activity of loped TiO_2 photocatalysts. TiO_2-based self-cleaning materials exhibited hydrophilic ind underwater superoleophobic properties, so these can be utilized in water man-igement, antifouling applications, and separation of oil in water emulsions.

The antifogging property finds extensive use in preventing fog formation in optical lenses, paint coatings, mirrors of bathrooms or dressing tables, rear view mirrors, and so on, while the self-cleaning property is helpful in efficient and rapid cleaning of headlights of vehicles, computer displays, signboards, and walls and lighting of tunnel. Rainfall will automatically assist in cleaning the photocatalytic surfaces in open areas like traffic signs, exterior windows and tiles, building stone and so on. Accelerated drying can also be achieved in photocatalytic surfaces present in bathroom, windows, toilets, and so on. The cleaning process can be made easy and time saving while cleaning tableware, kitchenware, showcase, kitchen, interior furnishings, and so on.

The superhydrophilicity can be applied for many other products utilizing secondary properties. A surface dries quickly utilizing this property because water cannot flatly spread on it. This property can be applied to prevent dewdrops from being formed inside a windowpane for the purpose of protecting vegetables from rotting by dewdrops.

TiO_2 is reported to kill not only bacteria but also viruses and cancer cells. Indoor uses of TiO_2 could greatly improve the quality of life, but it has a specific challenge in that indoor light is visible light but it needs UV light to become active. An efficient and affordable way to split water into hydrogen may find numerous uses and hydroponic agriculture is based on using TiO_2 to keep the plant water clean so that better plant growth is obtained.

Recently, the UV light emitting diode (LED) has attracted the attention of the scientific community for possible use in photocatalytic applications. These LED can be used in solar panel also as they need less power around 3.0 eV. It is hoped that some water purification system will come soon in the market based on this. Efforts are focused on energy-efficient sources and effective photocatalyst design so as to make photocatalysis a promising technology of future.

The photocatalytic efficiency of a material depends on the type of contaminant to be removed. A particular catalyst which is most active for one contaminant may or may not be active for some other pollutant. Therefore, there cannot be a universal catalyst for all the contaminants.

The most important step is either to develop lower band gap photocatalysts or to modify them so as to narrow their band gap. Efforts have been focused on reducing the band gaps of semiconductors through doping or alloying. A totally new type of visible light energy harvesting material is plasmonic nanostructures. Plasmonic nanostructures donate electrons to the attached cocatalysts. Overall water splitting has been observed by integrating plasmonic materials with hydrogen and oxygen cocatalysts. A major part of the solar spectrum can be absorbed by tuning the resonance of surface plasmon resonances.

The field of materials engineering, device design, and intricacies of mechanisms are developing at an ever-increasing pace. H_2 production from water splitting, conversion of CO_2, DSSCs, photocatalytic wastewater and gas treatment, and application of nanostructured materials are some of the major areas in this direction.

Photocatalysis is an area of research with immense promises and many more feathers will be added in its cap in years to come.

Index

A

Acetamiprid, photocatalytic degradation of, 18, 215
Adsorptivity of catalyst, 139
Advanced oxidation processes (AOPs), 17
Aeromonas hydrophila, 265
AFB photoreactor, *see* Annular fluidized bed photoreactor
AFBPR, *see* Anaerobic fluidized bed photoreactor
Ag/Ga$_2$O$_3$ photocatalysts, *see* Ag-loaded Ga$_2$O$_3$ photocatalysts
Ag(I) ions, doping energy level of, 82–83
Ag-loaded Ga$_2$O$_3$ (Ag/Ga$_2$O$_3$) photocatalysts, 231
Ag NPs–TiO$_2$ system, Schottky barrier in, 68
Ag–ZnO, *see* Silver–zinc oxide
Air purification, 317
Al-doped ZnO (AZO), 85
Anaerobic fluidized bed photoreactor (AFBPR), 301
Anatase phase, 224
Anionic doping, 3
Annular fluidized bed (AFB) photoreactor, 300–301
Annular reactor, 305–306
Annular slurry photoreactor (ASP) system, 310
Anthocyanin, 123
Antibacterial effects of photocatalysts, 5, 265
Anticancer activity, 273–275
Antifogging, 5, 284–287
Antifungal activity, 271–272
Antiviral activity, 272
AO10, *see* Monoazo dye acid orange 10
AOPs, *see* Advanced oxidation processes
APCVD, *see* Atmospheric pressure chemical vapor deposition
AR114, *see* Diazo dye acid red 114
Artificial photosynthesis, 242–243
CO$_2$ reduction and fuel production, 252–256
different approaches to, 256–257
goals and benefits, 258
limitations of, 258–259
natural photosystem and hybrid photosystem, 243–244
role of, 245–247
water splitting, 248–252
Artificial photosystem, 243–244
Artificial Z-scheme photocatalyst, 234
ASP system, *see* Annular slurry photoreactor system

Atmospheric pressure chemical vapor deposition (APCVD), 266, 271
Auxochrome, photosensitizer, 110
AZO, *see* Al-doped ZnO

B

Bacillus subtilis, 265, 267, 275
Band gap, of different semiconductors, 12
Barium titanate (BaTiO$_3$), 38–41
Batch reactor, 302–304
BaTiO$_3$, *see* Barium titanate
Bed reactor
annular fluidized bed photoreactor, 300–301
fixed bed photoreactor, 299–300
packed bed photoreactor, 302
Betalain, 122
BET surface area, *see* Brunauer–Emmett–Teller surface area
Bi$_{0.5}$Na$_{0.45}$Li$_{0.05}$K$_{0.5}$TiO$_3$– BaTiO$_3$ (BNKLBT), 41
Binary semiconductors; *see also* Oxides
zinc sulfide, 28–29
Biological safety cabinet (BSC), 274
Bismuth vanadate (BiVO$_4$), 42–44
BiVO$_4$, *see* Bismuth vanadate
BNKLBT, *see* Bi$_{0.5}$Na$_{0.45}$Li$_{0.05}$K$_{0.5}$TiO$_3$–BaTiO$_3$
Borosilicate glass irradiation tubes, 304
Brilliant green yellow, photocatalytic decolorization of, 20
Brunauer–Emmett–Teller (BET) surface area, 207
BSC, *see* Biological safety cabinet

C

Cadmium sulfide (CdS), 26–29, 71–74
CaFe$_2$O$_4$/TiO$_2$, 177
Cancer treatment, 5
Candida albicans, 271, 275
Carbon-based supports/natural products, 156–158
Carbon dioxide, reduction of
artificial photosynthesis, 252–256
nonoxides, *see* Nonoxide photocatalysts
photocatalytic, *see* Photocatalytic reduction, of carbon dioxide
Carbon-doped zinc oxide nanostructures, 92
Carbon nanotubeanatase titanium dioxide (CNT–TiO$_2$), 135–136
Carbon nanotube–based composites, 135–136

319

For Product Safety Concerns and Information please contact our EU
representative GPSR@taylorandfrancis.com
Taylor & Francis Verlag GmbH, Kaufingerstraße 24, 80331 München, Germany